Advances in Heterojunction
Photocatalysts

Advances in Heterojunction Photocatalysts

Editors

Yongming Fu
Qian Zhang

Basel • Beijing • Wuhan • Barcelona • Belgrade • Novi Sad • Cluj • Manchester

Editors

Yongming Fu
School of Physics and
Electronic Engineering
Shanxi University
Taiyuan
China

Qian Zhang
School of Materials
Sun Yat-Sen University
Shenzhen
China

Editorial Office
MDPI
St. Alban-Anlage 66
4052 Basel, Switzerland

This is a reprint of articles from the Special Issue published online in the open access journal *Catalysts* (ISSN 2073-4344) (available at: www.mdpi.com/journal/catalysts/special_issues/ advances_heterojunction_photocatalysts).

For citation purposes, cite each article independently as indicated on the article page online and as indicated below:

Lastname, A.A.; Lastname, B.B. Article Title. *Journal Name* **Year**, *Volume Number*, Page Range.

ISBN 978-3-7258-0852-6 (Hbk)
ISBN 978-3-7258-0851-9 (PDF)
doi.org/10.3390/books978-3-7258-0851-9

© 2024 by the authors. Articles in this book are Open Access and distributed under the Creative Commons Attribution (CC BY) license. The book as a whole is distributed by MDPI under the terms and conditions of the Creative Commons Attribution-NonCommercial-NoDerivs (CC BY-NC-ND) license.

Contents

About the Editors vii

Preface ix

Peng Chen, Xiu Li, Zeqian Ren, Jizhou Wu, Yuqing Li and Wenliang Liu et al.
Enhancing Photocatalysis of Ag Nanoparticles Decorated $BaTiO_3$ Nanofibers through Plasmon-Induced Resonance Energy Transfer Turned by Piezoelectric Field
Reprinted from: Catalysts **2022**, 12, 987, doi:10.3390/catal12090987 1

Zhanyong Gu, Mengdie Jin, Xin Wang, Ruotong Zhi, Zhenghao Hou and Jing Yang et al.
Recent Advances in $g-C_3N_4$-Based Photocatalysts for NO_x Removal
Reprinted from: Catalysts **2023**, 13, 192, doi:10.3390/catal13010192 12

Zihao Xia, Ting Cai, Xiangguo Li, Qian Zhang, Jing Shuai and Shenghua Liu
Recent Progress of Printing Technologies for High-Efficient Organic Solar Cells
Reprinted from: Catalysts **2023**, 13, 156, doi:10.3390/catal13010156 36

Huangzhaoxiang Chen, Qian Zhang, Aumber Abbas, Wenran Zhang, Shuzhou Huang and Xiangguo Li et al.
$BiVO_4$ Photoanodes Modified with Synergetic Effects between Heterojunction Functionalized $FeCoO_x$ and Plasma Au Nanoparticles
Reprinted from: Catalysts **2023**, 13, 1063, doi:10.3390/catal13071063 58

Muhammad Humayun, Ayesha Bahadur, Abbas Khan and Mohamed Bououdina
Exceptional Photocatalytic Performance of the $LaFeO_3/g-C_3N_4$ Z-Scheme Heterojunction for Water Splitting and Organic Dyes Degradation
Reprinted from: Catalysts **2023**, 13, 907, doi:10.3390/catal13050907 70

Ling Wang, Keyi Xu, Hongwang Tang and Lianwen Zhu
Vertical Growth of WO_3 Nanosheets on TiO_2 Nanoribbons as 2D/1D Heterojunction Photocatalysts with Improved Photocatalytic Performance under Visible Light
Reprinted from: Catalysts **2023**, 13, 556, doi:10.3390/catal13030556 84

Xiaohang Yang, Yulin Zhang, Jiayuan Deng, Xuyang Huo, Yanling Wang and Ruokun Jia
Fabrication of Porous Hydrophilic CN/PANI Heterojunction Film for High-Efficiency Photocatalytic H_2 Evolution
Reprinted from: Catalysts **2023**, 13, 139, doi:10.3390/catal13010139 100

Amit Imbar, Vinod Kumar Vadivel and Hadas Mamane
Solvothermal Synthesis of $g-C_3N_4/TiO_2$ Hybrid Photocatalyst with a Broaden Activation Spectrum
Reprinted from: Catalysts **2022**, 13, 46, doi:10.3390/catal13010046 112

Chenjing Sun, Kaiqing Zhang, Bingquan Wang and Rui Wang
Synergistic Effect of Amorphous Ti(IV)-Hole and Ni(II)-Electron Cocatalysts for Enhanced Photocatalytic Performance of Bi_2WO_6
Reprinted from: Catalysts **2022**, 12, 1633, doi:10.3390/catal12121633 123

Murugan Arunachalapandi, Thangapandi Chellapandi, Gunabalan Madhumitha, Ravichandran Manjupriya, Kumar Aravindraj and Selvaraj Mohana Roopan
Direct Z-Scheme $g-C_3N_5/Cu_3TiO_4$ Heterojunction Enhanced Photocatalytic Performance of Chromene-3-Carbonitriles Synthesis under Visible Light Irradiation
Reprinted from: Catalysts **2022**, 12, 1593, doi:10.3390/catal12121593 140

Tingting Ma, Zhen Li, Gan Wang, Jinfeng Zhang and Zhenghua Wang
Efficient Visible-Light Driven Photocatalytic Hydrogen Production by Z-Scheme $ZnWO_4/Mn_{0.5}Cd_{0.5}S$ Nanocomposite without Precious Metal Cocatalyst
Reprinted from: *Catalysts* **2022**, *12*, 1527, doi:10.3390/catal12121527 . . . . **156**

Zhen Li, Ligong Zhai, Tingting Ma, Jinfeng Zhang and Zhenghua Wang
Efficient and Stable Catalytic Hydrogen Evolution of ZrO_2/CdSe-DETA Nanocomposites under Visible Light
Reprinted from: *Catalysts* **2022**, *12*, 1385, doi:10.3390/catal12111385 . . . . **168**

Xuefeng Hu, Ting Luo, Yuhan Lin and Mina Yang
Construction of Novel Z-Scheme g-C_3N_4/AgBr-Ag Composite for Efficient Photocatalytic Degradation of Organic Pollutants under Visible Light
Reprinted from: *Catalysts* **2022**, *12*, 1309, doi:10.3390/catal12111309 . . . . **179**

Lian Sun, Qian Zhang, Qijie Liang, Wenbo Li, Xiangguo Li and Shenghua Liu et al.
α-Fe_2O_3/Reduced Graphene Oxide Composites as Cost-Effective Counter Electrode for Dye-Sensitized Solar Cells
Reprinted from: *Catalysts* **2022**, *12*, 645, doi:10.3390/catal12060645 . . . . **190**

Qiong Lu, Jing An, Yandong Duan, Qingzhi Luo, Yunyun Shang and Qiunan Liu et al.
Highly Efficient and Selective Carbon-Doped BN Photocatalyst Derived from a Homogeneous Precursor Reconfiguration
Reprinted from: *Catalysts* **2022**, *12*, 555, doi:10.3390/catal12050555 . . . . **200**

About the Editors

Yongming Fu

Yongming Fu is an Associate Professor and Master's Supervisor at the School of Physical and Electronic Engineering, Shanxi University. He serves as a Guest Editor and a member of TAP for Catalysts, as well as a Guest Editor for *Sustainability* and a Youth Editorial Board Member for the *Journal of Advanced Dielectrics*. Additionally, he holds the position of Executive Director on the Expert Committee of the China Opto-electronic Industry Platform. His research interests lie in condensed matter physics, nanomaterials, and the interdisciplinary field of optics, with a particular focus on the fundamental and applied aspects of multiphysical field coupling phenomena in piezoelectric nanomaterials. Yongming Fu has published over 70 papers in prestigious journals, including *Nano Energy*, *Advanced Functional Materials*, *Applied Catalysis B*, and *Nano-Micro Letters*, two of which are highly cited papers. His work has been cited more than 2300 times, earning him an h-index of 31. He has led or participated in several national and provincial projects, including key international cooperation research projects and youth fund projects sponsored by the National Natural Science Foundation of China. He has applied for five Chinese invention patents, two international patents, and holds two software copyrights. In 2023, he won the Second Prize in Natural Science Award from the government of Shanxi Province.

Qian Zhang

Qian Zhang is currently an Associate Professor at the School of Materials at Sun Yat-sen University. She received her PHD from University of Science and Technology Beijing. She then worked at the National University of Singapore as a Research Fellow. Her research interests are low-dimensional nanomaterials and multifunctional devices. So far, she has co-authored more than 40 papers in the international refereed journals, such as *Nat. Commun., Energ. Environ. Sci., Adv. Mater., Adv. Funct. Mater.*, and *ACS Nano*. Three of her papers have been recognized as ESI Highly Cited Papers. Her research on "shadow-effect generator" was featured on the cover of EES and received high praise in a featured review by Prof. Geoffery Ozin. The interesting "shadow-effect generator" has been reported by over 100 domestic and international outlets. She has filed for several international and national patents. She has frequently presented at domestic and international academic conferences. She is currently the principal investigator for projects funded by the Natural Science Foundation of Guangdong Province and the Shenzhen Science and Technology Program, among others. She has been asked to be an independent reviewer for over 20 international journals, including *Chem. Soc. Rev., Energ. Environ. Sci, Adv. Mater.*, and *ACS Nano*. In addition, she has served as a Guest Editor for the SCI journal *Catalysts*.

Preface

In the quest for sustainable solutions to global energy and environmental challenges, the innovative use of semiconductor-based photocatalysis has emerged as a beacon of hope. It is with immense pleasure and a deep sense of gratitude that I introduce "*Advances in Heterojunction Photocatalysts*", a reprint inspired by the Special Issue of the same name published in the esteemed Journal *Catalysts*. This Special Issue was meticulously organized by Professor Yongming Fu of Shanxi University and Professor Qian Zhang of Sun Yat-sen University. This reprint is a compilation of 15 seminal works, chosen from the contributions to the Special Issue, each representing a stride forward in the field of heterojunction photocatalysts.

It is important to recognize the collective endeavor of all the authors who contributed their research to the Special Issue of the Journal *Catalysts*. Their rigorous work and insightful findings have laid the groundwork for this publication. The diversity and depth of the research presented here are a testament to their expertise and commitment to advancing our understanding and application of heterojunction photocatalysts.

At the core of this compilation is an exploration of heterojunction photocatalysts. These materials are engineered to enhance the spatial separation of photogenerated electron–hole pairs, significantly boosting their efficiency in converting solar energy into solar fuels and in degrading pollutants.

This reprint proudly presents 13 original research articles and 2 comprehensive review papers—each a beacon of expertise in its own right. From the innovative synthesis of noble metal-modified semiconductors to the development of novel Z-Scheme heterojunctions, the research articles delve deeply into the latest advancements and experimental breakthroughs. Meanwhile, the review articles offer an expansive overview of their respective fields, synthesizing past achievements, current developments, and future prospects.

By offering both detailed findings and broad overviews, this book caters to a wide audience, from researchers to students keen on photocatalysis. It aims to highlight scientific achievements and the potential applications of heterojunction photocatalysts in environmental remediation and solar fuel generation, underscoring their versatility and contribution to a sustainable future.

As we turn the pages, we embark on a journey of discovery and inspiration, fueled by the collective wisdom and creativity of the scientific community. "*Advances in Heterojunction Photocatalysts*" is more than a compilation of current research; it is a clarion call to further exploration, innovation, and exploitation of the vast potential of photocatalytic materials for the benefit of society and the environment at large.

Yongming Fu and Qian Zhang
Editors

Article

Enhancing Photocatalysis of Ag Nanoparticles Decorated BaTiO₃ Nanofibers through Plasmon-Induced Resonance Energy Transfer Turned by Piezoelectric Field

Peng Chen [1,2], Xiu Li [1], Zeqian Ren [1], Jizhou Wu [1,2], Yuqing Li [1,2], Wenliang Liu [1,2], Peng Li [1], Yongming Fu [1,*] and Jie Ma [1,2,*]

[1] State Key Laboratory of Quantum Optics and Quantum Optics Devices, School of Physics and Electronic Engineering, Institute of Laser Spectroscopy, Shanxi University, Taiyuan 030006, China
[2] Collaborative Innovation Center of Extreme Optics, Shanxi University, Taiyuan 030006, China
* Correspondence: fuyongming@sxu.edu.cn (Y.F.); mj@sxu.edu.cn (J.M.)

Citation: Chen, P.; Li, X.; Ren, Z.; Wu, J.; Li, Y.; Liu, W.; Li, P.; Fu, Y.; Ma, J. Enhancing Photocatalysis of Ag Nanoparticles Decorated BaTiO₃ Nanofibers through Plasmon-Induced Resonance Energy Transfer Turned by Piezoelectric Field. *Catalysts* 2022, 12, 987. https://doi.org/10.3390/catal12090987

Academic Editor: Vincenzo Vaiano

Received: 13 July 2022
Accepted: 30 August 2022
Published: 1 September 2022

Publisher's Note: MDPI stays neutral with regard to jurisdictional claims in published maps and institutional affiliations.

Copyright: © 2022 by the authors. Licensee MDPI, Basel, Switzerland. This article is an open access article distributed under the terms and conditions of the Creative Commons Attribution (CC BY) license (https://creativecommons.org/licenses/by/4.0/).

Abstract: Revealing the charge transfer path is very important for studying the photocatalytic mechanism and improving photocatalytic performance. In this work, the charge transfer path turned by the piezoelectricity in Ag-BaTiO₃ nanofibers is discussed through degrading methyl orange. The piezo-photocatalytic degradation rate of Ag-BaTiO₃ is much higher than the photocatalysis of Ag-BaTiO₃ and piezo-photocatalysis of BaTiO₃, implying the coupling effect between Ag nanoparticle-induced localized surface plasmon resonance (LSPR), photoexcited electron-hole pairs, and deformation-induced piezoelectric field. With the distribution density of Ag nanoparticles doubling, the LSPR field increases by one order of magnitude. Combined with charge separation driven by the piezoelectric field, more electrons in BaTiO₃ nanofibers are excited by plasmon-induced resonance energy transfer to improve the photocatalytic property.

Keywords: photocatalysis; piezoelectric; plasmon; Ag nanoparticle; BaTiO₃ nanofiber; resonance energy transfer

1. Introduction

The increasing worsening of global water pollution seriously threatens human health and social development. Solar-induced photocatalytic degradation of pollutants based on semiconducting photocatalysts is an effective method for wastewater treatment [1–4]. Under light irradiation with applicable energy, the semiconductor photocatalysts are excited to generate free electron-hole pairs, which migrate to the solid/liquid interface and generate reactive oxygen species (ROSs) for participating in the degradation of various pollutants [5–8]. However, the electron-hole pairs are easy to recombine due to the conduction of the Coulomb force; only a few carriers can successfully migrate to the surface of semiconductor particles to participate in the photocatalytic reaction [9–13]. It is a key problem to effectively improve the separation of electron-hole pairs for photocatalysis.

Recently, a novel method, named "piezo-photocatalysis" has been developed based on piezoelectric photocatalysts by coupling the semiconductor, photoexcitation, and piezoelectric effect to achieve simple and efficient separation of electron hole pairs [14–17]. BaTiO₃ (BTO) is one of the most promising piezoelectric materials that can generate a strong internal field through crystal deformation under mechanical strain [18–20], which is considered a potential piezoelectric photocatalyst [21–24]. However, pure BTO can only absorb UV light due to its wide band gap (~3.6 eV). On the other hand, piezo-photocatalysis can be further improved by coupling with other photocatalytic enhancement methods, including heterojunctions, plasmons, and defect projects. In particular, the plasmons raised from metal nanoparticles loading on semiconductors with an appropriate amount are efficient to enhance the catalytic performance, which has been widely studied in photocatalytic degradation, nitrogen fixation, CO_2 reduction, and electrocatalysis [25–28].

In this work, BTO nanofibers are synthesized to study the photocatalytic and piezo-photocatalytic performances through degrading methyl orange (MO). Ag nanoparticles with different densities are coated on the BTO surface to further improve the catalytic performance. The BTO nanofibers with high Ag loading mass exhibit better photocatalytic performance, which is significantly improved by ultrasonication-induced piezoelectricity. To study the charge transfer paths, the BTO and Ag-BTO are excited by UV and visible light, respectively. Combining the simulation results, the energy transfer mechanism of piezo-photocatalysis of Ag-BTO is proposed.

2. Results and Discussion

2.1. Characterization of Ag-BTO Nanofibers

Figure 1a shows the SEM image of pure BTO nanofibers, illustrating the ultralong un-directional nanofibers with few fractures. Figure 1b is the SEM image of Ag-BTO-2 nanofibers, displaying a similar morphology with pure BTO. The low-resolution TEM image of Ag-BTO-2 nanofibers is shown in Figure 1c, depicting the uniform distribution of Ag nanoparticles on the whole surface of BTO nanofibers. The corresponding SAED pattern is shown in Figure 1d, where the diffraction rings depending on (110), (111), (200), (211), and (202) crystal planes of BTO are observed [29]. The HRTEM images of Ag-BTO-1 and Ag-BTO-2 are shown in Figure 1e,f, respectively. The distances of 0.235 and 0.422 nm belong to the (111) and (001) crystal planes of Ag and BTO, respectively. For Ag-BTO-2, the size of the Ag nanoparticle is equal to Ag-BTO-1, while the distribution density is double.

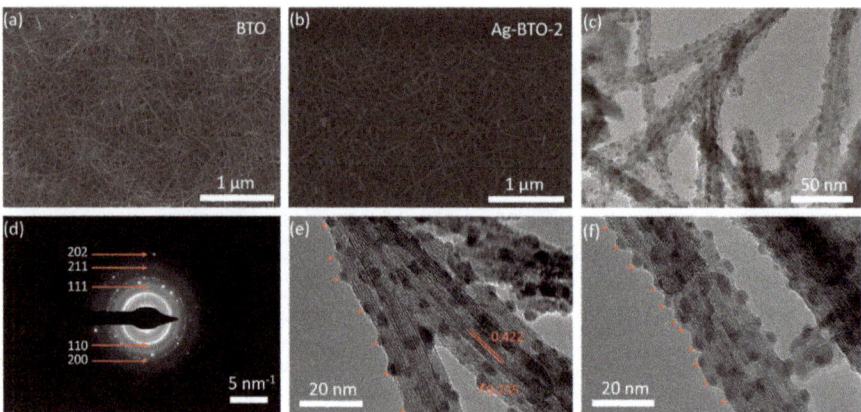

Figure 1. The characterizations of the samples. (**a,b**) SEM images of BTO and Ag-BTO nanofibers. (**c**) TEM image of Ag-BTO nanofibers. (**d**) SAED pattern of Ag-BTO nanofibers. (**e,f**) HRTEM images of Ag-BTO-1 and Ag-BTO-2.

The XRD and Raman curves of pure BTO, Ag-BTO-1, and Ag-BTO-2 are compared in Figure 2. As shown in Figure 2a, all the major diffraction peaks of the three samples are assigned to BTO (PDF No. 79-2265) [30]. Particularly, the peak splitting around 45° indicates the high purity of tetragonal BTO. The trace peak at 38.62° assigned to (111) plane of Ag crystal can only be observed in Ag-BTO-2, which is attributed to the low proportion of Ag nanoparticles. In Figure 2b, the three normalized Raman curves are basically the same, where the peaks at 265, 309, 517, and 716 cm^{-1} correspond to BTO [31]. A sharp peak appearing around 309 cm^{-1} further confirms the presence of tetragonal BTO. The DRS spectra of the three samples are shown in Figure 2c. Both the samples display two similar adsorption peaks in UV range around 200 and 300 nm. For Ag-BTO-1, a gentle peak appears around 500–700 nm due to the LSPR of Ag nanoparticles. For Ag-BTO-2, the intensity of the gentle peak triples, implying the LSPR of Ag nanoparticles drastically increases. To characterize the crystal structure of the synthesized BTO nanofibers in detail, the refined XRD pattern of pure BTO nanofibers is tested and analyzed by Rietveld refinement, as shown in Figure 3. The refined XRD pattern is indexed by tetragonal BaTiO$_3$ (piezoelectric, space group P4mm) and cubic BaTiO$_3$ (non-piezoelectric, space group Pm–3m) with a ratio of 2:1, as well as a small portion of BaCO$_3$. The refinement results are shown in Table 1, where the R_{wp} = 6.06%, R_p = 3.98%, and GOF = 3.32. The lattice parameters for tetragonal and cubic BTO are calculated to be a = 4.0150, c = 3.9984, and a = 4.0081 Å, respectively, which are very close to the standard data. These results suggest that the piezoelectric BaTiO$_3$ are successfully synthesized under a relatively low temperature.

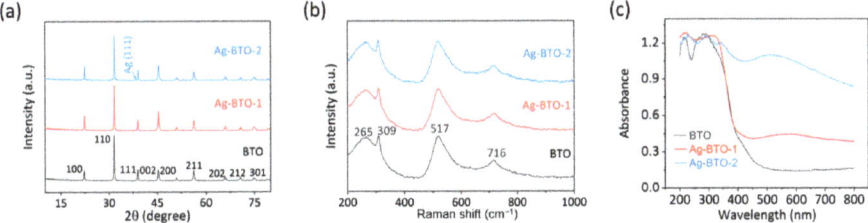

Figure 2. (**a**) XRD patterns, (**b**) Raman spectra, and (**c**) DRS spectra of the three samples.

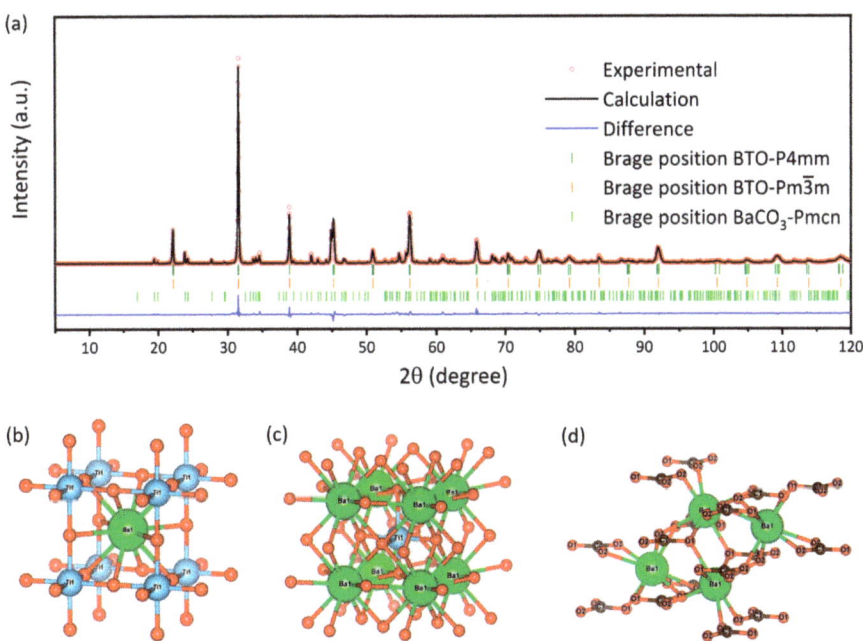

Figure 3. (**a**) Rietveld refined XRD pattern of the BTO nanofibers. (**b**–**d**) The refined crystal lattice structures of (**b**) BaTiO$_3$-Pm-3m, (**c**) BaTiO$_3$-P4mm, and (**d**) BaCO$_3$-Pmcn.

The XPS spectra of pure BTO and Ag-BTO-2 are almost the same, as shown in Figure 4. The survey spectra depict the presence of C, O, Ti, and Ba in pure BTO and C, O, Ti, Ba, and Ag in Ag-BTO, respectively (Figure 4a). The C 1s spectra are divided into three peaks around 284.80, 286.48, and 288.78 eV corresponding to the environmental C–C (or C–H) groups, CO_3^{2-} ions, and C–O groups, respectively (Figure 4b) [32]. The O 1s spectra are also divided into three peaks assigning BaTiO$_3$, CO_3^{2-} ions, and C–O groups, respectively

(Figure 4c). The Ba 3d spectra shows two peaks around 793.65 and 778.48 eV, which are assigned to Ba $3d_{3/2}$ and Ba $3d_{5/2}$, respectively. Furthermore, the peaks are well divided into two couple of peaks assigned to $BaTiO_3$ and $BaCO_3$, respectively (Figure 4d). Ti 2p photoelectron peaks reveal the purity of $BaTiO_3$ without TiO_2 or sodium titanates (Figure 4e). For Ag-BTO-2, the energy difference between Ag $3d_{3/2}$ (373.68) and $3d_{5/2}$ (367.67) is 6.01 eV, indicating the zero-valent state of Ag element (Figure 4f). These results imply that Ag nanoparticles are successfully loaded on the surface of BTO nanofibers without affecting the morphology, microstructure, and chemical state of BTO nanofibers.

Table 1. Rietveld refined cell parameters of the synthesized BTO nanofibers.

Cell Parameters	Tetragonal $BaTiO_3$	Cubic $BaTiO_3$	$BaCO_3$
Proportion	64.32%	32.20%	3.48%
Space Group	P4mm	Pm-3m	Pmcn
a (Å)	4.0150	4.0081	5.3130
b (Å)	4.0150	4.0081	8.9038
c (Å)	3.9984	4.0081	6.4361
α (°)	90	90	90
β (°)	90	90	90
γ (°)	90	90	90
Volume (Å^3)	64.45	64.392	304.47

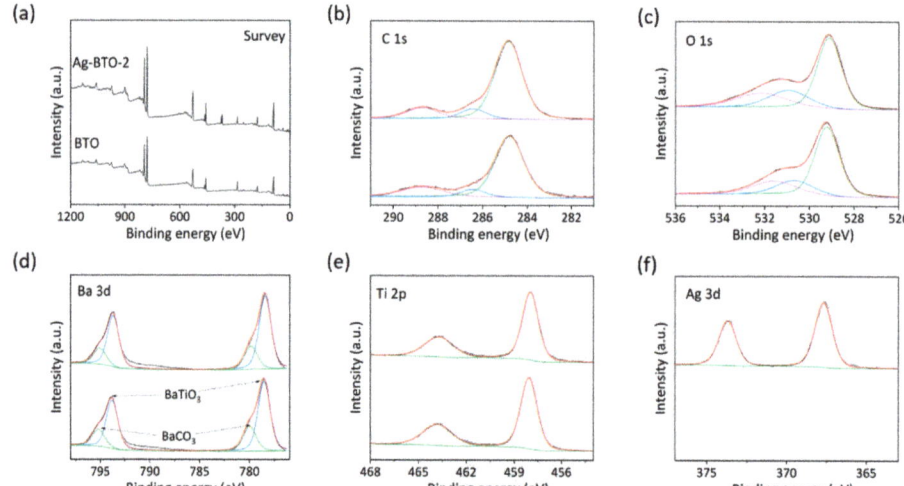

Figure 4. The XPS spectra of BTO and Ag-BTO. (**a**) Survey spectra. (**b**) C1s. (**c**) O 1s. (**d**) Ba 3d. (**e**) Ti 2p. (**f**) Ag 3d.

2.2. Piezo-Photocatalytic Property of Ag-BTO Nanofibers

The piezo-photocatalytic properties of the nanofibers are evaluated by degrading MO solution under ultrasonication and solar irradiation, as shown in Figure 5. Figure 5a–c show the MB degradation rates catalyzed by pure BTO, Ag-BTO-1, and Ag-BTO-2 under different conditions for 120 min, respectively. Under only ultrasonication, the degradation rates of the three samples are relatively low, indicating the ultrasound-induced piezoelectric field failed to efficiently degrade MO. Under light irradiation, the loading of Ag nanoparticles greatly improves the degradation rate, which further increases with the increasing Ag loading mass. The maximum degradation rate (82.7%) is achieved by Ag-BTO-2 under the cooperation of ultrasonication and light irradiation, suggesting the synergistic effect between piezoelectricity and photocatalysis. Figure 5d shows the degradation kinetic curves, indicating all the degradation processes follow the pseudo-first-order reaction kinetics.

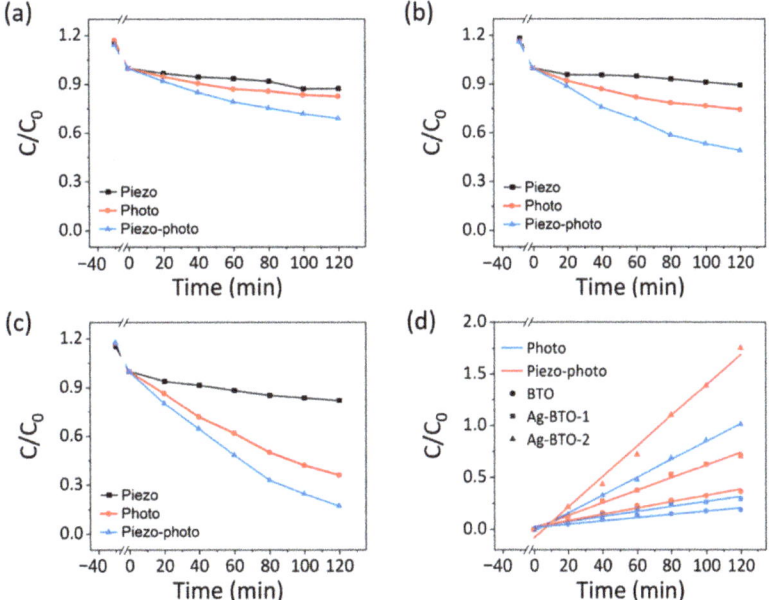

Figure 5. Catalytic performance comparison. (**a**–**c**) The piezocatalysis, photocatalysis, and piezo-photocatalysis of (**a**) BTO, (**b**) Ag-BTO-1, and (**c**) Ag-BTO-2. (**d**) The reaction kinetics.

To further study the function of the piezoelectricity on photocatalysis, the catalytic performances of BTO and Ag-BTO-2 samples are measured under UV and visible light, respectively, as shown in Figure 6. All the catalytic degradation rates of Ag-BTO-2 catalysts are higher than pure BTO. For pure BTO nanofibers (Figure 6a), the photocatalysis and piezo-catalysis under visible light are very weak (2.4% and 13.7%), and those under UV light are also limited (19.4% and 26.6%). As shown in Figure 6b, the UV-driven photocatalysis of the Ag-BTO-2 sample is moderately improved from 26.0% to 38.5% by piezoelectricity. In comparison, the visible-driven photocatalysis is developed from 32.4% to 51.7% by piezoelectricity, displaying a giant enhancement.

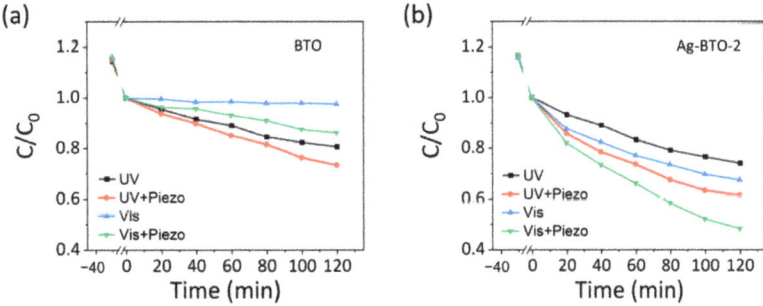

Figure 6. The photocatalysis and piezo-photocatalysis under UV and visible light of (**a**) BTO and (**b**) Ag-BTO-2.

The stability of the catalysts is shown in Figure 7. Figure 7a is a TEM image of Ag-BTO-2 after piezo-photocatalytic process, keeping the original morphology and nanoparticle-nanofiber structure. The corresponding XRD pattern is shown in Figure 7b, also showing similar data with the sample before piezo-photocatalysis. Particularly, the minor peak related to Ag nanoparticles is still obtained, and no Ag_2O peak is observed, indicating that the Ag nanoparticles maintain metallic state.

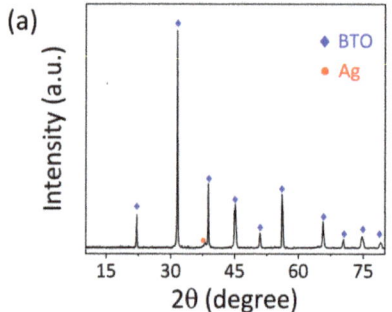 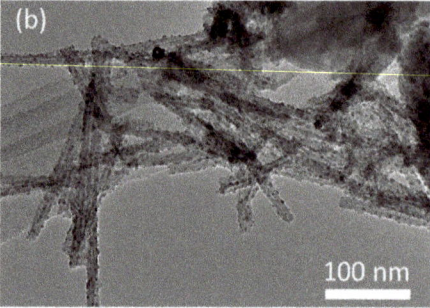

Figure 7. (**a**) XRD pattern and (**b**) TEM image of Ag-BTO-2 after piezo-photocatalysis.

2.3. Mechanism Analysis

Based on the above experimental results, the mechanisms of photocatalysis and piezo-photocatalysis of Ag-BTO nanofibers are developed. The possible photocatalytic mechanism of charge transfer in Ag-BTO is simply discussed. The generation and separation of photoexcited electron-hole pairs are the most important for photocatalysis. BTO nanofibers and Ag nanoparticles form Schottky contact with band offset (Figure 8a) [33]. BTO can absorb UV light to generate electron-hole pairs but cannot absorb visible light due to the wide bandgap (Figure 8b). Ag nanoparticles can absorb visible light through located surface plasmon resonance (LSPR) [34–36], but the UV absorption is very low due to the large detuning from resonance frequency (Figure 8c). As the Ag nanoparticles are synthesized on the BTO surface, there are three possible paths to enhance the solar-driven photocatalysis: the first is that the photogenerated electrons in BTO migrate to Ag nanoparticles due to the lower work function, leading to the separation of electron-hole pairs (Figure 8d) [37–39]; the second is that the LSPR-induced hot electrons in Ag nanoparticles migrate to BTO, increasing the number of free electrons in BTO (Figure 8e) [40–43]; the last is that the plasmons in Ag can transfer energy to BTO to generate electron-hole pairs, named "plasmon-induced resonance energy transfer" (PIRET) through dipole-dipole resonance (Figure 8f) [44–47]. In this case, the UV photocatalytic property of Ag-BTO is a little higher than that of BTO, ruling out the existence of the first path. On the other hand, the Ag-BTO samples exhibit good visible photocatalytic properties, indicating that the visible light is absorbed by Ag nanoparticles through LSPR and transfers hot electrons or energy to BTO, corresponding to the second or third path.

The LSPR electromagnetic fields of Ag-BTO-1 and Ag-BTO-2 are simulated by FEM, as shown in Figure 9a,b, respectively. As the distance of Ag nanoparticle arrays decreases from 10 to 5 nm, the average strength of the LSPR electromagnetic field at 550 nm increases by one order of magnitude. Thus, the photocatalysis and piezo-photocatalysis of Ag-BTO-2 are much higher than those of Ag-BTO-1. Under ultrasonication, BTO is deformed by ultrasound-induced cavitation blasting and generates a built-in piezoelectric field, which promotes the electron-hole separation in BTO and increases the Ag-BTO interface barrier (Figure 9c). As reported, the hot-electron injection path is significantly affected by the interface barrier, while the PERET path is a non-contact route rising superior to the interface barrier [48]. Under solar irradiation, the electron density of state around Ag nanoparticles is much higher than that in BTO, and the migrating of free electrons from BTO to Ag is restrained. The piezoelectric-enhanced interface barrier prevents the hot electrons from injecting into BTO. With a large number of Ag nanoparticles loading on the BTO surface, the first path and second path are both excluded during the piezo-photocatalysis process. Thus, the piezo-photocatalytic mechanism of Ag-BTO nanofibers is described below: under solar irradiation, BTO is excited by UV light to generate electron-hole pairs. The piezoelectric field contributes to the separation of electron-hole pairs and increases the Ag-BTO interface barrier. On the other hand, visible light is captured by high-density Ag nanoparticles through LSPR, transferring energy to further excite more electron-hole pairs for catalysis.

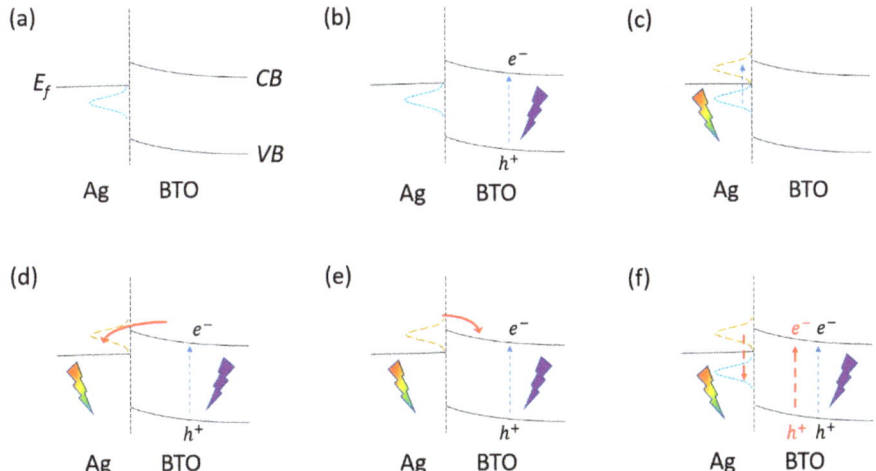

Figure 8. The schematic diagrams of charge transfer paths of Ag–BTO. (**a**) Schottky contact of Ag–BTO. (**b**) UV excitation. (**c**) Visible light excitation. (**d**–**f**) The transfer paths of (**d**) charge separation, (**e**) hot-electron injection, and (**f**) PIRET.

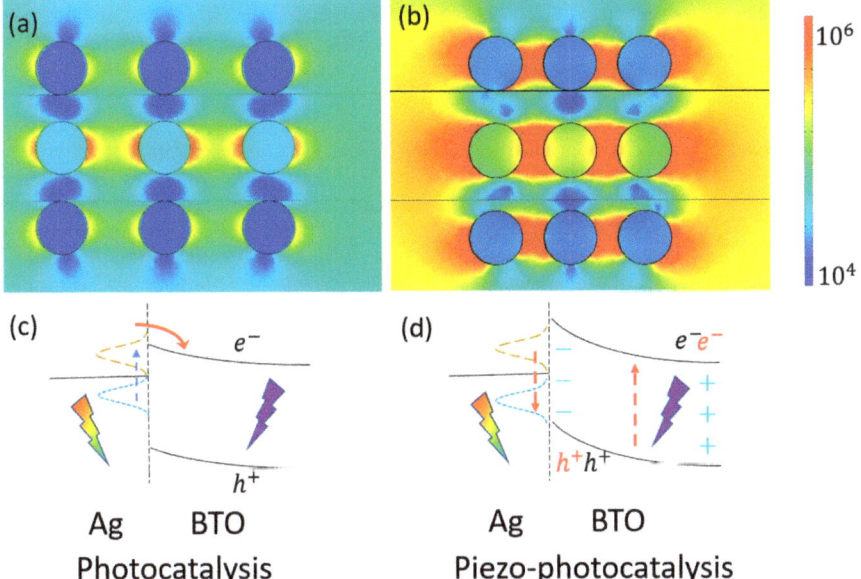

Figure 9. The proposed mechanism of piezo-photocatalysis. (**a**,**b**) The FEM simulation of the LSPR electromagnetic field of (**a**) Ag-BTO-1 and (**b**) Ag-BTO-2. (**c**) The hot-electron injection process in photocatalysis. (**d**) The PIRET process in piezo-photocatalysis.

3. Materials and Methods

3.1. Synthesis of BaTiO$_3$ Nanofibers

All reagents were analytical reagents and purchased from Sinophram Chemical Reagent, Shanghai, China. A piece of Ti foil (Haiyuan aluminum, Xining, China) was ultrasound-cleaned in deionized water/acetone for 10 min, heated to 650 °C with a heating rate of 10 °C min^{-1}, and maintained for 5 h in air. Then, the heat-treated Ti foil was immersed in 30 mL NaOH aqueous solution (12 M), sealed, and heated at 160 °C for 6 h, followed by immersion in Ba(OH)$_2$ aqueous solution (0.02 M) at 210 °C for 7 h. After the reaction, the white film was carefully peeled from the surface of Ti foil and ground in an agate mortar for 30 min. Finally, the powders were ultrasound-dispersed in deionized water for 10 min and centrifuged at 4000 rpm for 3 min. To synthesize Ag-BTO nanofibers, the sample was dispersed in 50 mL AgNO$_3$ ethanol solution (0.02 and 0.05 M) under 100 rpm stirring (JoanLab MS5s, Huzhou, China) and 300 W Hg lamp (PerfectLight CHF-XM 300, Beijing, China) irradiation for 2 h, labeled as Ag-BTO-1 and Ag-BTO-2, respectively.

3.2. Characterization and Measurements

The morphology and microstructure were obtained by scanning electron microscope (SEM, Hitachi S-8100, Tokyo. Japan), transmission electron microscope (TEM, FEI Tecnai F20, Portland, OR, USA), X-ray powder diffractometer (XRD, Rigaku SmartLab, Tokyo, Japan), Raman spectrometer (HORIBA LabRAM Nano, Montpellier, France), and X-ray photoelectron spectroscope (XPS, Thermo Scientific K-Alpha, Waltham, MA, USA). Conventional XRD patterns were measured with a sweeping rate of 10° min^{-1}. The refined XRD pattern of pure BTO nanofibers was tested in a slow rate of 0.5° min^{-1} with a wide sweeping range from 5° to 120°, and Rietveld refinement is employed by TOPAS academic. The photocatalysis and piezo-photocatalysis were evaluated by degrading MO in a mechano-photo reaction apparatus (homemade). A 300 W Xe lamp (PerfectLight, Microsolar 300, Beijing, China) was used to directly provide the light source. UV and visible light were obtained by short-pass and long-pass filters, respectively. The ultrasound was generated by a 120 W ultrasonic vibrator (DongSen DS-120STS, Shenzhen, China) with a frequency of 24 kHz. During the catalytic process, 10 mg catalyst was used to degrade 50 mL MO solution (10 mg L^{-1}). The MO concentration was determined by absorption spectroscopy at 464 nm.

4. Conclusions

In summary, the piezo-photocatalytic performance of BaTiO$_3$ nanofibers is enhanced by loading with Ag nanoparticles. FEM simulation indicates that the visible-light absorbance of Ag nanoparticles exponentially increases with the distribution density. With the loading of Ag nanoparticles, the piezo-photocatalytic MO degradation rate increases from 30.8% to 50.8%, and further increases to 82.7% with doubling Ag density. The UV-driven and visible-driven catalytic performances were individually measured to study the contribution of Ag nanoparticles and BTO nanofibers, displaying that the visible-driven catalysis dominates the enhancement during the piezo-photocatalytic process. Finally, the piezo-photocatalytic mechanism of Ag-BTO was discussed. Under solar irradiation, electron-hole pairs are generated in BTO by UV light, separated by the piezoelectric field, and enhanced by the high-density Ag nanoparticles through PIRET. This work develops the investigation of plasmon-improved piezo-photocatalysis.

Author Contributions: Conceptualization, Y.F. and J.M.; methodology, Z.R. and P.L.; investigation, P.C. and X.L.; writing—original draft preparation, P.C. and Y.L.; writing—review and editing, Y.F. and J.W.; visualization, P.C. and W.L.; supervision, Y.F. and J.M.; project administration, Y.F. and J.M.; funding acquisition, J.M. All authors have read and agreed to the published version of the manuscript.

Funding: This research was funded by National Key R&D Program of China (2017YFA0304203), the National Natural Science Foundation of China (62020106014, 62175140, 61901249, 92165106, 12104276); PCSIRT (IRT—17R70), 111 project (D18001), the Program for the Outstanding Innovative Teams of Higher Learning Institutions of Shanxi (OIT), the Applied Basic Research Project of Shanxi Province, China (201901D211191, 201901D211188), the Shanxi 1331 KSC, and the collaborative grant by the Russian Foundation for Basic Research and NSF of China (62011530047, 20-53-53025 in the RFBR classification).

Data Availability Statement: The data presented in this study are available on request from the corresponding author.

Acknowledgments: The authors would like to thank Zhenyu Duan from Shiyanjia Lab (www.shiyanjia.com) for the helps on material characterizations.

Conflicts of Interest: The authors declare no conflict of interest.

References

1. Chen, T.; Liu, L.Z.; Hu, C.; Huang, H.W. Recent advances on Bi_2WO_6-based photocatalysts for environmental and energy applications. *Chin. J. Catal.* **2021**, *42*, 1413–1438. [CrossRef]
2. Goktas, S.; Goktas, A. A comparative study on recent progress in efficient ZnO based nanocomposite and heterojunction photocatalysts: A review. *J. Alloys Compd.* **2021**, *863*, 158734. [CrossRef]
3. Kumar, A.; Raizada, P.; Hosseini-Bandegharaei, A.; Thakur, V.K.; Nguyen, V.; Singh, P. C-, N-Vacancy defect engineered polymeric carbon nitride towards photocatalysis: Viewpoints and challenges. *J. Mater. Chem. A* **2021**, *9*, 111–153. [CrossRef]
4. Li, Y.; Wu, Y.; Jiang, H.; Wang, H. In situ stable growth of Bi_2WO_6 on natural hematite for efficient antibiotic wastewater purification by photocatalytic activation of peroxymonosulfate. *Chem. Eng. J.* **2022**, *446*, 136704. [CrossRef]
5. Jin, P.; Wang, L.; Ma, X.; Lian, R.; Huang, J.; She, H.; Zhang, M.; Wang, Q. Construction of hierarchical $ZnIn_2S_4$@PCN-224 heterojunction for boosting photocatalytic performance in hydrogen production and degradation of tetracycline hydrochloride. *Appl. Catal. B Environ.* **2021**, *284*, 119762. [CrossRef]
6. Moradi, M.; Hasanvandian, F.; Isari, A.A.; Hayati, F.; Kakavandi, B.; Setayesh, S.R. CuO and ZnO co-anchored on g-C_3N_4 nanosheets as an affordable double Z-scheme nanocomposite for photocatalytic decontamination of amoxicillin. *Appl. Catal. B Environ.* **2021**, *285*, 119838. [CrossRef]
7. Wang, Z.; Jiang, L.; Wang, K.; Li, Y.; Zhang, G. Novel $AgI/BiSbO_4$ heterojunction for efficient photocatalytic degradation of organic pollutants under visible light: Interfacial electron transfer pathway, DFT calculation and degradation mechanism study. *J. Hazard. Mater.* **2021**, *410*, 124948. [CrossRef]
8. Bayan, E.M.; Lupeiko, T.G.; Pustovaya, L.E.; Volkova, M.G. Synthesis and photocatalytic properties of Sn-TiO_2 nanomaterials. *J. Adv. Dielectr.* **2020**, *10*, 2060018. [CrossRef]
9. Chen, J.; Li, G.; Lu, N.; Lin, H.; Zhou, S.; Liu, F. Anchoring cobalt single atoms on 2D covalent triazine framework with charge nanospatial separation for enhanced photocatalytic pollution degradation. *Mater. Today Chem.* **2022**, *24*, 100832. [CrossRef]
10. Hu, J.; Chen, Y.; Zhou, Y.; Zeng, L.; Huang, Y.; Lan, S.; Zhu, M. Piezo-enhanced charge carrier separation over plasmonic Au-BiOBr for piezo-photocatalytic carbamazepine removal. *Appl. Catal. B Environ.* **2022**, *311*, 121369. [CrossRef]
11. Idris, A.M.; Zheng, S.; Wu, L.; Zhou, S.; Lin, H.; Chen, Z.; Xu, L.; Wang, J.; Li, Z. A heterostructure of halide and oxide double perovskites $Cs_2AgBiBr_6/Sr_2FeNbO_6$ for boosting the charge separation toward high efficient photocatalytic CO_2 reduction under visible-light irradiation. *Chem. Eng. J.* **2022**, *446*, 137197. [CrossRef]
12. Liao, B.; Liao, X.; Xie, H.; Qin, Y.; Zhu, Y.; Yu, Y.; Hou, S.; Zhang, Y.; Fan, X. Built in electric field boosted photocatalytic performance in a ferroelectric layered material $SrBi_2Ta_2O_9$ with oriented facets: Charge separation and mechanism insights. *J. Mater. Sci. Technol.* **2022**, *123*, 222–233. [CrossRef]
13. Li, X.; Kang, B.; Dong, F.; Zhang, Z.; Luo, X.; Han, L.; Huang, J.; Feng, Z.; Chen, Z.; Xu, J.; et al. Enhanced photocatalytic degradation and H_2/H_2O_2 production performance of S-pCN/$WO_{2.72}$ S-scheme heterojunction with appropriate surface oxygen vacancies. *Nano Energy* **2021**, *81*, 105671. [CrossRef]
14. Dai, B.Y.; Biesold, G.M.; Zhang, M.; Zou, H.Y.; Ding, Y.; Wang, Z.L.; Lin, Z.Q. Piezo-phototronic effect on photocatalysis, solar cells, photodetectors and light-emitting diodes. *Chem. Soc. Rev.* **2021**, *50*, 13646–13691. [CrossRef] [PubMed]
15. Guo, L.X.; Chen, Y.D.; Ren, Z.Q.; Li, X.; Zhang, Q.W.; Wu, J.Z.; Li, Y.Q.; Liu, W.L.; Li, P.; Fu, Y.M.; et al. Morphology engineering of type-II heterojunction nanoarrays for improved sonophotocatalytic capability. *Ultrason. SonoChem.* **2021**, *81*, 105849. [CrossRef]
16. Zhang, C.X.; Lei, D.; Xie, C.F.; Hang, X.S.; He, C.A.X.; Jiang, H.L. Piezo-Photocatalysis over Metal-Organic Frameworks: Promoting Photocatalytic Activity by Piezoelectric Effect. *Adv. Mater.* **2021**, *33*, 2106308. [CrossRef]
17. Li, X.; Wang, W.; Dong, F.; Zhang, Z.; Han, L.; Luo, X.; Huang, J.; Feng, Z.; Chen, Z.; Jia, G.; et al. Recent Advances in Noncontact External-Field-Assisted Photocatalysis: From Fundamentals to Applications. *ACS Catal.* **2021**, *11*, 4739–4769. [CrossRef]
18. Marshall, J.; Walker, D.; Thomas, P. Bismuth zinc niobate: BZN-BT, a new lead-free $BaTiO_3$-based ferroelectric relaxor? *J. Adv. Dielectr.* **2020**, *10*, 2050033. [CrossRef]

19. Sidorenko, E.; Ngoc, C.T.B.; Prikhodko, G.; Natkhin, I.; Shloma, A.; Kharchenko, D. The constant electric field effect on the radio absorption of crystals BaTiO$_3$ and piezoceramics PCR-1. *J. Adv. Dielectr.* **2020**, *10*, 2060020. [CrossRef]
20. Yao, M.; Li, L.; Wang, Y.; Yang, D.; Miao, L.; Wang, H.; Liu, M.; Ren, K.; Fan, H.; Hu, D. Mechanical Energy Harvesting and Specific Potential Distribution of a Flexible Piezoelectric Nanogenerator Based on 2-D BaTiO$_3$-Oriented Polycrystals. *ACS Sustain. Chem. Eng.* **2022**, *10*, 3276–3287. [CrossRef]
21. Cheng, S.S.; Luo, Y.; Zhang, J.; Shi, R.; Wei, S.T.; Dong, K.J.; Liu, X.M.; Wu, S.L.; Wang, H.B. The highly effective therapy of ovarian cancer by Bismuth-doped oxygen-deficient BaTiO$_3$ with enhanced sono-piezocatalytic effects. *Chem. Eng. J.* **2022**, *442*, 136380. [CrossRef]
22. Fu, Y.M.; Ren, Z.Q.; Guo, L.X.; Li, X.; Li, Y.Q.; Liu, W.L.; Li, P.; Wu, J.Z.; Ma, J. Piezotronics boosted plasmonic localization and hot electron injection of coralline-like Ag/BaTiO$_3$ nanoarrays for photocatalytic application. *J. Mater. Chem. C* **2021**, *9*, 12596–12604. [CrossRef]
23. Liu, Q.; Li, Z.Y.; Li, J.; Zhan, F.Q.; Zhai, D.; Sun, Q.W.; Xiao, Z.D.; Luo, H.; Zhang, D. Three dimensional BaTiO$_3$ piezoelectric ceramics coated with TiO$_2$ nanoarray for high performance of piezo-photoelectric catalysis. *Nano Energy* **2022**, *98*, 107267. [CrossRef]
24. Tang, Q.; Wu, J.; Kim, D.; Franco, C.; Terzopoulou, A.; Veciana, A.; Puigmarti-Luis, J.; Chen, X.Z.; Nelson, B.J.; Pane, S. Enhanced Piezocatalytic Performance of BaTiO$_3$ Nanosheets with Highly Exposed {001} Facets. *Adv. Funct. Mater.* **2022**, *32*, 2202180. [CrossRef]
25. Sanchis-Gual, R.; Otero, T.F.; Coronado-Puchau, M.; Coronado, E. Enhancing the electrocatalytic activity and stability of Prussian blue analogues by increasing their electroactive sites through the introduction of Au nanoparticles. *Nanoscale* **2021**, *13*, 12676–12686. [CrossRef]
26. Li, X.; Jiang, H.P.; Ma, C.C.; Zhu, Z.; Song, X.H.; Wang, H.Q.; Huo, P.W.; Li, X.Y. Local surface plasma resonance effect enhanced Z-scheme ZnO/Au/g-C$_3$N$_4$ film photocatalyst for reduction of CO$_2$ to CO. *Appl. Catal. B-Environ.* **2021**, *283*, 119638. [CrossRef]
27. Li, S.J.; Wang, C.C.; Liu, Y.P.; Xue, B.; Jiang, W.; Liu, Y.; Mo, L.Y.; Chen, X.B. Photocatalytic degradation of antibiotics using a novel Ag/Ag$_2$S/Bi$_2$MoO$_6$ plasmonic p-n heterojunction photocatalyst: Mineralization activity, degradation pathways and boosted charge separation mechanism. *Chem. Eng. J.* **2021**, *415*, 128991. [CrossRef]
28. Chen, L.W.; Hao, Y.C.; Guo, Y.; Zhang, Q.H.; Li, J.N.; Gao, W.Y.; Ren, L.T.; Su, X.; Hu, L.Y.; Zhang, N.; et al. Metal-Organic Framework Membranes Encapsulating Gold Nanoparticles for Direct Plasmonic Photocatalytic Nitrogen Fixation. *J. Am. Chem. Soc.* **2021**, *143*, 5727–5736. [CrossRef]
29. Chen, T.; Meng, J.; Wu, S.; Pei, J.; Lin, Q.; Wei, X.; Li, J.; Zhang, Z. Room temperature synthesized BaTiO$_3$ for photocatalytic hydrogen evolution. *J. Alloys Compd.* **2018**, *754*, 184–189. [CrossRef]
30. On, D.V.; Vuong, L.D.; Chuong, T.V.; Quang, D.A.; Tuyen, H.V.; Tung, V.T. Influence of sintering behavior on the microstructure and electrical properties of BaTiO$_3$ lead-free ceramics from hydrothermal synthesized precursor nanoparticles. *J. Adv. Dielectr.* **2021**, *11*, 2150014. [CrossRef]
31. Charoonsuk, T.; Sriphan, S.; Nawanil, C.; Chanlek, N.; Vittayakorn, W.; Vittayakorn, N. Tetragonal BaTiO$_3$ nanowires: A template-free salt-flux-assisted synthesis and its piezoelectric response based on mechanical energy harvesting. *J. Mater. Chem. C* **2019**, *7*, 8277–8286. [CrossRef]
32. Miot, C.; Husson, E.; Proust, C.; Erre, R.; Coutures, J.P. Residual carbon evolution in BaTiO$_3$ ceramics studied by XPS after ion etching. *J. Eur. Ceram. Soc.* **1998**, *18*, 339–343. [CrossRef]
33. Nithya, P.M.; Devi, L.G. Effect of surface Ag metallization on the photocatalytic properties of BaTiO$_3$: Surface plasmon effect and variation in the Schottky barrier height. *Surf. Interfaces* **2019**, *15*, 205–215. [CrossRef]
34. Chen, Z.; Chen, M.; Yan, H.; Zhou, P.; Chen, X. Enhanced solar thermal conversion performance of plasmonic gold dimer nanofluids. *Appl. Therm. Eng.* **2020**, *178*, 115561. [CrossRef]
35. Liu, X.-D.; Chen, B.; Wang, G.-G.; Ma, S.; Cheng, L.; Liu, W.; Zhou, L.; Wang, Q.-Q. Controlled Growth of Hierarchical Bi$_2$Se$_3$/CdSe-Au Nanorods with Optimized Photothermal Conversion and Demonstrations in Photothermal Therapy. *Adv. Funct. Mater.* **2021**, *31*, 2104424. [CrossRef]
36. Sun, Y.; Hou, P.; Wu, S.; Yu, L.; Dong, L. The enhanced photocatalytic activity of Ag-Fe$_2$O$_3$-TiO$_2$ performed in Z-scheme route associated with localized surface plasmon resonance effect. *Colloids Surf. A* **2021**, *628*, 127304. [CrossRef]
37. Humayun, M.; Ullah, H.; Cheng, Z.-E.; Tahir, A.A.; Luo, W.; Wang, C. Au surface plasmon resonance promoted charge transfer in Z-scheme system enables exceptional photocatalytic hydrogen evolution. *Appl. Catal. B Environ.* **2022**, *310*, 121322. [CrossRef]
38. Xu, Q.; Knezevic, M.; Laachachi, A.; Franger, S.; Colbeau-Justin, C.; Ghazzal, M.N. Insight into Interfacial Charge Transfer during Photocatalytic H$_2$ Evolution through Fe, Ni, Cu and Au Embedded in a Mesoporous TiO$_2$@SiO$_2$ Core-shell. *ChemCatChem* **2022**, *14*, e202200102. [CrossRef]
39. Zhang, J.; Gu, H.; Wang, X.; Zhang, H.; Chang, S.; Li, Q.; Dai, W.-L. Robust S-scheme hierarchical Au-ZnIn$_2$S$_4$/NaTaO$_3$: Facile synthesis, superior photocatalytic H$_2$ production and its charge transfer mechanism. *J. Colloid Interface Sci.* **2022**, *625*, 785–799. [CrossRef]
40. Berdakin, M.; Soldano, G.; Bonafe, F.P.; Liubov, V.; Aradi, B.; Frauenheim, T.; Sanchez, C.G. Dynamical evolution of the Schottky barrier as a determinant contribution to electron-hole pair stabilization and photocatalysis of plasmon-induced hot carriers. *Nanoscale* **2022**, *14*, 2816–2825. [CrossRef]

41. Gai, Q.; Ren, S.; Zheng, X.; Liu, W.; Dong, Q. Enhanced photocatalytic performance of Ag/CdS by L-cysteine functionalization: Combination of introduced co-catalytic groups and optimized injection of hot electrons. *Appl. Surf. Sci.* **2022**, *579*, 151838. [CrossRef]
42. Manuel, A.P.; Shankar, K. Hot Electrons in TiO$_2$-Noble Metal Nano-Heterojunctions: Fundamental Science and Applications in Photocatalysis. *NanoMater* **2021**, *11*, 1249. [CrossRef] [PubMed]
43. Yuan, X.; Zhen, W.; Yu, S.; Xue, C. Plasmon Coupling-Induced Hot Electrons for Photocatalytic Hydrogen Generation. *Chem. Asian J.* **2021**, *16*, 3683–3688. [CrossRef] [PubMed]
44. Choi, Y.M.; Lee, B.W.; Jung, M.S.; Han, H.S.; Kim, S.H.; Chen, K.; Kim, D.H.; Heinz, T.F.; Fan, S.; Lee, J.; et al. Retarded Charge-Carrier Recombination in Photoelectrochemical Cells from Plasmon-Induced Resonance Energy Transfer. *Adv. Energy Mater.* **2020**, *10*, 2000570. [CrossRef]
45. Jia, H.; Wong, Y.L.; Wang, B.; Xing, G.; Tsoi, C.C.; Wang, M.; Zhang, W.; Jian, A.; Sang, S.; Lei, D.; et al. Enhanced solar water splitting using plasmon-induced resonance energy transfer and unidirectional charge carrier transport. *Opt. Express* **2021**, *29*, 34810–34825. [CrossRef]
46. Li, J.; Cushing, S.K.; Meng, F.; Senty, T.R.; Bristow, A.D.; Wu, N. Plasmon-induced resonance energy transfer for solar energy conversion. *Nat. Photonics* **2015**, *9*, 601–607. [CrossRef]
47. Ma, J.; Liu, X.; Wang, R.; Zhang, F.; Tu, G. Plasmon-induced near-field and resonance energy transfer enhancement of photodegradation activity by Au wrapped CuS dual-chain. *Nano Res.* **2022**, *15*, 5671–5677. [CrossRef]
48. Kohan, M.G.; You, S.; Camellini, A.; Concina, I.; Zavelani-Rossi, M.; Vomiero, A. Optical field coupling in ZnO nanorods decorated with silver plasmonic nanoparticles. *J. Mater. Chem. C* **2021**, *9*, 15452–15462. [CrossRef]

Review

Recent Advances in g-C₃N₄-Based Photocatalysts for NOₓ Removal

Zhanyong Gu [1,*], Mengdie Jin [1], Xin Wang [1], Ruotong Zhi [1], Zhenghao Hou [1], Jing Yang [1], Hongfang Hao [1], Shaoyan Zhang [1], Xionglei Wang [2], Erpeng Zhou [1,*] and Shu Yin [3,4,*]

[1] Shijiazhuang Key Laboratory of Low-Carbon Energy Materials, College of Chemical Engineering, Shijiazhuang University, Shijiazhuang 050035, China
[2] School of Material Science and Engineering, Hebei University of Engineering, Handan 056038, China
[3] Institute of Multidisciplinary Research for Advanced Materials (IMRAM), Tohoku University, 2-1-1, Katahira, Aoba-ku, Sendai 980-8577, Japan
[4] Advanced Institute for Materials Research (WPI-AIMR), Tohoku University, 2-1-1, Katahira, Aoba-ku, Sendai 980-8577, Japan
* Correspondence: gyz030201@163.com (Z.G.); zhouep@126.com (E.Z.); yin.shu.b5@tohoku.ac.jp (S.Y.)

Abstract: Nitrogen oxides (NOₓ) pollutants can cause a series of environmental issues, such as acid rain, ground-level ozone pollution, photochemical smog and global warming. Photocatalysis is supposed to be a promising technology to solve NOₓ pollution. Graphitic carbon nitride (g-C₃N₄) as a metal-free photocatalyst has attracted much attention since 2009. However, the pristine g-C₃N₄ suffers from poor response to visible light, rapid charge carrier recombination, small specific surface areas and few active sites, which results in deficient solar light efficiency and unsatisfactory photocatalytic performance. In this review, we summarize and highlight the recent advances in g-C₃N₄-based photocatalysts for photocatalytic NOₓ removal. Firstly, we attempt to elucidate the mechanism of the photocatalytic NOₓ removal process and introduce the metal-free g-C₃N₄ photocatalyst. Then, different kinds of modification strategies to enhance the photocatalytic NOₓ removal performance of g-C₃N₄-based photocatalysts are summarized and discussed in detail. Finally, we propose the significant challenges and future research topics on g-C₃N₄-based photocatalysts for photocatalytic NOₓ removal, which should be further investigated and resolved in this interesting research field.

Keywords: photocatalysis; g-C₃N₄; NOₓ; mechanism; modification strategy

1. Introduction

With the fast development of the economy and modern industrialization, environmental pollution and the energy crisis have become the two major challenges in the world [1–7]. Air pollution is one of the serious environmental problems [8,9]. There are a variety of air pollutants including sulfur oxides (SOₓ), nitrogen oxides (NOₓ), carbon monoxide (CO) and so on [10,11]. Air pollution would bring about a number of environmental issues: it can damage and corrode buildings and equipment; it is harmful to human beings, animals and vegetables; and it can lead to ecological deterioration of the environment. NOₓ is especially serious among air pollutants [12–18]. To be more specific, as shown in Figure 1, NOₓ pollutants result in ground-level ozone pollution, acid rain, photochemical smog, global warming and so on. Moreover, it could cause damage to human health and increase the risk of diseases such as emphysema, bronchitis and respiratory disease. If human beings want to realize sustainable development and have a bright future, we must solve air pollution as soon as possible.

An enormous amount of research work has been carried out to deal with NOₓ pollution [19–22]. For example, traditional adsorption, filtration and selective catalytic reduction technologies have been extensively used to solve NOₓ pollution [23–25]. However, these technologies have low efficiency to remove the low-concentration NOₓ pollutants and also have disposal and regeneration issues [26]. Recently, photocatalysis as a green technology has attracted a substantial amount of attention [8,27]. Compared with traditional physical or chemical methods, photocatalytic NO_x removal technology is more efficient and environment-friendly with a semiconductor as the photocatalyst and solar energy as the driving force. Photocatalytic technology can convert the low concentration of NO_x in the atmosphere into non-toxic products without disposal and regeneration issues. The greatest challenge in the photocatalytic process is to develop low-cost and highly efficient photocatalysts, which can make full use of solar energy to remediate environmental problems. Unveiling the photocatalytic mechanism is the premise of developing highly effective photocatalysts. Since Fujishima and Honda reported photoelectrochemical water splitting in 1972 [28], great progress has been made to unravel the photocatalysis mechanism [29–31]. Typically, as shown in Scheme 1, the photocatalytic process includes three consecutive steps [32–34]: (1) light harvesting using a semiconductor photocatalyst for charge carrier excitation; (2) photogenerated electron-hole pairs separation and migration and (3) surface reduction and oxidation reactions. It should be pointed out that a large number of photogenerated electrons and holes recombine during the charge carrier separation and migration process, i.e., volume recombination and surface recombination. Only the effective charge carriers, in other words, the remaining charge carriers, can attend the surface reduction and oxidation reactions [35–37]. Therefore, if we attempt to improve the photocatalytic performance of the photocatalysts, we can enhance the light-harvesting capability, promote the charge carrier separation efficiency and facilitate the surface reduction and oxidation reactions.

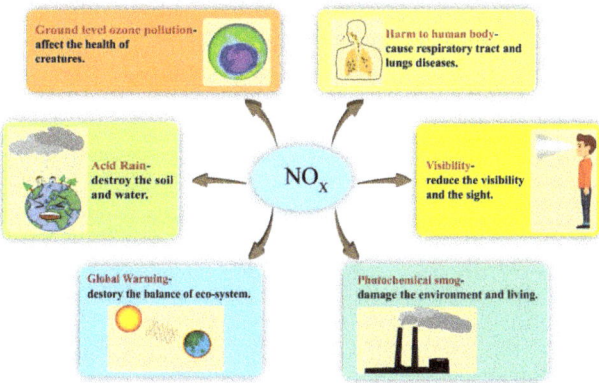

Figure 1. The environmental issues caused by NO_x pollutants.

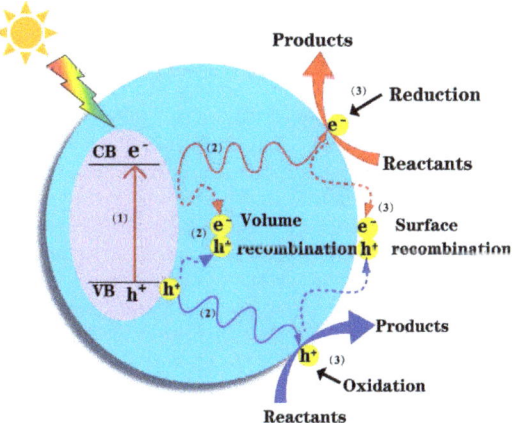

Scheme 1. The basic principle of photocatalysis over semiconductor photocatalyst.

As one of the typical photocatalytic processes, the reaction mechanism of photocatalytic NO_x removal has been carefully discussed. When the photocatalyst is irradiated by light with equal or greater energy than the bandgap energy of the photocatalyst, the electrons (e^-) can be excited from the valence band to the conduction band, resulting in positive holes (h^+) in the valance band [8,38]. It is well-understood that oxygen molecules are the second most abundant gases, consisting of 21% of the atmosphere. The oxygen molecules absorbed on the surface would be reduced, which brought about the superoxide anion radicals ($\cdot O_2^-$) [39]. At the same time, holes (h^+) in the valance band can oxide the water molecules and hydroxide ion (OH^-) to obtain hydroxyl radicals ($\cdot OH$) [10,39]. It should be noted that these radicals play significant roles during the photocatalytic NO_x removal process. To be more specific, the NO absorbed on the surface of a photocatalyst would be oxidized by superoxide radicals ($\cdot O_2^-$) and/or hydroxyl radicals ($\cdot OH$) and/or positive holes (h^+) to become the final products of the nitrate ion (NO_3^-), through the formation of intermediates related to NO_2 and HNO_2. These series of reactions are illustrated as follows [10,24,39–42]:

$$\text{Photocatalyst} + h\nu \rightarrow e^- + h^+ \tag{1}$$

$$O_2 + e^- \rightarrow \cdot O_2^- \tag{2}$$

$$h^+ + OH^- \rightarrow \cdot OH \tag{3}$$

$$NO_x + \cdot O_2^- \rightarrow NO_3^- \tag{4}$$

$$NO + \cdot OH \rightarrow HNO_2 \tag{5}$$

$$HNO_2 + \cdot OH \rightarrow NO_2 + H_2O \tag{6}$$

$$NO_2 + \cdot OH \rightarrow NO_3^- \tag{7}$$

It is universally acknowledged that the key point of photocatalysis is to develop highly effective and stable photocatalysts. Since the pioneering work of Fujishima and Honda in 1972 [28], a great number of photocatalysts have developed to be used in energy conversion and environmental remediation [1,21,43–50]. The reported photocatalysts include metal oxides such as Fe_2O_3 [51,52], WO_3 [53], TiO_2 [54] and ZnO [55]; metal sulfides such as ZnS [56], CdS [57] and CuS [58]; bismuth-based photocatalysts such as BiOX (X = I, Br, Cl) [59] and Bi_2MO_6 (M = Mo, W) [60]; metal-organic frameworks (MOFs) [61]; covalent organic frameworks (COFs) [62] and many more. Among these semiconductor photocatalysts, TiO_2 is considered the most classical photocatalyst owing to its many advantages such as non-toxicity, low cost, long-term stability and so on [63,64]. However, the wide bandgap of TiO_2 (3.0–3.2 eV) constrains the photocatalytic response to only ultraviolet (UV) light [65]. As we know, the full solar spectrum consists of a near-infrared region (52%), a visible-light region (43%) and a UV region (5%) [66]. Therefore, metal and non-metal doping are performed to reduce the bandgap to make the best use of solar energy [67,68].

Recently, graphitic carbon nitride (g-C_3N_4), as a metal-free semiconductor photocatalyst, has attracted much attention since Wang et al. reported it could be used for H_2 production in 2009 [31]. The g-C_3N_4 exhibited outstanding optical, electrical and structural properties, such as visible-light response due to the suitable bandgap (2.7 eV), nanoscale thickness and high surface-to-volume ratio owing to the two-dimensional (2D) structure, facile synthesis with a cheap precursor, such as melamine, dicyandiamide, urea thiourea and cyanamide (Figure 2a,b), high thermal and chemical stability and nontoxic nature [32,38]. What is more, the electronic structures of the 2D materials could be modulated by controlling the thickness or doping strategy. Thanks to its excellent advantages, g-C_3N_4 has been used in different fields including energy issues and environmental remediation. To be more specific, it has been reported in supercapacitors [69], electrocatalysis [70], photo-electro catalytic reactions [71], N_2 fixation [72], pollutant degradation [73], CO_2 reduction [74], water splitting [30], organic catalysis [75] and sensing [76].

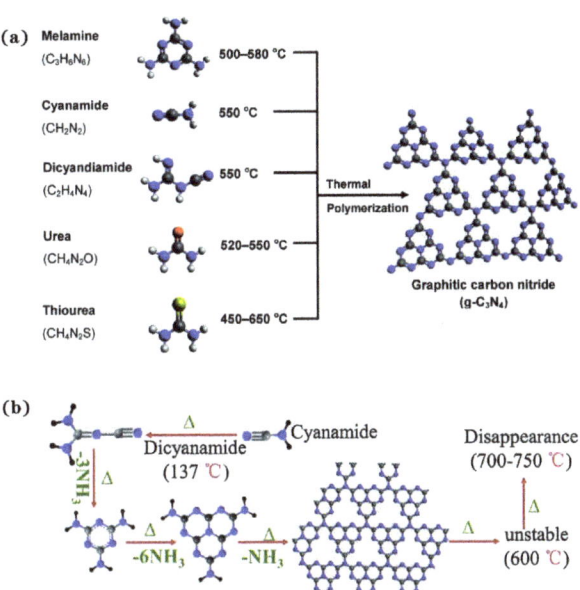

Figure 2. (**a**) Synthesis of g–C$_3$N$_4$ using thermal polymerization with different precursors [32]. Copyright 2016, American Chemical Society. (**b**) reaction pathway of g–C$_3$N$_4$ using cyanamide as a precursor [38]. Copyright 2017, Elsevier.

The long history of g-C$_3$N$_4$ could trace back to 1834 when Berzelius and Liebig prepared it by igniting mercuric thiocyanate [77,78]. Since then, g-C$_3$N$_4$ has become a hot research area. g-C$_3$N$_4$ is normally considered as three basic structures: triazine-based g-C$_3$N$_4$, heptazine-based g-C$_3$N$_4$ and triazine and heptazine mixed g-C$_3$N$_4$ [32,38]. Figure 3a,b shows the typical structures of triazine-based g-C$_3$N$_4$ and heptazine-based g-C$_3$N$_4$, respectively. However, it should be noted that heptazine-based g-C$_3$N$_4$ is the most stable phase at ambient conditions according to the first-principles density functional theory (DFT) calculations carried out by Kroke et al. [79]. Therefore, more and more researchers and scientists tend to recognize heptazine as the building block for the formation of g-C$_3$N$_4$.

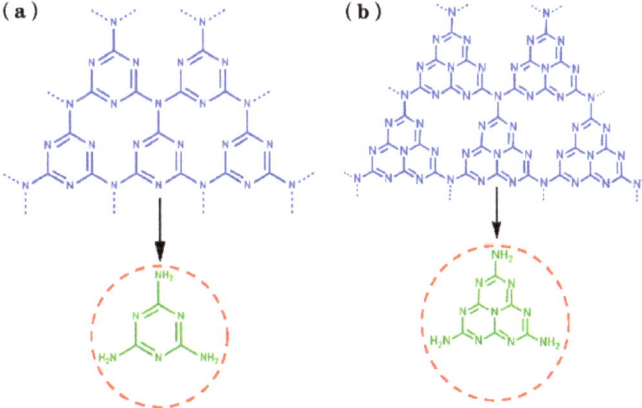

Figure 3. (**a**)Triazine and (**b**) tris–s–triazine (heptazine) structures of g–C$_3$N$_4$.

2. Modification Strategies of Pristine g-C_3N_4

g-C_3N_4, as the significant metal-free semiconductor photocatalyst, holds great potential in the application of the photocatalytic NO_x removal process due to its plentiful extraordinary advantages, such as visible light response properties, mild bandgap, low cost, facile preparation and high thermal stability. However, pristine g-C_3N_4 prepared using the traditional high-temperature solid reaction suffers from low specific surface areas and low crystallinity owing to kinetic hindrance, which results in small specific surface areas, few reactive sites, limited light-harvesting capacity, rapid recombination of photogenerated charge carriers and unsatisfactory photocatalytic NO_x removal performance. In order to improve the photocatalytic performance of pristine g-C_3N_4, a variety of modification strategies have been developed including metal doping, non-metal doping, defect engineering, crystallinity optimization, morphology controlling and heterojunction construction.

2.1. Morphology Controlling

Morphology controlling is considered a promising strategy to improve the photocatalytic performance of bulk g-C_3N_4 [80]. Since bulk g-C_3N_4 is synthesized using the high-temperature solid reaction, it suffers from low specific surface areas and few active sites, which is detrimental to photocatalytic performance. Moreover, bulk g-C_3N_4 exhibits a longer charge carrier migration distance, and thus the photogenerated electron and hole pairs achieve rapid charge recombination. In addition, bulk g-C_3N_4 is unfavorable for molecular mass transport, surface redox reactions and light harvesting in comparison with porous g-C_3N_4 [32,81]. In order to enlarge the specific surface areas, increase the active sites, promote the charge carrier separation efficiency and facilitate the molecular mass transport, much progress has been made such as the exfoliation of bulk g-C_3N_4 into nanosheets, template strategy including hard-template and soft-template and supramolecular preorganization method.

2.1.1. Nanosheets Structure

Inspired by the preparation of graphene nanosheets [82,83], scientific researchers attempt to exfoliate bulk g-C_3N_4 into nanosheets. Compared with bulk g-C_3N_4, g-C_3N_4 nanosheets exhibit a great deal of distinct benefits owing to morphology changes. It not only enlarges the specific surface areas and increases the active sites but also shortens the charge carrier's transport distance, improves the solubility and modifies the electronic structures owing to the famous quantum confine effect. More specifically, the photogenerated electrons and holes coming from the g-C_3N_4 nanosheets can easily migrate to the surface of the photocatalysts to attend the surface reactions through the shortened paths. This phenomenon is instrumental in facilitating the charge carrier separation efficiency to improve photocatalytic performance. In addition, the enlarged bandgap of g-C_3N_4 nanosheets leads to enhanced oxidation potential energy and reduced potential energy, which is useful for surface reactions.

Generally, the g-C_3N_4 nanosheet structures could be achieved using two different strategies, i.e., liquid exfoliation of bulk g-C_3N_4 and thermal exfoliation of bulk g-C_3N_4. Various solvents with suitable surface energy, such as water, methanol, ethanol, N-methylpyrrolidone (NMP), 1-isopropanol (IPA), acetone and their mixtures, have been used to overcome the weak van der Waals forces between the two adjacent layers of bulk g-C_3N_4 using facile sonication. For example, Xie et al. reported a green liquid exfoliation strategy to obtain ultrathin nanosheets using cheap and environmentally friendly water as the solvent (Figure 4) [84]. The thickness of the exfoliated nanosheet is about 2.5 nm in height (around seven layers) with the size distribution ranging from 70 nm to 160 nm. In addition, Zhu et al. reported a concentrated H_2SO_4 (98%) assisted liquid exfoliation strategy to fabricate a single atomic layer of g-C_3N_4 ultrathin nanosheets [85]. The intercalation of concentrated H_2SO_4 (98%) into the interplanar spacing of bulk g-C_3N_4 resulted in the graphene-like single-layer g-C_3N_4 structure with a small thickness of 0.4 nm and a large size of micrometers.

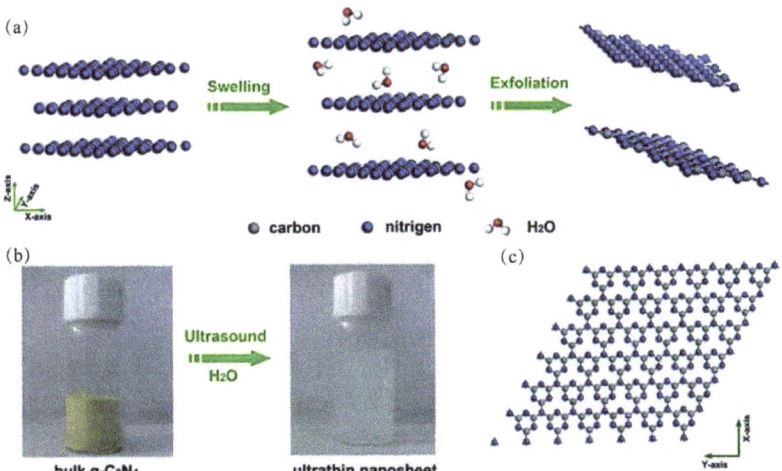

Figure 4. (a–c) Schematic illustration for the liquid-exfoliation process of bulk g–C_3N_4 [84]. Copyright 2012, American Chemical Society.

Compared with the liquid exfoliation strategy, thermal exfoliation is more facile and more environmentally friendly because it does not involve toxic solutions such as aqueous ammonia, hydrochloric acid and concentrated H_2SO_4 (98%). The thermal exfoliation approach is fast, low-cost and low-pollution. However, the largest drawback of the thermal exfoliation strategy is the low yield due to thermal oxidation and thermal etching. For instance, Niu et al. obtained g-C_3N_4 nanosheets with a thickness of 2 nm (about six to seven layers) using the thermal exfoliation strategy [86]. The synthesized g-C_3N_4 nanosheets exhibited enhanced photocatalytic H_2 evolution under simulated solar light irradiation. The excellent photocatalytic H_2 production of the obtained g-C_3N_4 nanosheets was ascribed to the large specific surface area, low sheet thickness, enlarged band gap, increased electron-transport ability and prolonged lifetime of the charge carriers. In addition, Gu et al. reported that bulk g-C_3N_4 could be exfoliated into nanosheets to increase the specific surface areas and active sites using facile post-thermal treatment [87]. At the same time, the electronic structure of bulk g-C_3N_4 was optimized during the calcination process. The valence band of g-C_3N_4 nanosheets was increased owing to the quantum confinement effect and nitrogen vacancy, which led to the higher thermodynamic driving force during the photocatalytic NO_x removal process. The g-C_3N_4 nanosheets showed about 3.0 times higher photocatalytic NO_x removal performance than pristine g-C_3N_4 owing to the enlarged specific surface areas and optimized electronic structure. In addition, the impact of calcination temperature, calcination time and sample amount on the photocatalytic NO_x removal performance and the yield of g-C_3N_4 nanosheets were systematically studied. This research work provides new insight into the thermal exfoliation approach for enhancing the photocatalytic NO_x removal performance.

2.1.2. Porous Structure

The template strategy is an effective approach to fabricating porous nanostructured g-C_3N_4, which can increase the specific surface areas and active sites of bulk g-C_3N_4 [88,89]. Moreover, the high porosity of nanostructured g-C_3N_4 is beneficial for mass and gas transport. In addition, the voluminous void space in nanostructured g-C_3N_4 can enhance the light absorption efficiency owing to the light trapping effect. These plentiful advantages can bring about outstanding photocatalytic performance in comparison with the bulk counterpart [89]. In general, the template method is based on the use of inorganic or organic nanostructures as a template, i.e., a hard template and a soft template [90].

A hard template is a controllable and precious strategy to prepare nanostructured g-C_3N_4. The hard template method, in other words, solid material nano-casting, is performed using a physical structure agent to control the porous nanostructured g-C_3N_4. Up to now, a large number of hard templates have been studied. For example, Zhang et al. used HCl-treated SBA-15 silica as a hard template to prepare ordered mesoporous g-C_3N_4 [91]. The obtained mesoporous g-C_3N_4 displayed significantly enlarged specific surface area and pore volume, which were 517 m^2 g^{-1} and 0.49 cm^3 g^{-1}, respectively. Similarly, Sun et al. reported hollow nanospheres of g-C_3N_4 when using a silica-based hard template [92]. The hollow nanospheres of g-C_3N_4 displayed excellent photocatalytic H_2 evolution performance due to the hollow sphere structure. However, the biggest drawback of the silica-based hard template involves toxic reagents, such as ammonium hydrogen difluoride (NH_4HF_2), when we remove the silica-based hard template. Recently, Zhang et al. demonstrated that low-cost calcium carbonate ($CaCO_3$) is a promising environmentally friendly hard template [93]. After the $CaCO_3$ is removed using hydrochloric acid treatment, porous g-C_3N_4 was successfully prepared.

Since hard templates involve hazardous fluoride-containing reagents, a tremendous amount of work has been completed on the soft template [94]. The key point of a soft template is the molecular self-assembly process, which can chemically tailor the porosity and morphology of pristine g-C_3N_4. Various templates, such as non-ionic surfactants and amphiphilic block polymers, could be chosen as soft templates. In addition, ionic liquids have demonstrated an effective soft template [32,38]. The soft templates provide a facile and more environmentally friendly strategy to prepare nanostructured g-C_3N_4 [95]. Very recently, supramolecular preorganization has become an interesting topic to prepare nanostructured g-C_3N_4. Figure 5 shows the formation of the self-assembled structures used to prepare nanostructured g-C_3N_4 [75]. To some extent, the supramolecular preorganization strategy is similar to the soft template strategy. However, this strategy is based on the supramolecular interactions of g-C_3N_4 monomers, including hydrogen bonds, the π–π bond and so on. For instance, Zhang et al. reported a solvent-assisted strategy to prepare porous g-C_3N_4 with enhanced visible-light photocatalytic NO removal performance (Figure 6a) [96]. g-C_3N_4 prepared with the addition of water and ethanol exhibited significantly improved visible-light photocatalytic performance, with NO removal percentages of 37.2% and 48.3%, respectively (Figure 6b,c). The enhanced photocatalytic NO_x removal performance was ascribed to the unique microstructure and prolonged lifetime of the charge carriers.

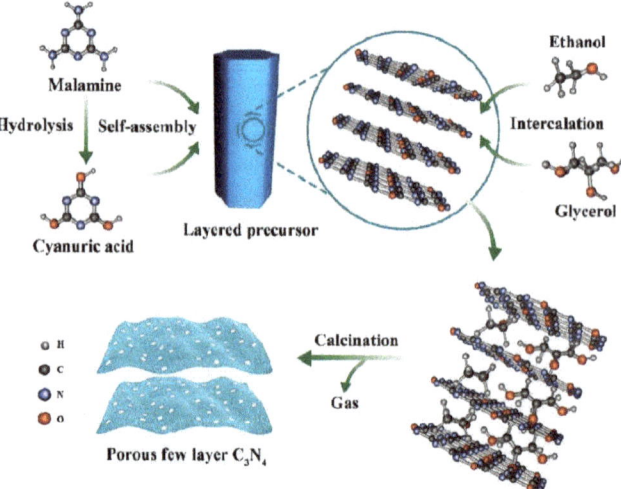

Figure 5. Schematic illustration for the self−assembled method to prepare few-layer porous structures of g−C_3N_4 [75]. Copyright 2019, American Chemical Society.

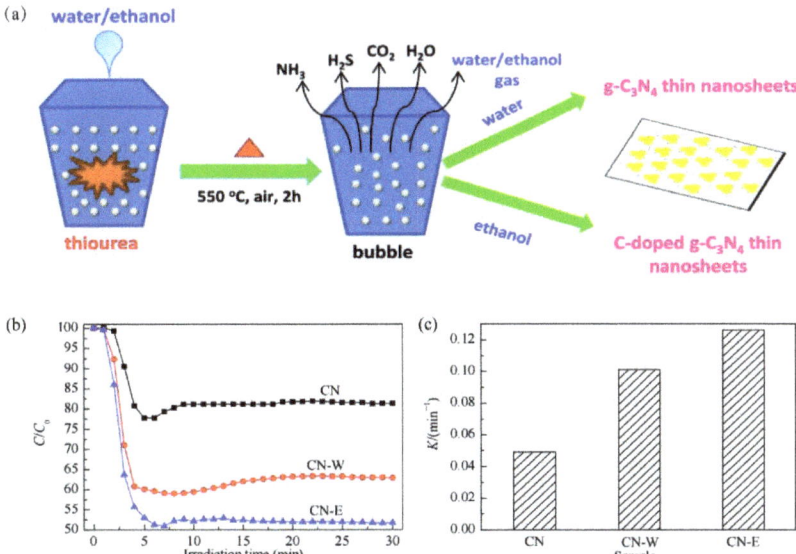

Figure 6. (a) Schematic illustration for preparation of porous $g-C_3N_4$ with the addition of water and ethanol. Photocatalytic performance (b) and Arrhenius rate constants (c) of $g-C_3N_4$, $g-C_3N_4-W$ and $g-C_3N_4-E$ for the removal of NO irradiated under visible light [96]. Copyright 2017, Elsevier.

2.2. Band Structure Engineering

The band structure of photocatalysts plays a crucial role in the photocatalytic process. The optimized band structure can absorb more solar energy to generate more electron-hole pairs; improve the charge carrier separation efficiency to obtain more effective electrons and holes for the surface reactions; and optimize the reaction sites and promote the adsorption of intermediates to improve the surface reactions. Up to now, tremendous efforts have been devoted to modulating the electronic structure of $g-C_3N_4$. The strategies of band structure engineering can be roughly divided into two categories: metal element doping and non-metal element doping [38].

2.2.1. Metal Element Doping

A series of metal cations have been used to modulate the band structure of pristine $g-C_3N_4$. There are two kinds of metal element doping related to $g-C_3N_4$, which are cave doping and interlayer doping. The metal cations can be introduced into the triangular pores of $g-C_3N_4$ between the heptazine structures [33]. The strong coordination interaction between the metal cations and $g-C_3N_4$ matrix and negatively charged nitrogen atoms can realize cave doping [38]. According to previous literature, the transition metal elements including Fe, Mn, Co, Ni and Zn have been demonstrated to be effective at optimizing the electronic structure [32,38]. For example, Wang et al. showed that the band gap could be reduced to enhance the visible-light harvesting capability using Fe and Zn doping into $g-C_3N_4$ [97]. Ding et al. also demonstrated that Fe, Mn, Co and Ni could be incorporated into the $g-C_3N_4$ framework to extend the visible-light absorption range and improve the separation efficiency of the photogenerated electrons and holes, which resulted in enhanced photocatalytic performance [98].

In addition, according to the first principle DFT calculation, Pan et al. predicted that the incorporation of Pt and Pd into the $g-C_3N_4$ framework could promote the charge carrier transport rate to improve the charge carrier separation efficiency and reduce the band gap to improve the light absorption, which played positive effects in improving the photocatalytic activity [99]. Recently, Dong et al. found that K atoms brought about interlayer doping instead of caving doping in the g-C$_3$N$_4$ matrix. Pristine g-C$_3$N$_4$ displayed a limited photocatalytic NO removal rate of 16%. The K-doped g-C$_3$N$_4$ exhibited approximately 2.3 times higher photocatalytic NO removal performance than pristine g-C$_3$N$_4$. The outstanding photocatalytic performance of K-doped g-C$_3$N$_4$ was ascribed to the benefits of K intercalation including bridging the layers, charge redistribution, facilitating the charge carrier separation and tuning band structure (Figure 7) [100]. In addition, Zhu et al. revealed that K doping could decrease the VB level of g-C$_3$N$_4$, leading to the promoted separation and transportation of photo-induced electrons and holes under visible light irradiation [101].

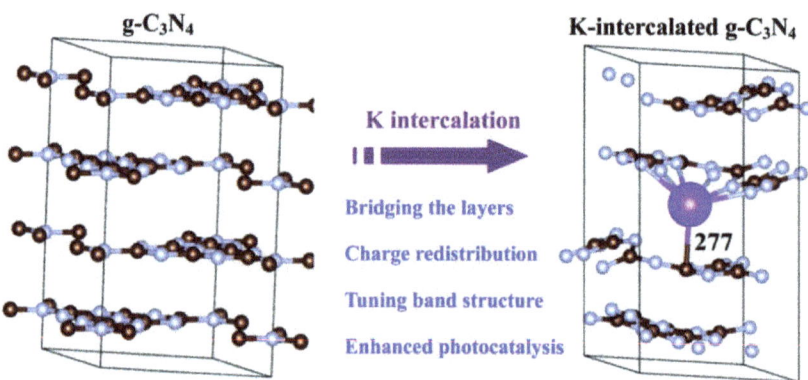

Figure 7. The proposed mechanism of K intercalation to improve photocatalytic performance of g−C$_3$N$_4$ [100]. Copyright 2016, American Chemical Society.

2.2.2. Non-Metal Element Doping

Compared with metal doping, the strategy of non-metal doping may be more popular because it not only tunes the electronic structure but also retains the metal-free property. So far, many non-metal elements such as S, P, B, O, C and I have been demonstrated to be effective for band-gap engineering using chemical substitution. As shown in Figure 8 [38], C atom self-doping can substitute the bridging N atoms while O, S and I atoms tend to replace the N atoms in the aromatic heptazine rings. Thanks to non-metal doping, the delocalization of the Π-conjugated electrons is enhanced to improve the conductivity, mobility and separation of the charge carriers, which is beneficial for improving the photocatalytic performance. As for the P and B atoms, they are inclined to substitute the C atoms. For instance, Wang et al. successfully synthesized B-doped g-C$_3$N$_4$ hollow tubes for improved photocatalytic NO$_x$ removal performance [102]. The B-doped g-C$_3$N$_4$ hollow tubes were fabricated by calcining the assembly supramolecular precursors, which were obtained using the self-conversion of melamine with the aid of boric acid (Figure 9a). The B-doped g-C$_3$N$_4$ hollow tubes displayed the best photocatalytic NO$_x$ removal performance (30.4%), which was 1.5 and 1.3 times higher than pristine g-C$_3$N$_4$ (20.8%) and g-C$_3$N$_4$ hollow tubes (22.9%), respectively (Figure 9b,c). The excellent photocatalytic NO$_x$ removal performance of B-doped g-C$_3$N$_4$ hollow tubes was attributed to the extended light-harvesting range and enhanced charge carrier efficiency (Figure 9d).

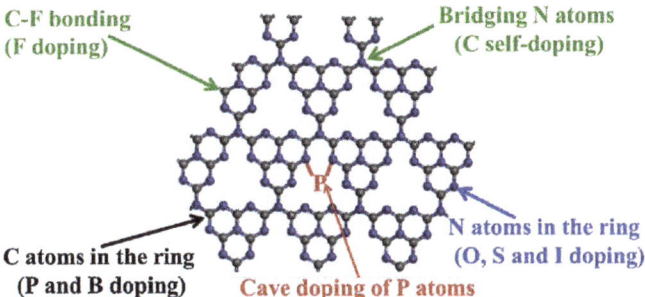

Figure 8. Schematic illustration of the non−metal doping of the g−C$_3$N$_4$ framework [38]. Copyright 2017, Elsevier.

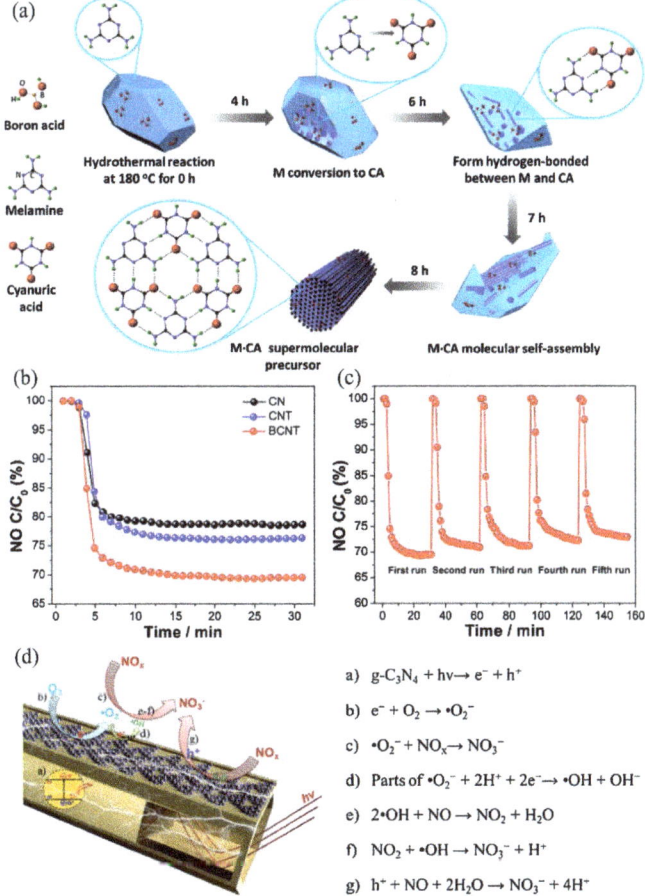

a) g-C$_3$N$_4$ + hv → e$^-$ + h$^+$
b) e$^-$ + O$_2$ → •O$_2^-$
c) •O$_2^-$ + NO$_x$ → NO$_3^-$
d) Parts of •O$_2^-$ + 2H$^+$ + 2e$^-$ → •OH + OH$^-$
e) 2•OH + NO → NO$_2$ + H$_2$O
f) NO$_2$ + •OH → NO$_3^-$ + H$^+$
g) h$^+$ + NO + 2H$_2$O → NO$_3^-$ + 4H$^+$

Figure 9. (**a**) Schematic illustration for the preparation of B−doped g−C$_3$N$_4$ hollow tubes. (**b**) Photocatalytic performance of bulk g−C$_3$N$_4$, g−C$_3$N$_4$ tubes and B−doped g−C$_3$N$_4$ tubes for the removal of NO irradiated under visible light. (**c**) Stability test of B−doped g−C$_3$N$_4$ tubes. (**d**) The proposed mechanism of B−doped g−C$_3$N$_4$ tubes to improve the photocatalytic performance of g−C$_3$N$_4$ [102]. Copyright 2018, Elsevier.

2.3. Defect Engineering

At the same time, defect engineering is also an effective strategy to improve the photocatalytic performance of pristine g-C_3N_4. The defect engineering strategy is premature to modify the electronic structures of pristine TiO_2, which may be due to the fact that TiO_2 is the most classical and fully-studied photocatalyst [103–106]. For example, the band structures and optical properties of pristine TiO_2 could be tuned by oxygen vacancies [107,108]. The oxygen vacancies-mediated TiO_2 can extend the visible-light range, enhance the charge carrier separation efficiency and improve the molecules to be adsorbed on the surface of the photocatalysts, which would result in excellent photocatalytic performance.

Inspired by the oxygen vacancies-mediated TiO_2, a defect engineering strategy is used to improve the photocatalytic performance of pristine g-C_3N_4. For example, Wang et al. reported the nitrogen vacancies-mediated g-C_3N_4 microtubes synthesized using a simple and green hydrothermal process (Figure 10a) [109]. The nitrogen vacancies-mediated g-C_3N_4 microtubes displayed significantly enhanced NO removal performance due to the enlarged specific surface areas and the curial roles of nitrogen vacancies. As shown in Figure 10b–g, the nitrogen vacancies-mediated g-C_3N_4 was beneficial for NO and O_2 adsorption, which contributed to attending the surface reactions. The enhanced surface reactions and increased active sites resulted in improved photocatalytic NO removal performance in comparison with pristine g-C_3N_4 under visible-light irradiation. Li et al. successfully synthesized carbon vacancies-modified g-C_3N_4 nanotubes by calcining the hydrolyzed melamine–urea mixture [110]. The EPR spectra confirmed the formation of carbon vacancies in g-C_3N_4 nanotubes because the EPR signal of carbon vacancies-modified g-C_3N_4 decreased significantly due to the fewer unpaired electrons.

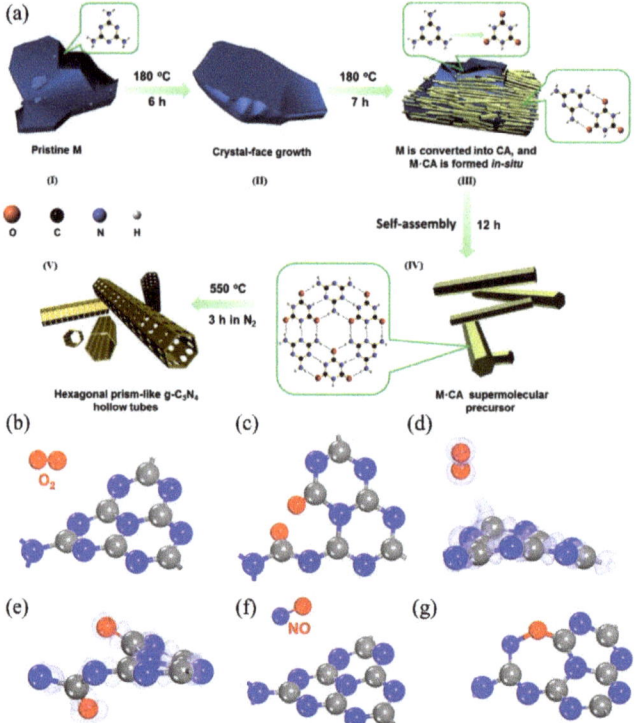

Figure 10. (a) Schematic illustration for the preparation of nitrogen vacancies–mediated g–C_3N_4 microtubes. (b–g) Schematic illustration for the nitrogen vacancies in g–C_3N_4 photocatalysts for the enhanced adsorption behavior [109]. Copyright 2019. American Chemical Society.

Gu et al. reported that the carbon vacancies and hydroxyls co-modified g-C$_3$N$_4$ were successfully prepared using a post-hydrothermal treatment [111]. Pristine g-C$_3$N$_4$ was first prepared using the thermally induced polymerization of melamine. Then a green hydrothermal treatment was employed to introduce the carbon vacancies and hydroxyls (Figure 11a). During the hydrothermal process, the water could induce the pristine g-C$_3$N$_4$ to partially hydrolyze, which introduced the carbon vacancies and hydroxyls into the pristine g-C$_3$N$_4$ simultaneously (Figure 11b). The obtained carbon vacancies and hydroxyls co-modified g-C$_3$N$_4$ showed 2.2 times higher photocatalytic NO removal activities than pristine g-C$_3$N$_4$. With the aid of DFT calculations and experimental calculations, Gu et al. revealed that carbon vacancies and hydroxyls played significant roles in enhancing the photocatalytic NO removal performance due to a synergistic effect. The carbon vacancies narrowed the band gap to extend the light-harvesting range and the hydroxyls could form the covalent bond acting as electron transport channels to facilitate the charge carrier separation efficiency (Figure 11c).

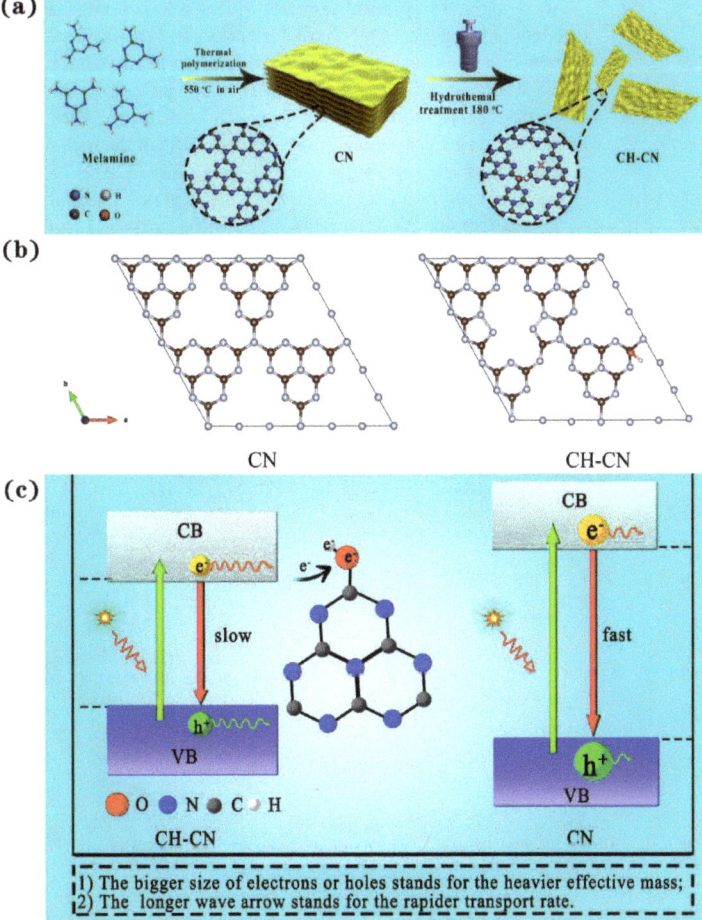

Figure 11. (a) Schematic illustration for the preparation of carbon vacancies and hydroxyls co−modified g−C$_3$N$_4$. (b) Schematic illustration for the g−C$_3$N$_4$ and carbon vacancies and hydroxyls co−modified g−C$_3$N$_4$. (c) The proposed mechanism of carbon vacancies and hydroxyls co−modified g−C$_3$N$_4$ for improved photocatalytic performance of g−C$_3$N$_4$ [111]. Copyright 2020, Elsevier.

2.4. Crystallinity Optimization

Recently, the crystallinity optimization strategy has attracted much attention for improving the photocatalytic activity of pristine g-C_3N_4 [78]. It is well-understood that kinetic hindrance is the major issue in the traditional high-temperature solid-state synthesis of pristine g-C_3N_4, which results in semi-crystalline or amorphous structures and limited photocatalytic performance. Since kinetic hindrance is a great problem in traditional high-temperature solid-state reactions, a novel liquid reaction synthesis technology was developed to solve this problem. Bojdys et al. first reported that triazine-based crystalline g-C_3N_4 was successfully synthesized with the ionothermal method using the eutectic mixture of LiCl/KCl as a high-temperature solvent [112].

Up to the present, a great deal of research work has been carried out to prepare crystalline g-C_3N_4 for enhancing photocatalytic performance. For example, Wang et al. prepared heptazine-based crystalline g-C_3N_4 with the molten salt method using preheated melamine as precursors. The melamine was first heated at 500 °C in a muffle furnace and the preheated melamine was mixed with KCl and LiCl [113]. Then, the mixtures were calcined again in a muffle furnace to prepare heptazine-based crystalline g-C_3N_4. Detailed experimental characterization and theoretical simulation showed that heptazine-based crystalline g-C_3N_4 displayed higher photocatalytic performance than triazine-based crystalline g-C_3N_4 owing to the enhanced light-harvesting property and increased mobility of photogenerated charge carriers. In addition, Wang et al. studied the crystallization process of g-C_3N_4 using different precursors of the melem-based oligomer and melon-based polymer with a molten salts method (Figure 12a) [114]. The melem-based oligomer and melon-based polymer represented different polymerization degrees of g-C_3N_4, which were calcined at 450 °C and 550 °C in a muffle furnace, respectively. Xiang et al. demonstrated that the crystallinity of crystalline g-C_3N_4 synthesized using the molten salts method could be further improved by hydrochloric acid treatment (Figure 12b) [115].

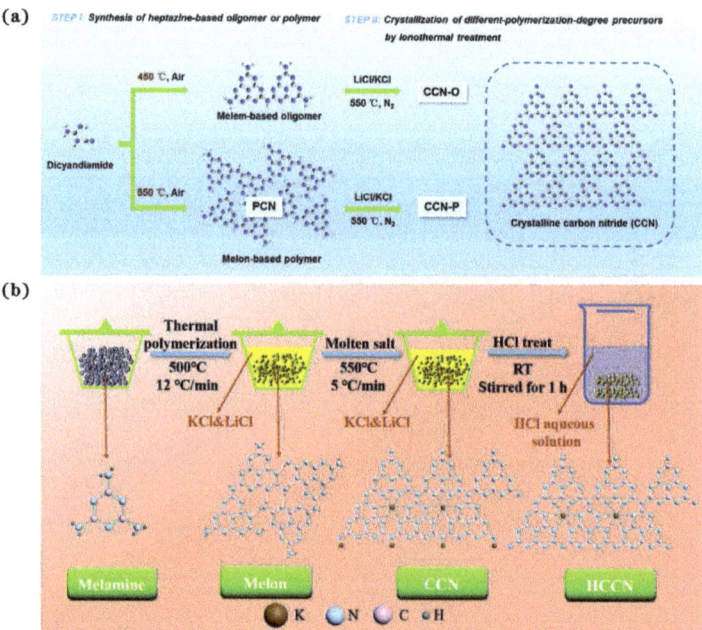

Figure 12. (a) Schematic illustration for the crystallization process of g—C_3N_4 [114]. Copyright 2020, Elsevier. (b) Schematic illustration for improving the crystallinity of g—C_3N_4 using hydrochloric acid treatment [115]. Copyright 2020, Elsevier.

It is important to point out that these reported molten salt methods were carried out under an inert gas atmosphere in a muffle furnace, which limited the large-scale production of crystalline g-C_3N_4. To solve this drawback, Gu et al. developed a modified molten salt method under ambient pressure using dicyanamide (DCDA) as the initial precursor (Figure 13a) [116]. The molten salts played two roles in the post-calcination process. One was improving the crystallinity of pristine g-C_3N_4 acting as the high-temperature solution, the other was protecting pristine g-C_3N_4 from contact with air since the pristine g-C_3N_4 was immersed in the solution. The crystalline g-C_3N_4 exhibited 3.0 times higher photocatalytic NO removal activity than pristine g-C_3N_4 under visible-light irradiation, with high stability under the cycling test (Figure 13b,c). The detailed experimental characterization and DFT calculation demonstrated that the optimized crystallinity played important roles in improving the photocatalytic NO removal activity of crystalline g-C_3N_4. The optimized crystallinity could decrease the band gap to extend the light-harvesting range, increase the conductivity to promote the photogenerated charge carrier separation efficiency and reduce the adsorption energy of NO and O_2 molecules to activate the surface reactions, which led to the significantly enhanced photocatalytic NO removal performance (Figure 13d–k).

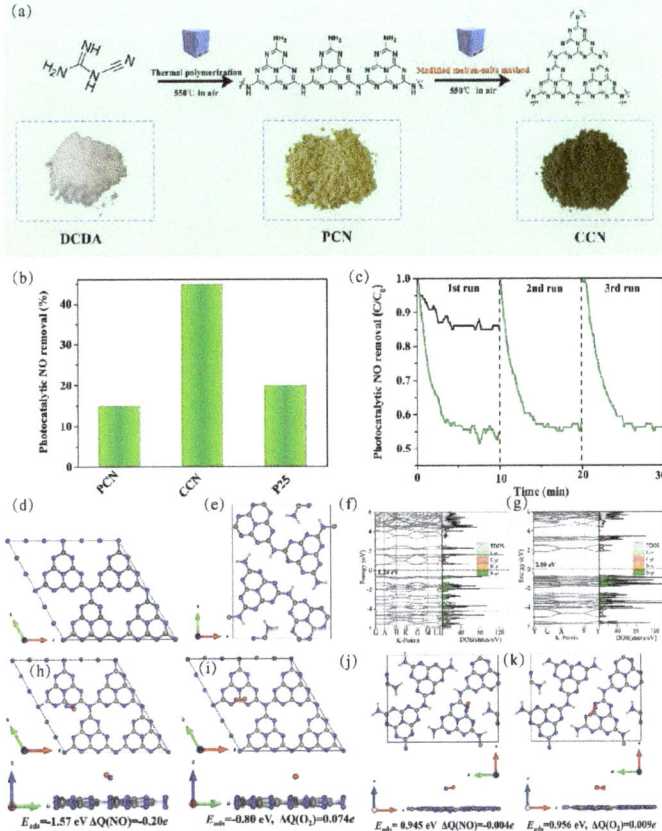

Figure 13. (a) Schematic illustration for the preparation of crystalline g−C_3N_4 using a modified molten salt method under ambient pressure. (b) Photocatalytic performance of pristine g−C_3N_4, crystalline g−C_3N_4 and P25 for the removal of NO irradiated under visible light. (c) Stability test of crystalline g−C_3N_4. (d–g) Schematic illustration for crystallinity in g−C_3N_4 photocatalysts for the enhanced light-harvesting properties. (h–k) Schematic illustration for crystallinity in g−C_3N_4 photocatalysts for the enhanced adsorption behavior [116]. Copyright 2021, Elsevier.

2.5. Heterojunction Construction

Charge carrier transport and separation is decisive in the photocatalytic process. A large number of photogenerated electrons and holes suffered from volume recombination and surface recombination, which result in unsatisfactory photocatalytic performance [117–120]. Constructing a g-C_3N_4-based heterojunction is an effective strategy to improve photocatalytic performance. The spatial separation of photogenerated electron-hole pairs can be achieved with efficient charge transfer across the interface between the two semiconductors. At the same time, the g-C_3N_4-based heterojunction can display the advantages of the counterpart. In other words, the g-C_3N_4- based heterojunction has both benefits of the two components. Up to the present, several types of g-C_3N_4-based heterojunction have attracted much attention including the traditional type-II heterojunction, all-solid-state Z-scheme heterojunction, step-scheme (S-scheme) heterojunction and g-C_3N_4/carbon heterojunction.

The traditional type-II heterojunction is facile constructed, and much progress has been made in this field. For example, Koci et al. reported that a series of TiO_2/g-C_3N_4 heterojunction photocatalysts were easily prepared using mechanical mixing of TiO_2 and g-C_3N_4 in a water suspension followed by calcination in a muffle furnace [121]. The TiO_2/g-C_3N_4 heterojunction with the optimal weight ratio of TiO_2 and g-C_3N_4 has shifted absorption edge energy towards longer wavelengths and decreased the recombination rate of charge carriers compared to pure g-C_3N_4.

Even though the traditional type-II heterojunction can improve the charge carrier separation efficiency to improve the photocatalytic performance, it sacrifices the oxidation potential energy and reduction potential energy. To overcome this drawback, the all-solid-state Z-scheme was developed inspired by the photosynthesis of plants [70]. The photosynthesis of plants consists of two isolated reactions of water oxidation and CO_2 reduction, which are linked together through redox mediators. Thanks to the unique structure, it keeps the strong redox ability, improves the charge carrier separation efficiency and results in enhanced photocatalytic performance. For instance, Zhang et al. reported an all-solid-state Z-scheme g-C_3N_4/Au/$ZnIn_2S_4$ heterojunction photocatalyst for enhanced photocatalytic NO removal performance [122]. The noble Au nanoparticles played an important role in the charge carrier transfer process, which acted as an electron acceptor and conductive channel for enhancing the charge carrier separation efficiency. Additionally, the all-solid-state Z-scheme heterojunction exhibited oxidation potential energy and reduction potential energy during the photocatalytic NO removal process. Therefore, the optimized Z-scheme g-C_3N_4/Au/$ZnIn_2S_4$ heterojunction photocatalyst showed photocatalytic NO removal efficiency of up to 59.7%.

Recently, direct Z-scheme heterojunction has become a research hotspot [37,123]. The direct Z-scheme heterojunction is totally different from the all-solid-state Z-scheme because there is no intermediate, either Au or Ag nanoparticles, used in the direct Z-scheme heterojunction [25,122,124]. To describe the photocatalytic mechanism of the direct Z-scheme heterojunction clearly and vividly, the Yu group first nominated the direct Z-scheme heterojunction for a step-scheme (S-scheme) heterojunction [125,126]. Yu et al., for the first time, reported the g-C_3N_4-TiO_2 direct Z-scheme heterojunction using the facile calcination method. Moreover, they found that the g-C_3N_4-TiO_2 direct Z-scheme heterojunction was largely dependent on the content of g-C_3N_4 [54]. To be more specific, if the surface of TiO_2 was partially covered by the g-C_3N_4, the g-C_3N_4-TiO_2 direct Z-scheme heterojunction would be obtained; if the content of g-C_3N_4 was too much, the traditional type-II heterojunction would be obtained. Lu et al. reported an α-Fe_2O_3/g-C_3N_4 direct Z-scheme heterojunction prepared using an impregnation–hydrothermal method (Figure 14a) [52]. The unique direct Z-scheme heterojunction brought about wide visible-light absorption and facilitated charge carrier separation efficiency (Figure 14b) [52]. The α-Fe_2O_3/g-C_3N_4 direct Z-scheme heterojunction displayed approximately 1.78 times higher photocatalytic NO removal performance than pristine g-C_3N_4. As for the S-scheme heterojunction, Yu et al., for the first time, designed and constructed the S-scheme heterojunction of WO_3/g-C_3N_4 using an electrostatic self-assembly strategy [126]. After that, a great number of

S-scheme heterojunctions have been reported for improving photocatalytic performance. For example, Zhang et al. reported a $Sb_2WO_6/g-C_3N_4$ S-scheme heterojunction prepared using an ultrasound-assisted strategy for improved visible-light photocatalytic NO removal performance. In addition, Cao et al. successfully reported a 2D/0D $g-C_3N_4/SnO_2$ S-scheme heterojunction using a hydrothermal and annealing strategy toward visible-light-driven NO degradation (Figure 15a) [127]. The S-scheme charge transfer mechanism was revealed using the Density-Functional Theory (DFT) calculation, trapping experiments and EPR spectra. Because of the unique structural features, the $g-C_3N_4/SnO_2$ S-scheme photocatalysts displayed a photocatalytic NO removal percentage of 40% irradiated under visible light (Figure 15b).

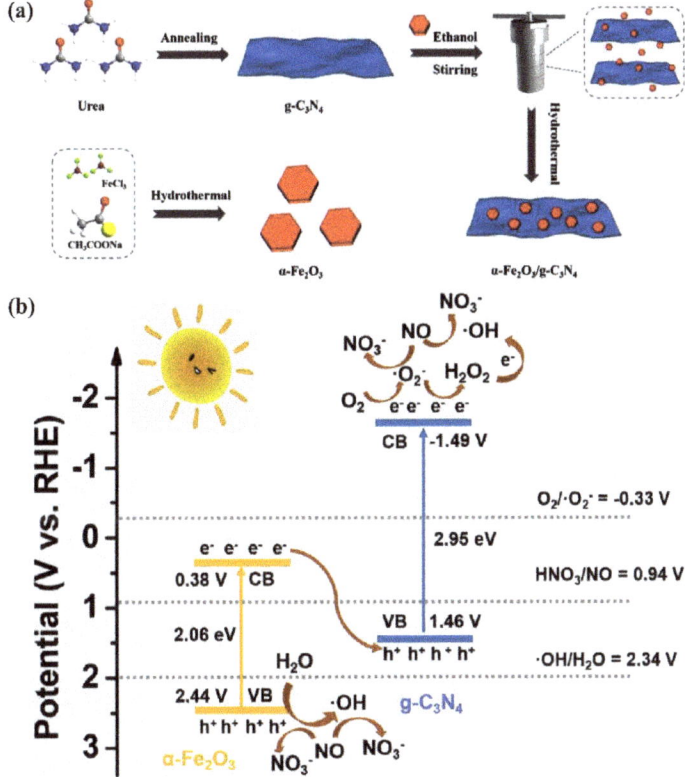

Figure 14. (a) Schematic illustration for the preparation of $\alpha-Fe_2O_3/g-C_3N_4$ direct Z−scheme heterojunction prepared using an impregnation−hydrothermal method. (b) The proposed mechanism of direct Z−scheme heterojunction to improve the photocatalytic performance of $\alpha-Fe_2O_3/g-C_3N_4$ [52]. Copyright 2021, Elsevier.

Carbon materials including graphene have brought about widespread attention in the field of photocatalysis owing to their outstanding physical and chemical properties including excellent electron conductivity, good light harvesting properties, large specific surface areas, low cost and high stability [128–130]. Constructing a $g-C_3N_4$/carbon heterojunction can take advantage of carbon materials to improve the photocatalytic performance of pristine $g-C_3N_4$. Firstly, coupling carbon materials with the $g-C_3N_4$ can significantly extend the light-harvesting range to near-infrared. Secondly, carbon materials can facilitate the photogenerated charge carrier separation efficiency since it acts as conductive channels for electron transfer. Lastly, the large specific surface areas of carbon materials can provide plentiful of supporting sites for $g-C_3N_4$. For example, Gu et al. reported an alkali-assisted hydrothermal method to prepare g-C_3N_4/reduced graphene oxide (g-C_3N_4/rGO) nanocomposites (Figure 16a) [131]. During the hydrothermal process, the NaOH could improve the reduction of GO to increase the conductivity of rGO and etch the pristine g-C_3N_4 into nanosheets to enlarge the specific surface areas. The g-C_3N_4/rGO nanocomposites displayed 2.7 times higher photocatalytic NO_x removal performance than pristine g-C_3N_4. The distinctly enhanced photocatalytic performance of g-C_3N_4/rGO nanocomposites is ascribed to the improved light-harvesting property, increased specific surface areas and active sites and facilitated charge carrier separation efficiency.

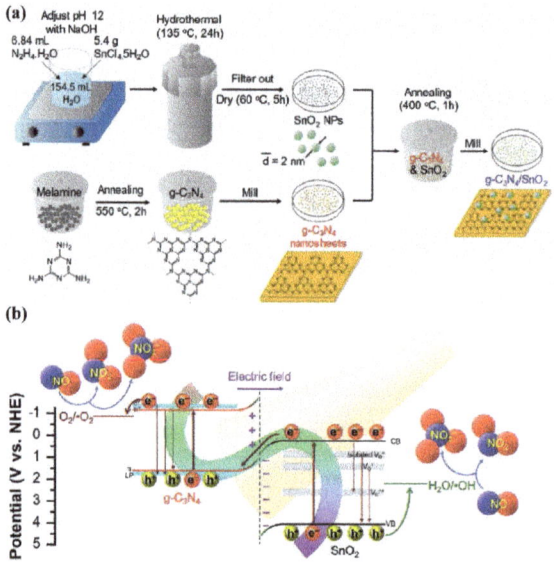

Figure 15. (a) Schematic illustration for the preparation of g–C_3N_4/SnO_2 S–scheme heterojunction using a hydrothermal and annealing strategy. (b) The proposed mechanism of S–scheme to improve the photocatalytic performance of g–C_3N_4/SnO_2 [127]. Copyright 2021, Elsevier.

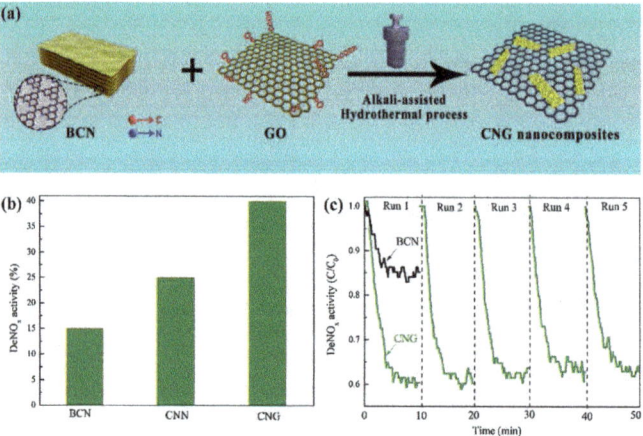

Figure 16. (a) Schematic illustration for the preparation of g–C_3N_4/rGO nanocomposites using an alkali–assisted hydrothermal method. (b) Photocatalytic performance of pristine g–C_3N_4 and g–C_3N_4 treated using an alkali-assisted process and g–C_3N_4/rGO for the removal of NO irradiated under visible light. (c) Stability test of g–C_3N_4/rGO [122]. Copyright 2020, Elsevier.

3. Conclusions and Prospects

In conclusion, photocatalysis is an environmentally friendly and low-cost technology to solve NO_x pollution. As a typical 2D metal-free semiconductor, the $g-C_3N_4$ photocatalyst has drawn great attention owing to its outstanding physical and chemical properties including visible light response, adjustable band structure, low cost, facile synthesis, high stability and so on. Therefore, the $g-C_3N_4$ photocatalyst possesses great potential for application in the photocatalytic NO_x removal process. However, pristine $g-C_3N_4$ synthesized using the traditional high-temperature solid reaction suffers from small specific surface areas and low crystallinity, which results in few reactive sites, limited light-harvesting capacity, rapid recombination of photogenerated charge carriers and unsatisfactory photocatalytic NO_x removal performance. In this review, we briefly summarize the recent advances in $g-C_3N_4$-based photocatalysts for the NO_x removal process. Various modification strategies are discussed including morphology controlling, band structure engineering, crystallinity optimization, defect engineering and heterojunction construction (Figure 17). The different modification strategies play different roles in improving the photocatalytic NO_x removal performance of pristine $g-C_3N_4$. Specifically, morphology controlling can not only enlarge the specific surface areas and increase the actives, but also shorten the charge carrier migration distance and thus suppress the photogenerated electron and hole pairs recombination. Band structure engineering can be achieved using metal and non-metal doping, which can reduce the bandgap and extend the light-harvesting range. As for defect engineering, it can modify the electronic structures and improve the molecules to be adsorbed on the surface of pristine $g-C_3N_4$. The crystallinity optimization strategy can increase the crystallinity of pristine $g-C_3N_4$ and decrease its band gap. In addition, it increases the conductivity to improve the photogenerated charge carrier separation efficiency. When it comes to the heterojunction construction, it mainly promotes charge carrier separation efficiency through efficient charge transfer across the interface between the two semiconductors. Up to now, great progress has been made to enhance the photocatalytic NO_x removal performance of $g-C_3N_4$. However, from our perspective, there are still some significant challenges for us to solve.

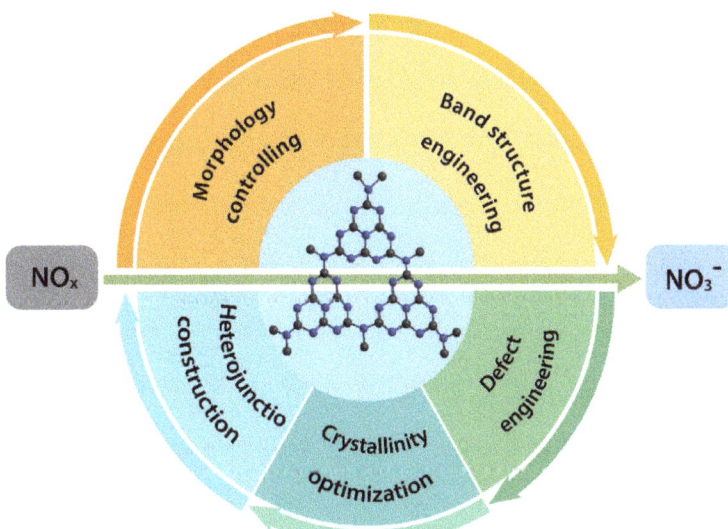

Figure 17. The different modification strategies to improve the photocatalytic NO_x removal performance of $g-C_3N_4$.

First of all, although a variety of modification strategies were developed to enhance the photocatalytic NO_x removal performance of $g-C_3N_4$, many of them are carried out under harsh conditions even involving toxic reagents. For example, dangerous hydrofluoric acid is used to remove the silica-based hard templates for preparing different nanostructured g-C_3N_4. Therefore, in the near future, the environmentally friendly and low-cost synthesis process which is suitable for large-scale production needs to be exploited for fabricating hollow or porous nanostructured g-C_3N_4.

Secondly, collaborative strategies should be adopted to modify pristine g-C_3N_4 for improving the photocatalytic NO_x removal performance. The majority of research works involve the single modification strategy, i.e., the element doping strategy, morphology control strategy, crystallinity optimization, defect engineering and heterojunction construction. The single modification strategy can enhance the photocatalytic activity of pristine g-C_3N_4 remarkably, but there is room for improvement. We can employ collaborative strategies to improve the photocatalytic NO_x removal performance of pristine g-C_3N_4. For instance, we can optimize the crystallinity and introduce nitrogen simultaneously to enhance the photocatalytic activity of pristine g-C_3N_4. Collaborative strategies extend the visible light response range, promote charge carrier separation efficiency and activate the reactants molecules, resulting in excellent photocatalytic performance.

Thirdly, g-C_3N_4-based photocatalysts with full-spectrum-activated photocatalytic activities from UV to near-infrared still remains a great challenge. It is well-understood that pristine g-C_3N_4 cannot harvest near-infrared (NIR) light, which accounts for 50% of the full solar spectrum. In order to make full use of solar energy, it is urgent to develop full-spectrum responsive g-C_3N_4-based photocatalysts for photocatalytic NO_x removal. Heterojunction construction with narrow optical materials, the combination of up-conversion materials and plasmonic materials and element doping collaboration with crystallinity optimization can be used to extend the light absorption of g-C_3N_4 to NIR light region.

Finally, the photocatalytic NO_x removal process is very complicated, and the related catalytic mechanism is not thoroughly studied or understood. Therefore, advanced characterization techniques and theoretical simulations are recommended for revealing the photocatalytic NO_x removal mechanism. Steady-state photoluminescence spectra and time-resolved photoluminescence spectra show the photogenerated charge carrier separation efficiency. The in situ infrared absorption spectroscopic analysis helps us to understand the mechanism of ROS generation and conversion. At the same time, the theoretical calculation should be widely used to analyze the effects of metal or non-metal doping on the band structure and the effects of carbon or nitrogen vacancies on the reactant molecule activation. Only when we clarify the photocatalytic NO_x removal mechanism comprehensively can we synthesize highly efficient g-C_3N_4-based photocatalysts for the photocatalytic NO_x removal process.

Author Contributions: Conceptualization, S.Y., E.Z. and Z.G.; Review and funding acquisition, Z.G. and X.W. (Xionglei Wang); Writing original draft preparation, Z.G., M.J., X.W. (Xin Wang), R.Z., Z.H., J.Y. and H.H.; writing review and editing, Z.G., S.Z. and X.W. (Xionglei Wang); supervision, S.Y., E.Z. and Z.G. All authors have read and agreed to the published version of the manuscript.

Funding: This work was supported by the Hebei Natural Science Foundation (E2022106008 and B2022402024) and Shijiazhuang University (22BS009). This work was also partly supported by the JSPS Grant-in-Aid for Scientific Research on Innovative Areas "Mixed anion" (No. 16H06439) Grant-in-Aid for Scientific Research (20H00297).

Acknowledgments: The authors acknowledge Ruotong Yang for her kind help.

Conflicts of Interest: The authors declare no conflict of interest.

References

1. Zhao, Z.; Fan, J.; Chang, H.; Asakura, Y.; Yin, S. Recent progress on mixed-anion type visible-light induced photocatalysts. *Sci. China Technol. Sci.* **2017**, *60*, 1447–1457. [CrossRef]
2. Guo, L.; Chen, Y.; Ren, Z.; Li, X.; Zhang, Q.; Wu, J.; Li, Y.; Liu, W.; Li, P.; Fu, Y. Morphology engineering of type-II heterojunction nanoarrays for improved sonophotocatalytic capability. *Ultrason. Sonochemistry* **2021**, *81*, 105849. [CrossRef]

3. Fu, Y.; Ren, Z.; Wu, J.; Li, Y.; Liu, W.; Li, P.; Xing, L.; Ma, J.; Wang, H.; Xue, X. Direct Z-scheme heterojunction of ZnO/MoS$_2$ nanoarrays realized by flowing-induced piezoelectric field for enhanced sunlight photocatalytic performances. *Appl. Catal. B Environ.* **2021**, *285*, 119785. [CrossRef]
4. Li, X.; Bi, W.; Zhang, L.; Tao, S.; Chu, W.; Zhang, Q.; Luo, Y.; Wu, C.; Xie, Y. Single-atom Pt as co-catalyst for enhanced photocatalytic H$_2$ evolution. *Adv. Mater.* **2016**, *28*, 2427–2431. [CrossRef]
5. Zhang, T.; Han, X.; Nguyen, N.T.; Yang, L.; Zhou, X. TiO$_2$-based photocatalysts for CO$_2$ reduction and solar fuel generation. *Chin. J. Catal.* **2022**, *43*, 2500–2529. [CrossRef]
6. Mamiyev, Z.; Balayeva, N.O. Metal Sulfide Photocatalysts for Hydrogen Generation: A Review of Recent Advances. *Catalysts* **2022**, *12*, 1316. [CrossRef]
7. Hejazi, S.; Killian, M.S.; Mazare, A.; Mohajernia, S. Single-Atom-Based Catalysts for Photocatalytic Water Splitting on TiO$_2$ Nanostructures. *Catalysts* **2022**, *12*, 905. [CrossRef]
8. Nikokavoura, A.; Trapalis, C. Graphene and g-C$_3$N$_4$ based photocatalysts for NO$_x$ removal: A review. *Appl. Surf. Sci.* **2018**, *430*, 18–52. [CrossRef]
9. Boyjoo, Y.; Sun, H.; Liu, J.; Pareek, V.K.; Wang, S. A review on photocatalysis for air treatment: From catalyst development to reactor design. *Chem. Eng. J.* **2017**, *310*, 537–559. [CrossRef]
10. Zhou, M.; Ou, H.; Li, S.; Qin, X.; Fang, Y.; Lee, S.C.; Wang, X.; Ho, W. Photocatalytic air purification using functional polymeric carbon nitrides. *Adv. Sci.* **2021**, *8*, 2102376. [CrossRef] [PubMed]
11. Pelaez, M.; Nolan, N.T.; Pillai, S.C.; Seery, M.K.; Falaras, P.; Kontos, A.G.; Dunlop, P.S.M.; Hamilton, J.W.J.; Byrne, J.A.; O'Shea, K. A review on the visible light active titanium dioxide photocatalysts for environmental applications. *Appl. Catal. B Environ.* **2012**, *125*, 331–349. [CrossRef]
12. Tran, H.H.; Bui, D.P.; Kang, F.; Wang, Y.-F.; Liu, S.-H.; Thi, C.M.; You, S.-J.; Chang, G.-M.; Pham, V.V. SnO$_2$/TiO$_2$ nanotube heterojunction: The first investigation of NO degradation by visible light-driven photocatalysis. *Chemosphere* **2018**, *215*, 323–332.
13. Zou, Y.; Xie, Y.; Yu, S.; Chen, L.; Cui, W.; Dong, F.; Zhou, Y. SnO$_2$ quantum dots anchored on g-C$_3$N$_4$ for enhanced visible-light photocatalytic removal of NO and toxic NO$_2$ inhibition. *Appl. Surf. Sci.* **2019**, *496*, 143630. [CrossRef]
14. Zhu, G.; Li, S.; Gao, J.; Zhang, F.; Liu, C.; Wang, Q.; Hojamberdiev, M. Constructing a 2D/2D Bi$_2$O$_2$CO$_3$/Bi$_4$O$_5$Br$_2$ heterostructure as a direct Z-scheme photocatalyst with enhanced photocatalytic activity for NO$_x$ removal. *Appl. Surf. Sci.* **2019**, *493*, 913–925. [CrossRef]
15. Zhao, L.; Li, C.; Li, S.; Du, X.; Zhang, J.; Huang, Y. Simultaneous removal of Hg0 and NO in simulated flue gas on transition metal oxide M (M = Fe$_2$O$_3$, MnO$_2$, and WO$_3$) doping on V$_2$O$_5$/ZrO$_2$-CeO$_2$ catalysts. *Appl. Surf. Sci.* **2019**, *483*, 260–269. [CrossRef]
16. Shen, S.; Liu, Y.; Zhai, D.; Qian, G. Electroplating sludge-derived spinel catalysts for NO removal via NH$_3$ selective catalysis reduction. *Appl. Surf. Sci.* **2020**, *528*, 146969. [CrossRef]
17. Li, Y.; Sun, Y.; Ho, W.; Zhang, Y.; Huang, H.; Cai, Q.; Dong, F. Highly enhanced visible-light photocatalytic NO$_x$ purification and conversion pathway on self-structurally modified g-C$_3$N$_4$ nanosheets. *Sci. Bull.* **2018**, *63*, 609–620. [CrossRef]
18. Duan, L.; Li, G.; Zhang, S.; Wang, H.; Zhao, Y.; Zhang, Y. Preparation of S-doped g-C$_3$N$_4$ with C vacancies using the desulfurized waste liquid extracting salt and its application for NO$_x$ removal. *Chem. Eng. J.* **2021**, *411*, 128551. [CrossRef]
19. Cheng, G.; Liu, X.; Song, X.; Chen, X.; Dai, W.; Yuan, R.; Fu, X. Visible-light-driven deep oxidation of NO over Fe doped TiO$_2$ catalyst: Synergic effect of Fe and oxygen vacancies. *Appl. Catal. B Environ.* **2020**, *277*, 119196. [CrossRef]
20. Guo, C.; Wu, X.; Yan, M.; Dong, Q.; Yin, S.; Sato, T.; Liu, S. The visible-light driven photocatalytic destruction of NO$_x$ using mesoporous TiO$_2$ spheres synthesized via a "water-controlled release process". *Nanoscale* **2013**, *5*, 8184–8191. [CrossRef]
21. Wu, X.; Yin, S.; Dong, Q.; Liu, B.; Wang, Y.; Sekino, T.; Lee, S.W.; Sato, T. UV, visible and near-infrared lights induced NO$_x$ destruction activity of (Yb, Er)-NaYF$_4$/C-TiO$_2$ composite. *Sci. Rep.* **2013**, *3*, 2918. [CrossRef] [PubMed]
22. Wang, H.; Sun, Y.; Jiang, G.; Zhang, Y.; Huang, H.; Wu, Z.; Lee, S.; Dong, F. Unraveling the mechanisms of visible light photocatalytic NO purification on earth-abundant insulator-based core–shell heterojunctions. *Environ. Sci. Technol.* **2018**, *52*, 1479–1487. [CrossRef]
23. Wu, X.; Yin, S.; Dong, Q.; Guo, C.; Li, H.; Kimura, T.; Sato, T. Synthesis of high visible light active carbon doped TiO$_2$ photocatalyst by a facile calcination assisted solvothermal method. *Appl. Catal. B Environ.* **2013**, *142*, 450–457. [CrossRef]
24. Lasek, J.; Yu, Y.-H.; Wu, J.C. Removal of NO$_x$ by photocatalytic processes. *J. Photochem. Photobiol. C Photochem. Rev.* **2013**, *14*, 29–52. [CrossRef]
25. Rhimi, B.; Padervand, M.; Jouini, H.; Ghasemi, S.; Bahnemann, D.W.; Wang, C. Recent progress in NO$_x$ photocatalytic removal: Surface/interface engineering and mechanistic understanding. *J. Environ. Chem. Eng.* **2022**, *10*, 108566. [CrossRef]
26. Li, N.; Wang, C.; Zhang, K.; Lv, H.; Yuan, M.; Bahnemann, D.W. Progress and prospects of photocatalytic conversion of low-concentration NO$_x$. *Chin. J. Catal.* **2022**, *43*, 2363–2387. [CrossRef]
27. Wu, H.; Chen, D.; Li, N.; Xu, Q.; Li, H.; He, J.; Lu, J. Hollow porous carbon nitride immobilized on carbonized nanofibers for highly efficient visible light photocatalytic removal of NO. *Nanoscale* **2016**, *8*, 12066–12072. [CrossRef]
28. Fujishima, A.; Honda, K. Electrochemical photolysis of water at a semiconductor electrode. *Nature* **1972**, *238*, 37. [CrossRef]
29. Yoon, T.P.; Ischay, M.A.; Du, J. Visible light photocatalysis as a greener approach to photochemical synthesis. *Nat. Chem.* **2010**, *2*, 527. [CrossRef]
30. Lin, L.; Lin, Z.; Zhang, J.; Cai, X.; Lin, W.; Yu, Z.; Wang, X. Molecular-level insights on the reactive facet of carbon nitride single crystals photocatalysing overall water splitting. *Nat. Catal.* **2020**, *3*, 649–655. [CrossRef]

31. Wang, X.; Maeda, K.; Thomas, A.; Takanabe, K.; Xin, G.; Carlsson, J.M.; Domen, K.; Antonietti, M. A metal-free polymeric photocatalyst for hydrogen production from water under visible light. *Nat. Mater.* **2009**, *8*, 76–80. [CrossRef]
32. Ong, W.J.; Tan, L.L.; Yun, H.N.; Yong, S.T.; Chai, S.P. Graphitic Carbon Nitride (g-C_3N_4)-Based Photocatalysts for Artificial Photosynthesis and Environmental Remediation: Are We a Step Closer To Achieving Sustainability? *Chem. Rev.* **2016**, *116*, 7159–7329. [CrossRef]
33. Cao, S.; Low, J.; Yu, J.; Jaroniec, M. Polymeric photocatalysts based on graphitic carbon nitride. *Adv. Mater.* **2015**, *27*, 2150–2176. [CrossRef] [PubMed]
34. Zhang, L.; Zhang, J.; Yu, H.; Yu, J. Emerging S-scheme photocatalyst. *Adv. Mater.* **2022**, *34*, 2107668. [CrossRef]
35. Fu, J.; Yu, J.; Jiang, C.; Cheng, B. g-C_3N_4-Based Heterostructured Photocatalysts. *Adv. Energy Mater.* **2018**, *8*, 1701503. [CrossRef]
36. Marschall, R. Semiconductor composites: Strategies for enhancing charge carrier separation to improve photocatalytic activity. *Adv. Funct. Mater.* **2014**, *24*, 2421–2440. [CrossRef]
37. Low, J.; Yu, J.; Jaroniec, M.; Wagheh, S.; Al-Ghamdi, A.A. Heterojunction Photocatalysts. *Adv. Mater.* **2017**, *29*, 1601694. [CrossRef] [PubMed]
38. Wen, J.; Xie, J.; Chen, X.; Li, X. A review on g-C_3N_4-based photocatalysts. *Appl. Surf. Sci.* **2017**, *391*, 72–123. [CrossRef]
39. Nosaka, Y.; Nosaka, A.Y. Generation and detection of reactive oxygen species in photocatalysis. *Chem. Rev.* **2017**, *117*, 11302–11336. [CrossRef]
40. Rehman, Z.U.; Butt, F.K.; Balayeva, N.O.; Idrees, F.; Hou, J.; Tariq, Z.; Rehman, S.U.; Haq, B.U.; Alfaify, S.; Ali, S. Two dimensional graphitic carbon nitride Nanosheets as prospective material for photocatalytic degradation of nitrogen oxides. *Diam. Relat. Mater.* **2021**, *120*, 108650. [CrossRef]
41. Balayeva, N.O.; Fleisch, M.; Bahnemann, D.W. Surface-grafted WO_3/TiO_2 photocatalysts: Enhanced visible-light activity towards indoor air purification. *Catal. Today* **2018**, *313*, 63–71. [CrossRef]
42. Khanal, V.; Balayeva, N.O.; Günnemann, C.; Mamiyev, Z.; Dillert, R.; Bahnemann, D.W.; Subramanian, V.R. Photocatalytic NO_x removal using tantalum oxide nanoparticles: A benign pathway. *Appl. Catal. B Environ.* **2021**, *291*, 119974. [CrossRef]
43. Xue, Y.; Yin, S. Element doping: A marvelous strategy for pioneering the smart applications of VO_2. *Nanoscale* **2022**, *14*, 11054–11097. [CrossRef] [PubMed]
44. Yin, S.; Hasegawa, T. Morphology Control of Transition Metal Oxides by Liquid-Phase Process and Their Material Development. *KONA Powder Part. J.* **2023**, *40*, 94–108. [CrossRef]
45. Yin, S. Creation of advanced optical responsive functionality of ceramics by green processes. *J. Ceram. Soc. Jpn.* **2015**, *123*, 823–834. [CrossRef]
46. Noda, C.; Asakura, Y.; Shiraki, K.; Yamakata, A.; Yin, S. Synthesis of three-component C_3N_4/rGO/C-TiO_2 photocatalyst with enhanced visible-light responsive photocatalytic deNO_x activity. *Chem. Eng. J.* **2020**, *390*, 124616. [CrossRef]
47. Komatsuda, S.; Asakura, Y.; Vequizo, J.J.M.; Yamakata, A.; Yin, S. Enhanced photocatalytic NO_x decomposition of visible-light responsive F-TiO_2/(N, C)-TiO_2 by charge transfer between F-TiO_2 and (N, C)-TiO_2 through their doping levels. *Appl. Catal. B Environ.* **2018**, *238*, 358–364. [CrossRef]
48. Cao, J.; Hasegawa, T.; Asakura, Y.; Sun, P.; Yang, S.; Li, B.; Cao, W.; Yin, S. Synthesis and color tuning of titanium oxide inorganic pigment by phase control and mixed-anion co-doping. *Adv. Powder Technol.* **2022**, *33*, 103576. [CrossRef]
49. Wang, Y.; Vogel, A.; Sachs, M.; Sprick, R.S.; Wilbraham, L.; Moniz, S.J.; Godin, R.; Zwijnenburg, M.A.; Durrant, J.R.; Cooper, A.I. Current understanding and challenges of solar-driven hydrogen generation using polymeric photocatalysts. *Nat. Energy* **2019**, *4*, 746–760. [CrossRef]
50. Cheng, C.; He, B.; Fan, J.; Cheng, B.; Cao, S.; Yu, J. An inorganic/organic S-scheme heterojunction H_2-production photocatalyst and its charge transfer mechanism. *Adv. Mater.* **2021**, *33*, 2100317. [CrossRef]
51. Wei, J.; Zhang, J. In Fe_2O_3/TiO_2 nanocomposite photocatalyst prepared by supercritical fluid combination technique and its application in degradation of acrylic acid. *IOP Conf. Ser. Mater. Sci. Eng.* **2017**, *167*, 012037. [CrossRef]
52. Geng, Y.; Chen, D.; Li, N.; Xu, Q.; Li, H.; He, J.; Lu, J. Z-Scheme 2D/2D α-Fe_2O_3/g-C_3N_4 heterojunction for photocatalytic oxidation of nitric oxide. *Appl. Catal. B Environ.* **2021**, *280*, 119409. [CrossRef]
53. Lei, B.; Cui, W.; Chen, P.; Chen, L.; Li, J.; Dong, F. C-Doping Induced Oxygen-Vacancy in WO_3 Nanosheets for CO_2 Activation and Photoreduction. *ACS Catal.* **2022**, *12*, 9670–9678. [CrossRef]
54. Yu, J.; Wang, S.; Low, J.; Xiao, W. Enhanced photocatalytic performance of direct Z-scheme g-C_3N_4–TiO_2 photocatalysts for the decomposition of formaldehyde in air. *Phys. Chem. Chem. Phys.* **2013**, *15*, 16883–16890. [CrossRef]
55. Wu, X.; Yin, S.; Xue, D.; Komarneni, S.; Sato, T. A $Cs_{(x)}WO_3$/ZnO nanocomposite as a smart coating for photocatalytic environmental cleanup and heat insulation. *Nanoscale* **2015**, *7*, 17048. [CrossRef] [PubMed]
56. Guan, L.; Wu, A.L.; Gu, T.K.; Yu, W.W. *Study on the Enhanced Visible Photocatalysis Activity in Transition Metal Doped ZnS*; Advanced Materials Research; Trans Tech Publications: Wollerau, Switzerland, 2013; pp. 2351–2355.
57. Tang, Z.R.; Han, B.; Han, C.; Xu, Y.J. One dimensional CdS based materials for artificial photoredox reactions. *J. Mater. Chem. A* **2016**, *5*, 2387–2410. [CrossRef]
58. Wu, J.; Liu, B.; Ren, Z.; Ni, M.; Li, C.; Gong, Y.; Qin, W.; Huang, Y.; Sun, C.Q.; Liu, X. CuS/RGO hybrid photocatalyst for full solar spectrum photoreduction from UV/Vis to Near-Infrared light. *J. Colloid Interface Sci.* **2017**, *517*, 80–85. [CrossRef]

59. Wu, X.; Zhang, K.; Zhang, G.; Yin, S. Facile preparation of BiOX (X = Cl, Br, I) nanoparticles and up-conversion phosphors/BiOBr composites for efficient degradation of NO gas: Oxygen vacancy effect and near infrared light responsive mechanism. *Chem. Eng. J.* **2017**, *325*, 59–70. [CrossRef]
60. Jing, T.; Dai, Y.; Wei, W.; Ma, X.; Huang, B. Near-infrared photocatalytic activity induced by intrinsic defects in Bi_2MO_6 (M=W, Mo). *Phys. Chem. Chem. Phys.* **2014**, *16*, 18596–18604. [CrossRef]
61. Ma, X.; Wang, L.; Zhang, Q.; Jiang, H.L. Switching on the Photocatalysis of Metal–Organic Frameworks by Engineering Structural Defects. *Angew. Chem.* **2019**, *131*, 12303–12307. [CrossRef]
62. Yusran, Y.; Guan, X.; Li, H.; Fang, Q.; Qiu, S. Postsynthetic functionalization of covalent organic frameworks. *Natl. Sci. Rev.* **2020**, *7*, 170–190. [CrossRef] [PubMed]
63. Mitoraj, D.; Kisch, H. The nature of nitrogen-modified titanium dioxide photocatalysts active in visible light. *Angew. Chem. Int. Ed.* **2008**, *47*, 9975–9978. [CrossRef] [PubMed]
64. Roy, P.; Berger, S.; Schmuki, P. TiO_2 nanotubes: Synthesis and applications. *Angew. Chem. Int. Ed.* **2011**, *50*, 2904–2939. [CrossRef]
65. Nowotny, J.; Alim, M.A.; Bak, T.; Idris, M.A.; Ionescu, M.; Prince, K.; Sahdan, M.Z.; Sopian, K.; Teridi, M.A.M.; Sigmund, W. Defect chemistry and defect engineering of TiO_2-based semiconductors for solar energy conversion. *Chem. Soc. Rev.* **2015**, *44*, 8424–8442. [CrossRef]
66. Chen, X.; Liu, L.; Huang, F. Black titanium dioxide (TiO_2) nanomaterials. *Chem. Soc. Rev.* **2015**, *44*, 1861–1885. [CrossRef]
67. Di Valentin, C.; Pacchioni, G.; Selloni, A. Theory of carbon doping of titanium dioxide. *Chem. Mater.* **2005**, *17*, 6656–6665. [CrossRef]
68. Nah, Y.C.; Paramasivam, I.; Schmuki, P. Doped TiO_2 and TiO_2 nanotubes: Synthesis and applications. *ChemPhysChem* **2010**, *11*, 2698–2713. [CrossRef]
69. Bai, L.; Huang, H.; Yu, S.; Zhang, D.; Huang, H.; Zhang, Y. Role of transition metal oxides in g-C_3N_4-based heterojunctions for photocatalysis and supercapacitors. *J. Energy Chem.* **2022**, *64*, 214–235. [CrossRef]
70. Zheng, Y.; Jiao, Y.; Zhu, Y.; Cai, Q.; Vasileff, A.; Li, L.H.; Han, Y.; Chen, Y.; Qiao, S.-Z. Molecule-level g-C_3N_4 coordinated transition metals as a new class of electrocatalysts for oxygen electrode reactions. *J. Am. Chem. Soc.* **2017**, *139*, 3336–3339. [CrossRef]
71. Shakeel, M.; Arif, M.; Yasin, G.; Li, B.; Khan, H.D. Layered by layered Ni-Mn-LDH/g-C_3N_4 nanohybrid for multi-purpose photo/electrocatalysis: Morphology controlled strategy for effective charge carriers separation. *Appl. Catal. B Environ.* **2019**, *242*, 485–498. [CrossRef]
72. Nguyen, V.-H.; Mousavi, M.; Ghasemi, J.B.; Van Le, Q.; Delbari, S.A.; Asl, M.S.; Shokouhimehr, M.; Mohammadi, M.; Azizian-Kalandaragh, Y.; Namini, A.S. In situ preparation of g-C_3N_4 nanosheet/FeOCl: Achievement and promoted photocatalytic nitrogen fixation activity. *J. Colloid Interface Sci.* **2021**, *587*, 538–549. [CrossRef] [PubMed]
73. Liu, X.; Ma, R.; Zhuang, L.; Hu, B.; Chen, J.; Liu, X.; Wang, X. Recent developments of doped g-C_3N_4 photocatalysts for the degradation of organic pollutants. *Crit. Rev. Environ. Sci. Technol.* **2021**, *51*, 751–790. [CrossRef]
74. Lu, Q.; Eid, K.; Li, W.; Abdullah, A.M.; Xu, G.; Varma, R.S. Engineering graphitic carbon nitride (g-C_3N_4) for catalytic reduction of CO_2 to fuels and chemicals: Strategy and mechanism. *Green Chem.* **2021**, *23*, 5394–5428. [CrossRef]
75. Xiao, Y.; Tian, G.; Li, W.; Xie, Y.; Jiang, B.; Tian, C.; Zhao, D.; Fu, H. Molecule self-assembly synthesis of porous few-layer carbon nitride for highly efficient photoredox catalysis. *J. Am. Chem. Soc.* **2019**, *141*, 2508–2515. [CrossRef]
76. Das, D.; Shinde, S.; Nanda, K. Temperature-dependent photoluminescence of g-C_3N_4: Implication for temperature sensing. *ACS Appl. Mater. Interfaces* **2016**, *8*, 2181–2186. [CrossRef]
77. Liebig, J. Uber einige Stickstoff-Verbindungen. *Ann. Der Pharm.* **1834**, *10*, 1–47. [CrossRef]
78. Lin, L.; Yu, Z.; Wang, X. Crystalline carbon nitride semiconductors for photocatalytic water splitting. *Angew. Chem.* **2019**, *131*, 6225–6236. [CrossRef]
79. Kroke, E.; Schwarz, M.; Horath-Bordon, E.; Kroll, P.; Noll, B.; Norman, A.D. Tri-s-triazine derivatives. Part I. From trichloro-tri-s-triazine to graphitic C_3N_4 structures. *New J. Chem.* **2002**, *26*, 508–512. [CrossRef]
80. Wang, S.; Zhang, J.; Li, B.; Sun, H.; Wang, S.; Duan, X. Morphology-dependent photocatalysis of graphitic carbon nitride for sustainable remediation of aqueous pollutants: A mini review. *J. Environ. Chem. Eng.* **2022**, *10*, 107438. [CrossRef]
81. Hu, C.; Lin, Y.R.; Yang, H.C. Recent development of g-C_3N_4-based hydrogels as photocatalysts: A minireview. *ChemSusChem* **2018**, *12*, 1794–1806.
82. Dimiev, A.M.; Tour, J.M. Mechanism of graphene oxide formation. *ACS Nano* **2014**, *8*, 3060–3068. [CrossRef]
83. Zhang, H.; Lv, X.; Li, Y.; Wang, Y.; Li, J. P25-Graphene Composite as a High Performance Photocatalyst. *ACS Nano* **2010**, *4*, 380. [CrossRef] [PubMed]
84. Zhang, X.; Xie, X.; Wang, H.; Zhang, J.; Pan, B.; Xie, Y. Enhanced photoresponsive ultrathin graphitic-phase C3N4 nanosheets for bioimaging. *J. Am. Chem. Soc.* **2012**, *135*, 18–21. [CrossRef]
85. Xu, J.; Zhang, L.; Shi, R.; Zhu, Y. Chemical exfoliation of graphitic carbon nitride for efficient heterogeneous photocatalysis. *J. Mater. Chem. A* **2013**, *1*, 14766–14772. [CrossRef]
86. Niu, P.; Zhang, L.; Liu, G.; Cheng, H.M. Graphene-like carbon nitride nanosheets for improved photocatalytic activities. *Adv. Funct. Mater.* **2012**, *22*, 4763–4770. [CrossRef]
87. Gu, Z.; Asakura, Y.; Yin, S. High yield post-thermal treatment of bulk graphitic carbon nitride with tunable band structure for enhanced deNOx photocatalysis. *Nanotechnology* **2019**, *31*, 114001. [CrossRef]

88. Sun, S.; Liang, S. Recent advances in functional mesoporous graphitic carbon nitride (mpg-C_3N_4) polymers. *Nanoscale* **2017**, *9*, 10544–10578. [CrossRef]
89. Prieto, G.; Tüysüz, H.; Duyckaerts, N.; Knossalla, J.; Wang, G.-H.; Schüth, F. Hollow nano-and microstructures as catalysts. *Chem. Rev.* **2016**, *116*, 14056–14119. [CrossRef]
90. Shi, Y.; Wan, Y.; Zhao, D. Ordered mesoporous non-oxide materials. *Chem. Soc. Rev.* **2011**, *40*, 3854–3878. [CrossRef]
91. Zhang, J.; Guo, F.; Wang, X. An optimized and general synthetic strategy for fabrication of polymeric carbon nitride nanoarchitectures. *Adv. Funct. Mater.* **2013**, *23*, 3008–3014. [CrossRef]
92. Sun, J.; Zhang, J.; Zhang, M.; Antonietti, M.; Fu, X.; Wang, X. Bioinspired hollow semiconductor nanospheres as photosynthetic nanoparticles. *Nat. Commun.* **2012**, *3*, 1139. [CrossRef]
93. Wang, J.; Zhang, C.; Shen, Y.; Zhou, Z.; Yu, J.; Li, Y.; Wei, W.; Liu, S.; Zhang, Y. Environment-friendly preparation of porous graphite-phase polymeric carbon nitride using calcium carbonate as templates, and enhanced photoelectrochemical activity. *J. Mater. Chem. A* **2015**, *3*, 5126–5131. [CrossRef]
94. Yang, Z.; Zhang, Y.; Schnepp, Z. Soft and hard templating of graphitic carbon nitride. *J. Mater. Chem. A* **2015**, *3*, 14081–14092. [CrossRef]
95. Barrio, J.; Shalom, M. Rational Design of Carbon Nitride Materials by Supramolecular Preorganization of Monomers. *ChemCatChem* **2018**, *10*, 5573–5586. [CrossRef]
96. Zhang, W.; Zhao, Z.; Dong, F.; Zhang, Y. Solvent-assisted synthesis of porous g-C_3N_4 with efficient visible-light photocatalytic performance for NO removal. *Chin. J. Catal.* **2017**, *38*, 372–378. [CrossRef]
97. Wang, X.; Chen, X.; Thomas, A.; Fu, X.; Antonietti, M. Metal-containing carbon nitride compounds: A new functional organic–metal hybrid material. *Adv. Mater.* **2009**, *21*, 1609–1612. [CrossRef]
98. Ding, Z.; Chen, X.; Antonietti, M.; Wang, X. Synthesis of transition metal-modified carbon nitride polymers for selective hydrocarbon oxidation. *ChemSusChem* **2011**, *4*, 274–281. [CrossRef]
99. Pan, H.; Zhang, Y.-W.; Shenoy, V.B.; Gao, H. Ab initio study on a novel photocatalyst: Functionalized graphitic carbon nitride nanotube. *ACS Catal.* **2011**, *1*, 99–104. [CrossRef]
100. Xiong, T.; Cen, W.; Zhang, Y.; Dong, F. Bridging the g-C_3N_4 interlayers for enhanced photocatalysis. *ACS Catal.* **2016**, *6*, 2462–2472. [CrossRef]
101. Zhang, M.; Bai, X.; Liu, D.; Wang, J.; Zhu, Y. Enhanced catalytic activity of potassium-doped graphitic carbon nitride induced by lower valence position. *Appl. Catal. B Environ.* **2015**, *164*, 77–81. [CrossRef]
102. Wang, Z.; Chen, M.; Huang, Y.; Shi, X.; Zhang, Y.; Huang, T.; Cao, J.; Ho, W.; Lee, S.C. Self-assembly synthesis of boron-doped graphitic carbon nitride hollow tubes for enhanced photocatalytic NOx removal under visible light. *Appl. Catal. B Environ.* **2018**, *239*, 352–361. [CrossRef]
103. Huang, Y.; Yu, Y.; Yu, Y.; Zhang, B. Oxygen Vacancy Engineering in Photocatalysis. *Sol. RRL* **2020**, *4*, 2000037. [CrossRef]
104. Morgan, B.J.; Watson, G.W. A DFT+ U description of oxygen vacancies at the TiO_2 rutile (1 1 0) surface. *Surf. Sci.* **2007**, *601*, 5034–5041. [CrossRef]
105. Linh, N.H.; Nguyen, T.Q.; Diño, W.A.; Kasai, H. Effect of oxygen vacancy on the adsorption of O_2 on anatase TiO_2 (001): A DFT-based study. *Surf. Sci.* **2015**, *633*, 38–45. [CrossRef]
106. Ran, L.; Hou, J.; Cao, S.; Li, Z.; Zhang, Y.; Wu, Y.; Zhang, B.; Zhai, P.; Sun, L. Defect engineering of photocatalysts for solar energy conversion. *Sol. RRL* **2020**, *4*, 1900487. [CrossRef]
107. Chen, X.; Liu, L.; Yu, P.Y.; Mao, S.S. Increasing solar absorption for photocatalysis with black hydrogenated titanium dioxide nanocrystals. *Science* **2011**, *331*, 746–750. [CrossRef] [PubMed]
108. Wang, G.; Ling, Y.; Li, Y. Oxygen-deficient metal oxide nanostructures for photoelectrochemical water oxidation and other applications. *Nanoscale* **2012**, *4*, 6682–6691. [CrossRef]
109. Wang, Z.; Huang, Y.; Chen, M.; Shi, X.; Zhang, Y.; Cao, J.; Ho, W.; Lee, S.C. Roles of N-Vacancies over Porous g-C_3N_4 Microtubes during Photocatalytic NO x Removal. *ACS Appl. Mater. Interfaces* **2019**, *11*, 10651–10662. [CrossRef]
110. Li, Y.; Gu, M.; Shi, T.; Cui, W.; Zhang, X.; Dong, F.; Cheng, J.; Fan, J.; Lv, K. Carbon vacancy in C_3N_4 nanotube: Electronic structure, photocatalysis mechanism and highly enhanced activity. *Appl. Catal. B Environ.* **2020**, *262*, 118281. [CrossRef]
111. Gu, Z.; Cui, Z.; Wang, Z.; Qin, K.S.; Asakura, Y.; Hasegawa, T.; Tsukuda, S.; Hongo, K.; Maezono, R.; Yin, S. Carbon vacancies and hydroxyls in graphitic carbon nitride: Promoted photocatalytic NO removal activity and mechanism. *Appl. Catal. B Environ.* **2020**, *279*, 119376. [CrossRef]
112. Bojdys, M.J.; Müller, J.O.; Antonietti, M.; Thomas, A. Ionothermal synthesis of crystalline, condensed, graphitic carbon nitride. *Chem. -A Eur. J.* **2008**, *14*, 8177–8182. [CrossRef] [PubMed]
113. Lin, L.; Ou, H.; Zhang, Y.; Wang, X. Tri-s-triazine-based crystalline graphitic carbon nitrides for highly efficient hydrogen evolution photocatalysis. *ACS Catal.* **2016**, *6*, 3921–3931. [CrossRef]
114. Ren, W.; Cheng, J.; Ou, H.; Huang, C.; Anpo, M.; Wang, X. Optimizing the Crystallization Process of Conjugated Polymer Photocatalysts to Promote Electron Transfer and Molecular Oxygen Activation. *J. Catal.* **2020**, *389*, 636–645. [CrossRef]
115. Li, Y.; Zhang, D.; Feng, X.; Xiang, Q. Enhanced photocatalytic hydrogen production activity of highly crystalline carbon nitride synthesized by hydrochloric acid treatment. *Chin. J. Catal.* **2020**, *41*, 21–30. [CrossRef]
116. Gu, Z.; Cui, Z.; Wang, Z.; Chen, T.; Sun, P.; Wen, D. Synthesis of crystalline carbon nitride with enhanced photocatalytic NO removal performance: An experimental and DFT theoretical study. *J. Mater. Sci. Technol.* **2021**, *83*, 113–122. [CrossRef]

117. Liu, D.; Yao, J.; Chen, S.; Zhang, J.; Li, R.; Peng, T. Construction of rGO-coupled C_3N_4/C_3N_5 2D/2D Z-scheme heterojunction to accelerate charge separation for efficient visible light H_2 evolution. *Appl. Catal. B Environ.* **2022**, *318*, 121822. [CrossRef]
118. Xu, Q.; Ma, D.; Yang, S.; Tian, Z.; Cheng, B.; Fan, J. Novel $g-C_3N_4/g-C_3N_4$ S-scheme isotype heterojunction for improved photocatalytic hydrogen generation. *Appl. Surf. Sci.* **2019**, *495*, 143555. [CrossRef]
119. Lin, F.; Zhou, S.; Wang, G.; Wang, J.; Gao, T.; Su, Y.; Wong, C.-P. Electrostatic self-assembly combined with microwave hydrothermal strategy: Construction of 1D/1D carbon nanofibers/crystalline $g-C_3N_4$ heterojunction for boosting photocatalytic hydrogen production. *Nano Energy* **2022**, *99*, 107432. [CrossRef]
120. Zhao, D.; Wang, Y.; Dong, C.-L.; Huang, Y.-C.; Chen, J.; Xue, F.; Shen, S.; Guo, L. Boron-doped nitrogen-deficient carbon nitride-based Z-scheme heterostructures for photocatalytic overall water splitting. *Nat. Energy* **2021**, *6*, 388–397. [CrossRef]
121. Kočí, K.; Reli, M.; Troppová, I.; Šihor, M.; Kupková, J.; Kustrowski, P.; Praus, P. Photocatalytic decomposition of N_2O over $TiO_2/g-C_3N_4$ photocatalysts heterojunction. *Appl. Surf. Sci.* **2017**, *396*, 1685–1695. [CrossRef]
122. Zhang, G.; Zhu, X.; Chen, D.; Li, N.; Xu, Q.; Li, H.; He, J.; Xu, H.; Lu, J. Hierarchical Z-scheme $g-C_3N_4/Au/ZnIn_2S_4$ photocatalyst for highly enhanced visible-light photocatalytic nitric oxide removal and carbon dioxide conversion. *Environ. Sci. Nano* **2020**, *7*, 676–687. [CrossRef]
123. Li, Y.; Zhou, M.; Cheng, B.; Shao, Y. Recent advances in $g-C_3N_4$-based heterojunction photocatalysts. *J. Mater. Sci. Technol.* **2020**, *56*, 1–17. [CrossRef]
124. Li, G.; Guo, J.; Hu, Y.; Wang, Y.; Wang, J.; Zhang, S.; Zhong, Q. Facile synthesis of the Z-scheme graphite-like carbon nitride/silver/silver phosphate nanocomposite for photocatalytic oxidative removal of nitric oxides under visible light. *J. Colloid Interface Sci.* **2021**, *588*, 110–121. [CrossRef] [PubMed]
125. Xu, Q.; Zhang, L.; Cheng, B.; Fan, J.; Yu, J. S-scheme heterojunction photocatalyst. *Chem* **2020**, *6*, 1543–1559. [CrossRef]
126. Fu, J.; Xu, Q.; Low, J.; Jiang, C.; Yu, J. Ultrathin 2D/2D $WO_3/g-C_3N_4$ step-scheme H_2-production photocatalyst. *Appl. Catal. B Environ.* **2019**, *243*, 556–565. [CrossRef]
127. Van Pham, V.; Mai, D.-Q.; Bui, D.-P.; Van Man, T.; Zhu, B.; Zhang, L.; Sangkaworn, J.; Tantirungrotechai, J.; Reutrakul, V.; Cao, T.M. Emerging 2D/0D $g-C_3N_4/SnO_2$ S-scheme photocatalyst: New generation architectural structure of heterojunctions toward visible-light-driven NO degradation. *Environ. Pollut.* **2021**, *286*, 117510. [CrossRef]
128. Cheng, L.; Zhang, H.; Li, X.; Fan, J.; Xiang, Q. Carbon-graphitic carbon nitride hybrids for heterogeneous photocatalysis. *Small* **2021**, *17*, 2005231. [CrossRef]
129. Zhu, B.; Cheng, B.; Fan, J.; Ho, W.; Yu, J. $g-C_3N_4$-based 2D/2D composite heterojunction photocatalyst. *Small Struct.* **2021**, *2*, 2100086. [CrossRef]
130. Ai, L.; Shi, R.; Yang, J.; Zhang, K.; Zhang, T.; Lu, S. Efficient Combination of $g-C_3N_4$ and CDs for Enhanced Photocatalytic Performance: A Review of Synthesis, Strategies, and Applications. *Small* **2021**, *17*, 2007523. [CrossRef]
131. Gu, Z.; Zhang, B.; Asakura, Y.; Tsukuda, S.; Kato, H.; Kakihana, M.; Yin, S. Alkali-assisted hydrothermal preparation of $g-C_3N_4/rGO$ nanocomposites with highly enhanced photocatalytic NO_x removal activity. *Appl. Surf. Sci.* **2020**, *521*, 146213. [CrossRef]

Disclaimer/Publisher's Note: The statements, opinions and data contained in all publications are solely those of the individual author(s) and contributor(s) and not of MDPI and/or the editor(s). MDPI and/or the editor(s) disclaim responsibility for any injury to people or property resulting from any ideas, methods, instructions or products referred to in the content.

Review

Recent Progress of Printing Technologies for High-Efficient Organic Solar Cells

Zihao Xia, Ting Cai, Xiangguo Li, Qian Zhang *, Jing Shuai * and Shenghua Liu *

School of Materials, Shenzhen Campus of Sun Yat-sen University, No. 66, Gongchang Road, Guangming District, Shenzhen 518107, China
* Correspondence: zhangqian6@mail.sysu.edu.cn (Q.Z.); shuaij3@mail.sysu.edu.cn (J.S.); liushengh@mail.sysu.edu.cn (S.L.)

Abstract: Organic solar cells (OSCs), as a renewable energy technology that converts solar energy into electricity, have exhibited great application potential. With the rapid development of novel materials and device structures, the power conversion efficiency (PCE) of non-fullerene OSCs has been increasingly enhanced, and over 19% has currently been achieved in single-junction devices. Compared with rigid silicon cells, OSCs have the characteristics of low cost, high flexibility, lightweight, and their inherent solution processability, which enables the devices to be manufactured by using printing technology for commercial applications. In recent years, to maximize the device performance of OSCs, many efforts have been devoted to improving the morphologies and properties of the active layer through various novel printing technologies. Herein, in this review, the recent progress and applications of several popular printing technologies to fabricate high-efficient OSCs are summarized, including blade-coating, slot-die coating, gravure printing, screen printing, inkjet printing, etc. The strengths and weaknesses of each printing technology are also outlined in detail. Ultimately, the challenges and opportunities of printing technology to fabricate OSC devices in industrial manufacturing are also presented.

Keywords: organic solar cells; printing technology; blade-coating; slot-die coating; inkjet printing

Citation: Xia, Z.; Cai, T.; Li, X.; Zhang, Q.; Shuai, J.; Liu, S. Recent Progress of Printing Technologies for High-Efficient Organic Solar Cells. *Catalysts* **2023**, *13*, 156. https://doi.org/10.3390/catal13010156

Academic Editor: Bruno Fabre

Received: 12 December 2022
Revised: 27 December 2022
Accepted: 28 December 2022
Published: 9 January 2023

Copyright: © 2023 by the authors. Licensee MDPI, Basel, Switzerland. This article is an open access article distributed under the terms and conditions of the Creative Commons Attribution (CC BY) license (https://creativecommons.org/licenses/by/4.0/).

1. Introduction

The extreme climate issue caused by resource depletion and excessive carbon emissions has become increasingly serious. To achieve a global carbon-neutral demand, it is urgent to replace the utilization of traditional fossil fuels with sustainable renewable energy. Solar energy, as the largest renewable energy on earth, is the key to satisfying the future energy demand. Currently, many researchers are focused on the development of high-efficient solar cells with low cost and high stability. Nowadays, the current photovoltaic market is dominated by traditional silicon-based solar cells owing to their ultrahigh conversion efficiency and long-term stability [1]. However, the complex manufactural processes, high cost, and insufficient flexibility of the device limit their further application in portable energy devices and wearable electronics.

The third-generation solar cells, including dye-sensitized solar cells, perovskite solar cells (PSCs), and organic solar cells (OSCs), have been developed rapidly to overcome the drawbacks of traditional solar cell technologies. The OSCs, as one of the representatives, exhibit great application potential due to their mechanical flexibility, semitransparency, and wearability [2–8]. Recently, great efforts have been made over the past few decades to synthesize novel acceptors and to substitute the fullerene systems with PCE enhancements simultaneously. Fortunately, a fused ring non-fullerene acceptor (NFA) called ITIC was first reported by Zhan's group [9]. These NFAs exhibit wide light absorption ranges and tunable bandgaps and are easily modified. Since then, lots of NFAs have been synthesized, and great achievements have been obtained in the efficiency breakthrough of OSCs. The record of power conversion efficiency (PCEs) in NFAs-OSCs has increased rapidly and has reached over 18% [10–15].

Meanwhile, research on large-scale and flexible OSCs has also attracted much attention due to their practical applications. In 2021, Huang's group [16] adopted a self-organization method by adding 2PACz into the active layer to process flexible and large-area OSCs (device area 1.0 cm^2) and achieved the highest PCE of 15.8%. More recently, Xie et al. [17]. reported 21 cm^2 flexible organic modules with an AgNWs-polymer transparent film as the top electrode, delivering an impressive PCE of 12.3%. However, these devices were mainly fabricated by the spin-coating method. Commonly, the spin-coating technique has been used to deposit high-quality thin films in OSCs with small areas [18–22]. The thickness of the film can be adjusted by controlling the spin-cast speed and solution viscosity. The solution was deposited on the substrate and dried quickly to avoid the risk of donor or acceptor aggregation in the wet film. Nevertheless, when it comes to large-scale substrates, the spin-coating method is inappropriate for controlling the film thickness and uniformity accurately. In addition, only a few materials remain on the substrate, most of which are spun off, resulting in a high percentage of material waste. In order to meet the requirements of future applications and industrial manufacturing, other fabrication technologies are urgently needed. Particularly, printing technologies are desirable and highly suitable for manufacturing large-area OSCs, such as blade-coating, slot-die coating, gravure printing, screen printing, inkjet printing et al. [23–27]. Recently, Wei et al. [28]. employed the slot-die method to fabricate 1 cm^2 flexible OSCs. With the fine-tuning of active layer morphology and flexible substrate properties, an efficiency of 12.16% was achieved, which was very close to the spin-coated rigid device (PCE of 12.37%). Furthermore, the same group studied the behaviors of film-dry kinetics during the slot-die process [29]. The PCE of 13.70% was obtained for 1 cm^2 large-area flexible OSCs. The 30 cm^2 flexible cells also delivered an impressive PCE of 12.20% due to the high tolerance of the film thickness. However, at present, less attention has been paid to the printing methods relative to spin coating. The performance of the lab-scale OSCs, based on printing technologies, is still lower than that of spin-coating. These printing technologies are more suitable to be adopted for the scalability and mass production of OSCs, which need to be further improved. Therefore, in this review, we summarized the recent progress of printing technologies, including blade-coating, slot-die coating, gravure printing, screen printing, inkjet printing et al. We first discussed the organic active layer and interfacial layer materials used for printing OSCs. The process, characteristics, and application of the above printing methods to fabricate the OSCs are also emphasized, including the strengths and weaknesses of each printing technology, which are outlined. In addition, the perspective for printing large-scale and flexible OSCs is also presented. We expect that this review can provide new strategies to accelerate the fabrication of OSCs with printing technology.

2. Device Structure and Materials in Organic Solar Cells

2.1. Device Architecture

The architecture of OSCs could be categorized into several types: a single-layer Schottky structure, bilayer planar heterostructure, bulk heterostructure (BHJ), and layer-by-layer (LBL) structure. For the simple single-layer structures studied in the early OSC, it is difficult to decompose the exciton into free electrons and holes due to the lack of a built-in electric field. Subsequently, a bilayer planar heterojunction solar cell (PHJ), as revealed in Figure 1a, was developed. The exciton in a PHJ device can generate charge transfer at the donor and acceptor interface, making exciton dissociation more effective. Then, when the concept of BHJ solar cells was introduced by Heeger et al. in 1995 [30], OSCs developed rapidly. The conventional structure of BHJ OSCs is shown in Figure 1b and includes a photoactive layer, hole transporting layers (HTLs), and electronic transporting layers (ETLs) coupled with electrodes. In the active layer of BHJ, the bi-continuous interpenetrating networks are formed with the mix of donor and acceptor materials, and the increasing D-A interfacial area facilitates the charge generation and separation effectively, leading to an improvement in PCE of OSCs [31]. For now, the BHJ devices have been a major research focus in the OSC field. The solution processability of active layer materials makes it possible to form electronic organic ink, which can be adopted for efficient printing technologies [32–37].

More recently, a sequential layer-by-layer (LBL) (Figure 1c) processed solar cells emerged and provided high device efficiency [38]. Such a strategy can form a pseudo bimolecular layer (p-i-n) structure in the active layer, in which appropriate vertical phase separation can be formed to promote exciton dissociation and optimize charge transport at the corresponding electrode to reduce energy losses. The first successful examples of highly efficient OSCs based on an LBL structure were reported simultaneously by two groups in 2018. Friend et al. [39] fabricated conventional LBL OSCs based on ITO/PEDOT: PSS/NCBDT/PBDB-T/PDINO/Al. They optimize photovoltaic performance by controlling the thickness of each active layer only, and the device achieved a high PCE of 10.19%, equivalent to the 10.04% PCE obtained by BHJ OSCs. Hou et al. [40] selected PBDB-TFS1 as the polymer donor and IT-4F as the non-fullerene acceptor, where the THF solvent was used for processing IT-4F on top of the layer. It was found that when the THF solvent was treated on PBDB-TFS1 films, the quality of the film could be maintained effectively. In order to control the inter-diffusion between PBDB-TFS1 and IT-4F, O-dichlorobenzene was introduced into THF as a cosolvent. During the film preparation, o-DCB can induce IT-4F molecules to penetrate into the bottom layer successfully; therefore, the vertical phase distribution can be adjusted by changing the amount of o-DCB. As a result, researchers found that as the amount of o-DCB increases, the device efficiency of LBL OSCs gradually increases from 8.11% to 13.0%. This result demonstrated that the optimized efficiency of the LBL device is higher than that of traditional BHJ OSCs (PCE: 11.8%), which has led to the rapid development of OSCs based on an LBL structure since then.

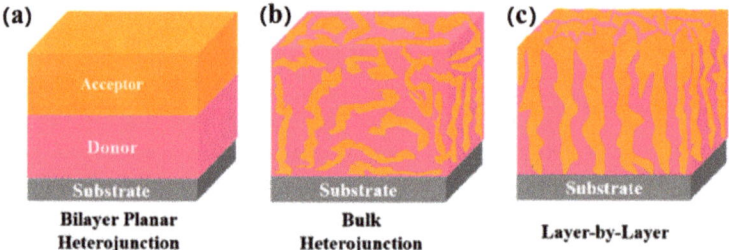

Figure 1. Schematic architecture of (**a**) Bilayer planar heterojunction, (**b**) Bulk heterojunction, and (**c**) Layer-by-layer. Adapted with permission from Ref. [41]. Copyright 2021 American Chemical Society.

2.2. Active Layer Materials

The active layer materials have a direct impact on the photovoltaic performance of OSCs. Generally speaking, the active layer of heterojunction devices consists of two parts, namely the donor and acceptor material. Among them, the energy level differences between the donor and acceptor provide a certain driving force for exciton separation, which requires the cascaded bandgap of the donor and the acceptor (the LUMO energy level of the donor material is at least 0.3~0.5 eV higher than that of the acceptor). Thus, the excitons can be separated at the donor/acceptor interface effectively. In addition, to satisfy the requirements of industrial fabrication, it is important to develop materials that can be prepared with an environmentally friendly solvent without toxicity. In addition, designing the active layer materials with a thickness insensitive is also valuable for large-area device fabrication [42–45]. Table 1 summarizes the recent progress in fabricating NFA-based OSCs with various printing technologies. According to the active layer materials and fabrication techniques, we can observe that blade coating and slot-die coating are the main printing methods to fabricate the high-efficient NFA OSCs with a maximum PCE beyond 17%, although the device areas are generally less than 1 cm^2, which should be further improved.

Table 1. The performance of NFA-based OSCs by printing technologies.

Device Structure	Active Layer Materials	Processing Method	Device Area [cm^2]	PCE [%]	Year	Ref.
ITO/ZnO/BHJ/MoO$_3$/Al	PBDB-TF:IT-4F	Blade coating	1.04	9.22	2018	[46]
ITO/ZnO/BHJ/MoO$_3$/Al	PBDB-TF:IT-4F	Blade coating	0.12	12.88	2018	[46]
ITO/PEDOT:PSS/BHJ/PFN–Br/Al	PBDB-T-SF:IT-4F	Slot-die coating	0.1	12.9	2019	[47]
ITO/PEDOT:PSS/BHJ/PFN–Br/Al	PBDB-T-SF:IT-4F	Slot-die coating	0.2	12.32	2019	[47]
ITO/PEDOT:PSS/BHJ/ZrAcac/Al	PM6:IT-4F	Blade coating	0.04	13.64	2019	[48]
ITO/PEDOT:PSS/BHJ/ZrAcac/Al	PM6:IT-4F	Blade coating	0.56	11.39	2019	[48]
ITO/ZnO/BHJ/MoO$_3$/Al	PBDB-T:i-IEICO-4F	Blade coating	0.04	11.6	2019	[49]
ITO/ZnO/BHJ/MoO$_3$/Al	PBDB-T:i-IEICO-4F	Slot-die coating	0.04	12.5	2019	[49]
ITO/PEDOT:PSS/LBL/PDINO/Al	J71:ITC6-IC	Blade coating	0.04	11.47	2019	[50]
ITO/PEDOT:PSS/LBL/PNDIT-F3N-Br/Al	PM6:Y6	Blade coating	0.04	16.35	2020	[51]
ITO/PEDOT:PSS/LBL/PNDIT-F3N-Br/Al	PM6:Y6	Blade coating	1	15.23	2020	[51]
ITO/ZnO/BHJ/MoO$_3$/Al	PM7:IT4F	Slot-die coating	0.04	13.2	2020	[52]
ITO/PEDOT:PSS/BHJ/PDINO/Al	PBDB-TF:BTP-4Cl-12	Blade coating	0.81	15.5	2020	[20]
ITO/ZnO NPs/BHJ/MoO$_3$/Ag	PTB7-Th:EH-IDTBR	Slot-die coating	1	9.43	2020	[53]
ITO/ZnO/BHJ/MoO$_3$/Ag	PM6:Y6	Slot-die coating	5.6	15.6	2020	[54]
ITO/PEDOT:PSS/BHJ/AZO/Ag	PTB7-Th:IEICO-4F	Inkjet printing	0.1	9.5	2020	[55]
ITO/ZnO/BHJ/MoO$_X$/Al	PM6:Y6	Slot-die coating	0.04	15.93	2021	[56]
ITO/ZnO/BHJ/MoO$_X$/Al	PM6:Y6	Slot-die coating	0.56	13.91	2021	[56]
ITO/PEDOT:PSS/BHJ/PFN-Br/Ag	PBDB-T:PYT	Blade coating	0.04	15.01	2021	[57]
ITO/PEDOT:PSS/BHJ/PFN-Br/Ag	PM6:BTP-eC9	Blade coating	0.04	16.77	2021	[58]
ITO/PEDOT:PSS/BHJ/PDINO/Al	PM6:BTP-eC9	Blade coating	0.04	16.58	2022	[59]
ITO/ZnO/BHJ/MoO$_3$/Al	D18:Y6	Slot-die coating	0.04	17.13	2022	[60]
ITO/ZnO/LBL/MoO$_3$/Al	PM6:BTP-BO-4Cl	Inkjet printing	0.04	13.09	2022	[61]
ITO/ZnO/BHJ/MoO$_3$/Al	D18:BTR–Cl:Y6	Slot-die coating	0.04	17.2	2022	[62]
ITO/ZnO/BHJ/MoO$_3$/Al	D18:BTR–Cl:Y6	Slot-die coating	1	16.3	2022	[62]
ITO/AZO/LBL/MoO$_3$/Al	PM6:Y6	Blade coating	0.1	16.26	2022	[63]

For donor materials, conjugated polymers are widely used due to their wide light absorption wavelength and excellent molecular packing structure [64–66]. The design of the printable donor materials should meet the following requirements. Firstly, the materials should have strong and wide absorption in the visible and near-infrared regions, which is crucial to the improvement of the device current. Secondly, the donor material should possess higher and more balanced carrier mobility to ensure that sufficient photogenerated excitons and charges are transported to the electrodes with less recombination. Thirdly, the material should exhibit good solubility and miscibility, which is conducive to the film morphology and avoid excess aggregation in the solvent. At present, plenty of donor materials show good photovoltaic performance. Among these, the materials based on benzodithiophene are the most excellent donors, such as PM6, PM7, D18, and their derivatives, as shown in Figure 2.

The development of the acceptors has meant that a transition has been experienced from fullerene to non-fullerene materials. Figure 3 lists several frequently used NFAs. For the requirement of acceptor materials, the matchable energy levels between donor and acceptor materials are necessary. Additionally, the crystallinity and solubility of the acceptor materials should also be emphasized, which are beneficial to the formation of the appropriate domain size and phase separation in the active layer. In 2019, Zou et al. reported a new non-fullerene acceptor, Y6 [67]. This molecule replaces the sp3 hybrid carbon atom in the acceptor with the nitrogen atom in the pyrrole ring, thus, reducing energy loss and improving electron mobility effectively. Y6 and its derivatives are currently the best non-fullerene acceptors for photovoltaic performance, which greatly promotes the development of OSCs, making the energy conversion efficiency reach 15~19%. Currently, the modification of the Y series is a hotspot in the research of OSCs. Yan et al. [68] further studied the effect of the position of alkyl chain branches on the photovoltaic performance of the device. It was found that the change in the position of alkyl chain branches could make a significant difference between the molecular stack and the morphology of the blend film. Sidechain engineering has made great achievements in controlling molecular crystallization properties, which results in the formation of high-quality films in the printing process.

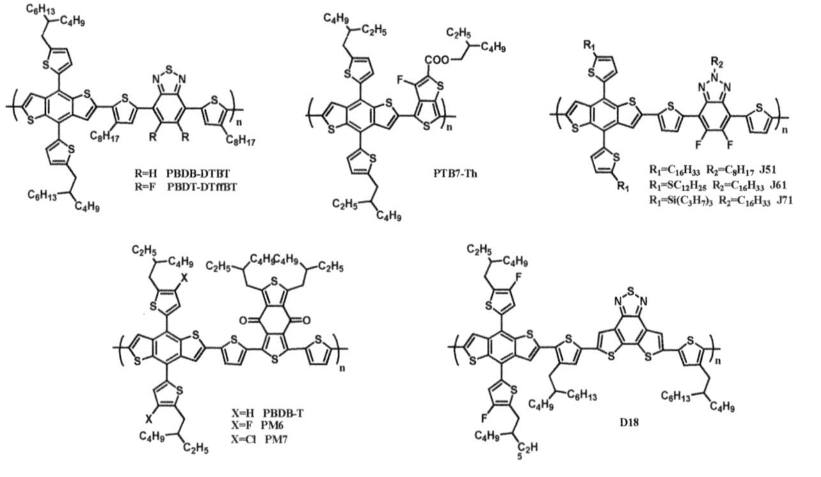

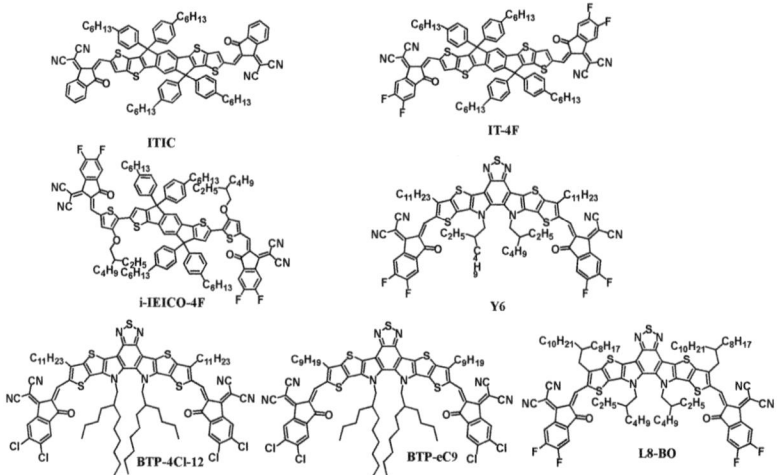

Figure 2. High-efficient printable donor materials listed in Table 1.

Figure 3. High efficiency printable acceptor materials listed in Table 1.

2.3. Interfacial Layer Materials

The charge transporting layers are usually called the buffer layers or the modification layers and are sandwiched between the electrode and active layer, playing a critical role in facilitating charge extraction, trap passivation, and carrier transport. Introducing organic or inorganic interface materials to realize the ohmic contact and optimize charge transport between the active layer and electrode is an effective way to improve the performance of OSCs.

As for the conventional structure of OSCs, a PEDOT:PSS aqueous solution is the most widely used hole transporting layer material because of its excellent electrical conductivity with a high adjustability and surface wettability solution deposition [69–72]; the molecular structures of this are presented in Figure 4a. Many strategies have been devoted to modifying the properties of such conductive solutions. Recently, Howells et al. [73] introduced a polymeric fluorinated additive into PEDOT:PSS, which avoided water diffusion into the hygroscopic HTLs and protected the Al electrode from deteriorating. In addition, the fluorinated additive also allowed the HTLs to achieve suitable built-in electric field distributions to obtain a faster charge extraction. Xu's groups [74] added a multifunctional organic material (2,3-dihydroxypyridine, DOH) into the PEDOT:PSS aqueous solution as an additive, which enhanced the conductivity and hole mobility of HTLs. They found that DOH doping can facilitate the phase separation between the PEDOT and PSS chains by inducing the conformational transformation of the PEDOT chain (Figure 4b). As a result, the device efficiency based on DOH incorporation was enhanced by 20%, with a significant improvement in thermal and air stability. Kee et al. [75] added different ionic liquids into HTLs to regulate the molecular ordering of PEDOT:PSS. By controlling the counter-ion exchange between the ionic liquids and PEDOT:PSS, the molecular packing of PEDOT was rearranged (Figure 4e,f). Due to the planar and rigid molecular structures, PSS and PEDOT molecules were reassembled through a strong π–π interaction, leading to a highly ordered nanofilm of PEDOT, making the charge transfer more effective. Hou's group [76] incorporated WO_x nanoparticles into HTLs, improving the interfacial properties and device performance successfully. After blending WO_x with PEDOT:PSS, the film showed better transparency, leading to a higher short current of the OSCs. According to morphological analysis, the physical crosslinking was formed due to the fine interaction between WO_x and PEDOT:PSS. Therefore, the surface free energy and phase deviation of the thin film increased significantly when compared to the pure WO_x and pristine PEDOT:PSS as the HTL. As a result, a raising fill factor of 80.79% and PCE of 14.57% were achieved based on the PBDB-TF:IT-4F BHJ device. In addition, some other inorganic HTLs materials, including NiO_x, MoO_x, VO_x, and Ag NWs, have also been introduced into the PEDOT:PSS solution as additives to tune the work function and enhance the device performance effectively [77–80].

The modified layer between the active layer and the cathode is called the electron transporting layer. Zinc oxide (ZnO) is a wide bandgap metal oxide with high electron mobility used as the ETLs material. Hou et al. developed a high-quality printable ZnO layer through sol-gel technology. They found that the OSCs with a ZnO precursor synthesized by an n-propylamine (PA) Lewis base exhibited the best performance; it could suppress the bumps and coffee rings during the blade coating process. Thus, the 1 cm^2 flexible OSC fabricated with PA-ZnO ETLs showed excellent photostability and obtained a high PCE of 16.71%. Although many efforts have demonstrated that devices with ZnO ETLs show excellent photovoltaic performance [46,81–83], the amine residues from the ZnO preparation process affect the active layer materials, deteriorating the device's efficiency [84,85]. Moreover, the fluidic characteristics of ZnO ETLs in the printing process are still unclear and need further research in the application of large-area OSCs [86]. As ZnO alternatives, some alcohol-soluble organic ETL materials with high conductivity could afford a suitable energy level alignment at the electrode interface. The naphthalene diimide (NDI), perylene diimide (PDI), and their derivatives are also most representative of ETLs. Zhang et al. [87] reported an aliphatic amino-functionalized PDI derivative named PDINN. It not only has excellent electronic transmission performance but also has high crystallinity.

When introducing PDINN as the ETLs, the charge transfer performance of the device is effectively improved, and the OSC based on PM6:Y6 achieved a high PCE of 17.23%. Liu et al. [88] introduced imidazole functional groups into the molecules to substitute the amine terminal group. Two novel interlayers, molecular NDI-M and PDI-Ms, were synthesized successfully by a condensation reaction. Compared to the molecular with amine groups, the imidazole-group-based molecules of NDI-M and PDI-M have deeper energy levels, as shown in Figure 4c, which facilitated the electron extraction and enhanced the charge transport at the interface. The devices based on D18-Cl:Y6:PC$_{71}$BM with the PDI-M as ETLs obtained a high PCE of 17.98%. Figure 4d–f shows the versatility and applicability of these imidazole-functionalized small molecules in different OSCs.

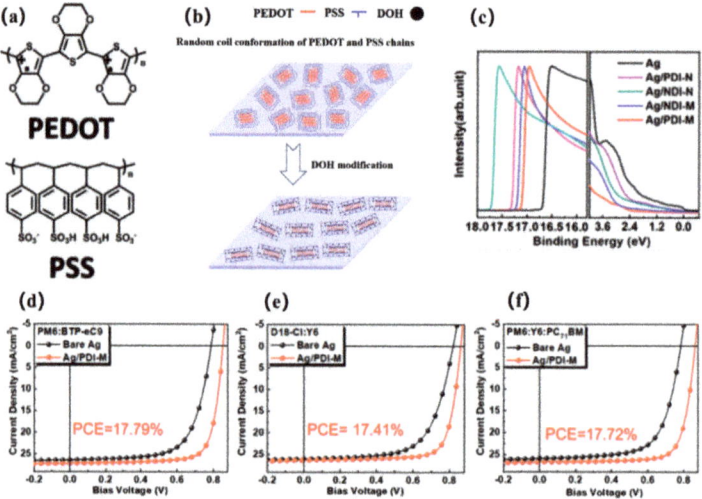

Figure 4. (a) Molecular structures of PEDOT and PSS. (b) The conformational transformation of PEDOT and PSS chains. (c) UPS of various ETLs (d–f), *J–V* curves of different OSCs with and w/o PDI-M as ETL. Adapted with permission from Ref. [88]. Copyright 2021 American Chemical Society.

3. Methods for Printing Technologies

Recently, a variety of film-forming technologies for solution-processed OSCs have been developed. The high-efficiency OSCs are basically manufactured by the spin-coating method. Despite the excellent behaviors of lab-scaling devices with spin-casts, recently, this method has not been favorable for upscale production [89]. When it comes to large substrates, the limitation of the spin process leads to inhomogeneous thickness at the edge of the blend films, affecting the device's performance. In addition, a large amount of solution wasted in the spin coating process further increases the costs. Therefore, for the mass and scalability production of OSCs, upscale printing fabrication technologies are urgently needed as shown in Figure 5. One type of printing method is when the mixed solution is transferred to the substrate by pouring, spraying, or casting. There is no contact between the coating head and the substrate. These methods can be divided into blade coating, slot-die coating, spray coating et al. On the other hand, the gravure printing, screen printing, flexographic printing, et al. methods are stamping processes in which the solution and substrate contact directly and can form two-dimensional patterns. In addition, inkjet printing has favorable applications for producing complex patterns. At present, fabrication technologies allow mechanical flexibility and solution-processable materials to be printed on flexible or rigid substrates, which are simple, efficient, and environmentally friendly methods for film fabrication and can also be integrated with Roll-to-Roll processing [90–92].

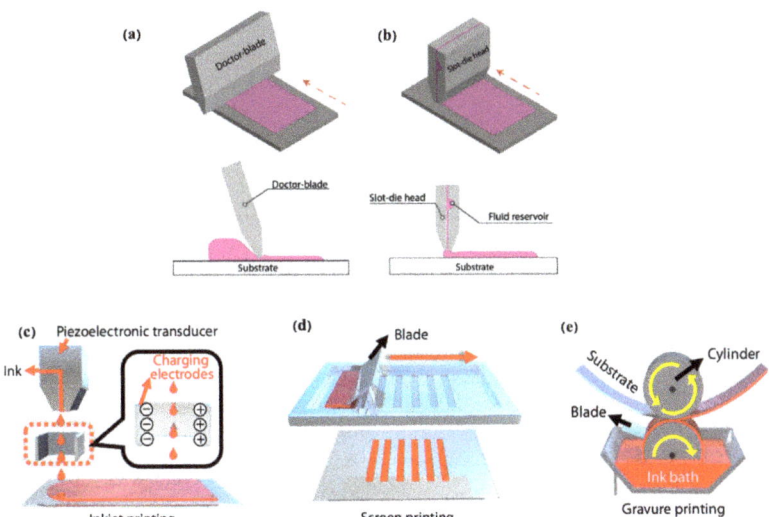

Figure 5. Schematic representations of: (**a**) Blade coating and (**b**) Slot-die coating. Adapted with permission from Ref. [23]. Copyright 2022 Wiley–VCH (**c**) Inkjet printing, (**d**) Screen printing, and (**e**) Gravure printing. Adapted with permission from Ref. [16]. Copyright 2022 Chinese Chemical Society.

3.1. Blade Coating

Blade-coating, known as doctor-blade, is a promising printing technique for preparing large-scale films. It is also a continuous fabrication process and is wildly used in laboratory preparation with the advantages of equipment simplicity, sufficient material utilization, and tunable parameters [93]. During the coating process, the precursor solution is deposited in front of the blade on the heating plate then the blade moves parallel to the substrate at a certain rate to disperse the solution and form a wet film. In order to obtain high-quality active layers, the properties of the film can be controlled by adjusting the parameter of the coating speed, solution concentration, solution species, plate temperature et al. The thickness (d) of the dry film can be calculated from the following empirical formula:

$$d = \frac{1}{2}\left(g\frac{c}{\rho}\right) \quad (1)$$

where "g" is the distance between the blade and substrate, "c" is the concentration of the solution, and "ρ" is the density of the final dry film. It is noteworthy that after coating, the wet film requires a long time to solidify. During the phase transition process, the solution tends to aggregate or self-assemble, especially for polymer materials, which leads to poor film morphology [51,63,94,95]. Therefore, great efforts have been devoted to the study of a mechanism of film formation in the blade coating process. This printing technique was first applied to the fullerene system by Mens et al. [96] to study the crystallization and phase separation of the blend film based on MDMO-PPV and PCBM. According to the results of solid-state NMR spectroscopy, they found that PCBM exhibited higher crystallinity in the blade coating method than in spin coating, which was related to the slower solvent evaporation rate in the blade coating process. In 2018, Ma et al. [46] developed the blade coating method for preparing inverted non-fullerene solar cells. The device structure with ITO/ZnO/PBDB-TF:IT-4F/MoO$_3$/Al delivered a PCE of 12.88% by improving the surface morphology of the ZnO buffer layer. Considering the differences in the spreading force and drying dynamics between spin-coating and blade coating, particularly in the preparation of large-area devices. Thus, they fabricated a smoother ZnO layer successfully to enhance the interfacial contact with the active layer, and consequently, led to a higher performance for the blade coating device. In the same years, Zhao et al. [97] also used the blade coating method to fabricate PBTA-TF:IT-M-based and PBDB-TF:IT-4F-based solar cells via a vacuum-assisted annealing (VAA) strategy, as depicted in Figure 6a. A similar phase separation of the OSCs was observed in both blade coating and spin-casting methods, with the maximum PCEs of 10.72% and 13.55%, respectively (Figure 6b). Due to the effect of the VAA process, the unfavorable morphology caused by the prolonged drying process was suppressed, and the large-area OSCs module with a 12.6 cm^2 large area based on PBDB-TF: IT-4F attained a PCE of 10.21% with a V_{oc} of 2.56 V, J_{sc} of 6.23 mA cm^{-2} and FF of 64.02%, which proved that the VAA method is a feasible way for blade coating large-area modules. On the basis of these previous studies, attention has been paid to the defects in film-forming during the coating process, such as inhomogeneous phase distribution, self-aggregation, and large phase separation. Strategies have been put forward to improve fluid flow and adjust film wettability, such as selecting solvents, adding additives, or controlling temperature et al. [56,87,94,97–99]. For example, Li et al. [97] developed a green solvent O-xylene for the eco-friendly printing of OSCs via blade-coating under high-temperature conditions. The excessive aggregation of the Y6 acceptor was inhibited effectively under 90 °C during the coating process, which shortened the drying period of the wet film. They also introduced 1,2-dimethylnaphthalene as the solvent additive to facilitate the crystallinity of the blend films. Contributing to enhanced photon absorption and reduced energy loss, the device based on PM6:Y6 obtained a PCE of 15.51%. In addition, large-area solar cells with 1.00 cm^2 were fabricated in the air, delivering a high PCE of 13.87%. More recently, Yuan et al. [56] designed the micro cylinder arrays patterned blade in order to control the fluid flow and optimize the morphology of the PM6:Y6 blend films, as represented in Figure 6c–e. They discovered that the arrays of the patterned blade had changed the fluid flow into a stable, unidirectional, and external flow type, which enhanced the rate of the extensional and shear strain. Thus, the polymer chains of PM6 were effectively stretched and aligned, leading to favorable phase separation, as shown in Figure 6f. As a result, the blend films used a patterned blade coating method via a lower coating speed and exhibited enhanced crystallinity and optimized morphology when compared to the normal blade coating. In addition, this novel strategy improved the exciton dissociation and charge transport efficiency, which was also applied in the fabrication of large-area devices successfully.

Recently, the sequential solution deposition of the layer-by-layer structure has been demonstrated as a promising way to achieve high-performance OSCs, and blade coating is also widely applied in the fabrication of this type of device. Sun et al. [50] reported an LBL processing approach using a sequential blade coating method to investigate the differences with the BHJ structure and studied morphology, photophysical dynamics, and device performance systematically (Figure 7a). They reported that the vertical distribution of the donor and acceptor brought by LBL processing is more advantageous than that of BHJ, which not only facilitated the charge transfer but also enhanced the stability of the devices. Several non-fullerene blend systems fabricated by this technique all exhibited higher efficiency and lower voltage loss, which indicated the excellent universality of sequential blade coating and its compatibility with different active layer materials. It is worth noting that the interpenetrating network structure formed in BHJ is generated by a spontaneous nanophase separation, which is metastable. When the morphology is adjusted to the thermodynamic equilibrium state, a large-scale phase separation can be observed. As a result, the device performance, especially for large-area solar cells, deteriorates [38]. Under this consideration, Min et al. [51] fabricated LBL devices based on PM6:Y6 with a sequential blade coating method, achieving a high efficiency of 16.35% for a small area (0.04 cm^2). The OSCs, as shown in Figure 7b,c, was better than the BHJ-bladed devices. They found that when introducing the LBL processing, the blend films could exhibit higher absorption and an obviously enhanced charge transport ability. In order to further explore the universality of this printing technique, other non-fullerene systems, PM6:Y6-

2Cl, PTQ10:Y6, and PM6:Y6-C2, were also selected. Benefiting from the physical dynamics to form proper surface uniformity for the sequential blade coating films, they applied this strategy to fabricate large-scale solar modules (Figure 7d). The 11.52 cm² module delivered an impressive PCE of 11.86% with a V_{oc} of 3.20 V, J_{sc} of 6.41 mA cm^{-2}, and an FF of 57.85%, which is the highest efficiency of large-area OSCs. These results demonstrate that the LBL printing technique shows great potential for the high performance of OSCs with mass production and a decrease in the PCE roll-off effect. Additionally, Li et al. [63] deeply studied the mechanism of film formation by adopting a reversible and LBL deposition method with sequential twice forward/reverse blade-coating (Figure 8a). The viscosity of PM6 in chloroform is related to the shear rate of the fluid, showing the feature of non-Newtonian fluid. During the coating process, the inhomogeneous viscosity of PM6 leads to poor mass distribution on the substrates. Thus, they developed an RS-LBL strategy to compensate for PM6 mass loss during printing, resulting in more uniform film thickness and higher light absorption. Benefiting from the uniform phase distribution, the active layer exhibited excellent face-on stacking and a PCE of 13.47% with an enhanced V_{oc} of 9.90 V, J_{sc} of 1.93 mA cm^{-2}, and FF of 70.53% was achieved in the large-area (36 cm²) solar modules (seen in Figure 8b–e).

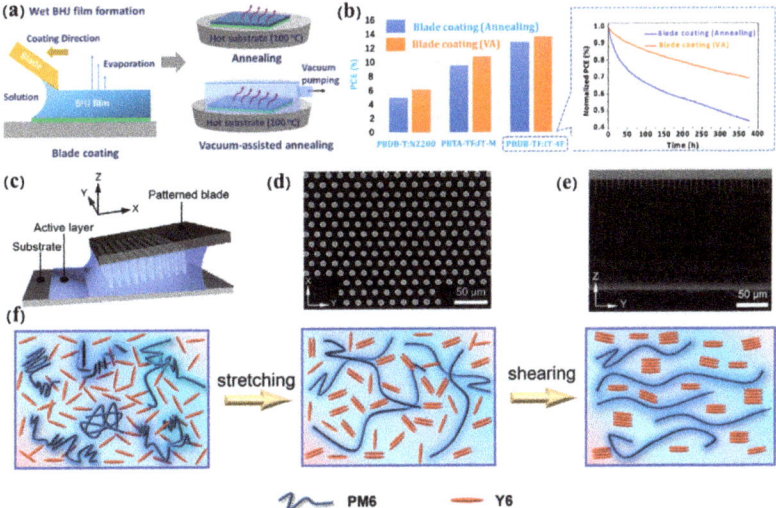

Figure 6. (a) Schematic illustration of blade-coated OSCs and vacuum-assisted annealing method. (b) PCE of various non-fullerene systems via different post-treatment methods. Adapted with permission from Ref. [97]. Copyright 2019 The Royal Society of Chemistry. (c) Schematic illustration of patterned blade coating. (d) Typical scanning electron microscopy (SEM) image of the circular patterned coating blade. (e) Cross-sectional SEM image of the circular patterned coating blade. (f) Schematic diagram of conformational changes in polymer and non-fullerene small molecules in stretching and shearing field Adapted with permission from Ref. [56]. Copyright 2021 Wiley–VCH.

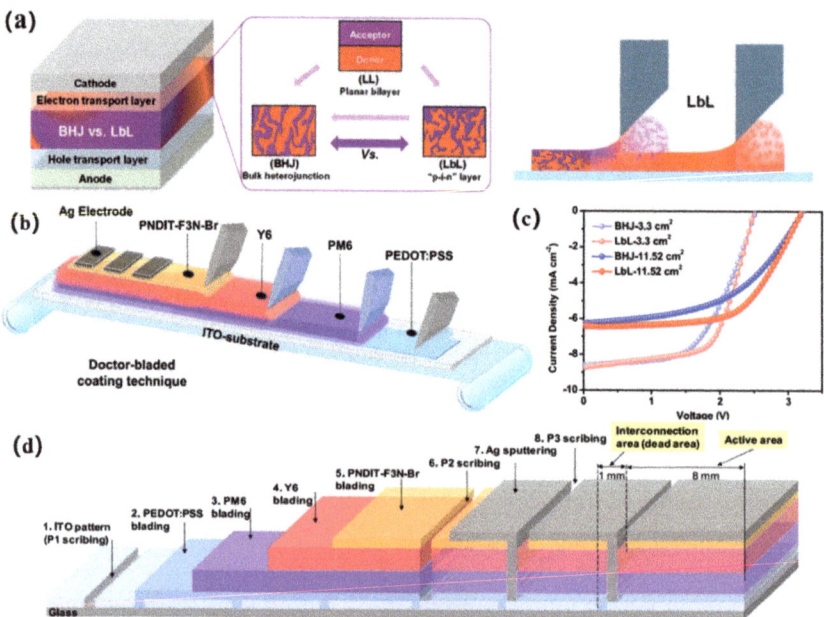

Figure 7. (**a**) Schematic of device architecture with different active layers and sequential blade coating method. Adapted with permission from Ref. [50]. Copyright 2019 the Royal Society of Chemistry. (**b**) Schematic illustration of the LBL blade coating and device architecture of OSCs [87], (**c**) *J*–*V* curves of BHJ and LBL devices using blade coating method. (**d**) Large-area solar modules for blade coating process. Adapted with permission from Ref. [51]. Copyright 2019 Elsevier Inc.

3.2. Slot-Die Coating

Slot-die coating is another printing technique that forms stripe patterns by controlling ink properties with excellent reproducibility, which is very suitable for printing multi-layer OSCs. The slot-die coating head is an ink reservoir consisting of two movable metal blades, a meniscus guide, and a gasket to support the pressurization of the ink. During the coating process, ink is provided through the coating head with a pump continuously and is deposited between the tip of the head and the substrate. Then, the coating head simply moves along the substrate direction to print a wet film. Similar to the film formation mechanism in the blade coating, the film thickness can be accurately determined by controlling the ink feeding rate, or the speed of the slot die head, which can be calculated as follows:

$$d = \frac{f}{Sw}\frac{c}{\rho} \tag{2}$$

where "f" is the flow rate of the ink, "S" is the coated speed, "w" is the coated width, "c" is the concentration of the ink, and "ρ" is the density of the final dry film. In addition, the slot-die coating is a closed feeding system, and it can efficiently prevent ink pollution and excessive evaporation during coating, which is critical for future industrial preparations [100–102]. Researchers are devoted to improving film quality and developing high-performance devices by the applied slot-die coating method. Zhao et al. [54] reported a slot-die coating process to fabricate hydrocarbon-based solvent solar cells with the device inverted structure of ITO/ZnO/PM6:Y6/MoO$_3$/Al. The aggregation behaviors of PM6 and Y6 in each solvent solution were different and exhibited temperature dependence (Figure 9a–c). For the coated films prepared in chlorobenzene, 1,2,4-trimethylbenzene, and ortho-xylene as solvents, they all exhibited higher crystallinity and appropriate morphology, resulting in enhanced and balanced charge transport along with reduced nonradiative losses. By optimizing the process temperature of slot-die coating, they achieved an impressive PCE of 15.2%, 15.4%, and 15.6%, respectively, which proved this scalable printing technique to be suitable for large-scale and industrial and environmentally friendly OSC production. To further understand the influence of solvents during the coating process, Yang et al. [103] used a sequential slot-die coating approach to prepare PM6:Y6 blend films by adopting two different kinds of boiling point solvents: chlorobenzene and chloroform. The donor layer was coated using high-boiling chlorobenzene first, while the acceptor dissolved in chloroform was slot-die-coated on the donor layer, in which the vertical phase separation was formed. They discovered that the film formation process of different solvents accompanied by different drying kinetics affected the morphology of the active layer (Figure 9d–e). The high-boiling solvent with less evaporation remained in the donor layer when the acceptor Y6 was deposited sequentially. The well blend of the PM6:Y6 film achieved an optimized vertical phase distribution and impressive interpenetrating network structure. As a result, the sequential slot-die coated device with two different solvents delivered a PCE of up to 14.42%, which is higher than the ones in traditional single solvents. Recently, Wei et al. [29] studied the film-dry kinetic during the slot-die coating process systematically by selecting two different NFAs: Qx-1 and Qx-2. Through in situ UV-vis absorption measurements, they found that Qx-1 maintained an appropriate aggregation with suitable crystallization during the drying process, while Qx-2 exhibited excessive aggregation, resulting in poor device performance, as shown in Figure 9d,e. The benefit from the desirable domain size of QX-1 and high thickness tolerance of blend films was that the 1 cm^2 flexible OSC based on PM6: Qx-1 via a slot-die coating method delivered a PCE of 13.70% and FF of 71%. Furthermore, they also fabricated 30 cm^2 large-area OSC modules along with an outstanding PCE of 12.20% and superior storage stability of over 6000 h. Despite the high materials utilization and low waste of this printing technique, large amounts of ink are needed to fill in the slot before coating. Therefore, it is an inappropriate method for preparing high-cost ink coating [104].

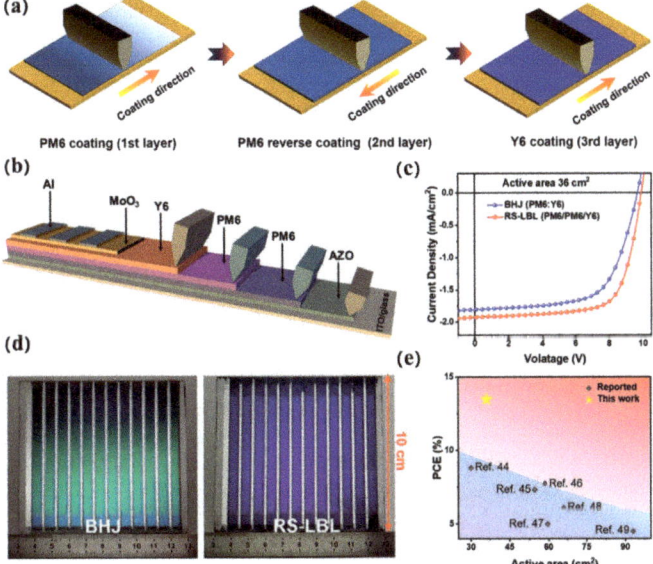

Figure 8. Schematic diagram of (**a**) RS–LBL fabrication. (**b**) Large-area OSC modules. (**c**) The *J–V* curves of the OSC; (**d**) The photograph of the blade-coated 36 cm^2 OSC modules on a 10 × 10 cm^2 substrate. (**e**) The reported PCEs of binary OSC modules with an active area over 30 cm^2. Adapted with permission from Ref. [63]. Copyright 2022 Wiley–VCH.

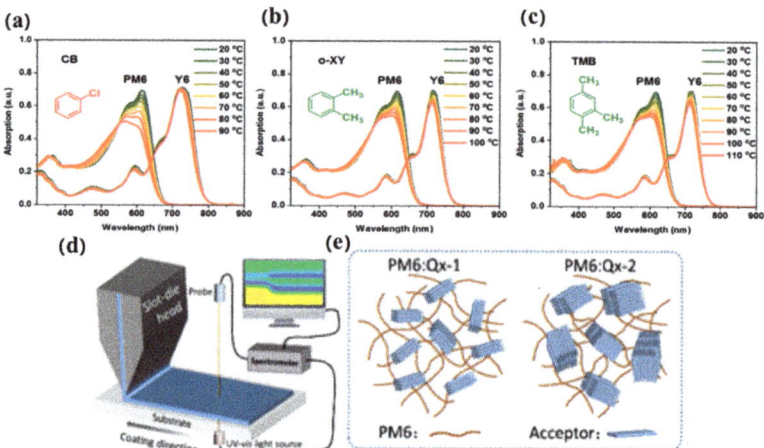

Figure 9. Temperature-dependent UV–vis absorption spectra of PM6 and Y6 in (**a**) CB, (**b**) o–XY, and (**c**) TMB. Adapted with permission from Ref. [54]. Copyright 2020 Wiley–VCH. (**d**) Schematic illustration of in situ UV–vis absorption measurement and slot-die coating process, (**e**) Diagram of the morphology of PM6: Qx-1 and PM6: Qx-2. Adapted with permission from Ref. [29]. Copyright 2022 Wiley–VCH.

3.3. Inkjet Printing

Inkjet printing is a digitally controlled printing technique and is widely used to fabricate the functional layer in OSCs with the advantages of low material waste, patternable preparation, and a maskless and contact-free process. The operating principle can be divided into the formation and spraying of liquid drops, the position, and diffusion of drops on the substrate, the evaporation of the solvent, and the formation of dry film. The ink is deposited on the substrate through a nozzle. Additionally, the size of the droplet can be well controlled by heating or mechanical compression. When the substrate moves forward, the wet film is prepared by moving the printing head in the transverse direction. There are two common inkjet printing models used to generate droplets, which are continuous inkjet printing (CIJ) and drop-on-demand (DoD)' [105,106], as shown in Figure 10a,b, respectively. In CIJ, the ink is sprayed from the nozzle, and the continuous droplets are charged by the electrode. When passing through the deflection plate, the electrostatic field deflects the charged ink droplets to a certain angle and then deposits them on the substrate selectively. In the DoD mode, it usually requires several printing heads, in which the ink droplets are formed by the pressure pulse of the piezoelectric stimulation or a thermal inkjet bubble. The droplets are sprayed from the nozzle only when demanded. Although CIJ is a fast-printing method, its complex operation and low printing resolution limit its application in large-area device fabrications. DoD technology is favored for the industrial production of functional materials. The DoD inkjet printing is stable for growing droplets to avoid the secondary droplets or satellite spots that influence the film properties. The DoD inkjet is a nearly no-waste process with high material utilization. The behaviors of ink droplets are related to the fluid properties, such as the viscosity, density, or surface tension of the ink, which should be adjusted to satisfy the printing requirements, and it can be characterized as Z, which is a dimensionless inverse Oh number as described in the following equation:

$$Z = \frac{1}{Oh} = \frac{\sqrt{\rho d \gamma}}{\eta} \qquad (3)$$

where "ρ" is density, "γ" is the surface tension, "d" is the nozzle diameter and "η" is the viscosity of the solution. Generally, the ink with a Z value in the range of $1 < Z < 10$ can form stable droplets, which is suitable for inkjet printing [107–110]. For example, Corzo et al. [111] reported that they developed the DoD inkjet printing to fabricate the P3HT:O-IDTBR BHJ device, yielding an impressive PCE of up to 6.47%, which was the first time the inkjet employed a printing technique into the NFA system for OSCs. According to rheological properties, they dissolved the P3HT:O-IDTBR blend in chlorobenzene-based and hydrocarbon-based ink, respectively, to achieve consistent jetting and optimize the droplet spacing and deposition temperature to form uniform films. In addition, they fabricated a 2.2 cm^2 large-area free-form solar cell by digital inkjet printing, delivering a PCE of 4.76%. Later, the same group yielded a PCE of up to 12.4% for opaque devices and 9.5% for semitransparent devices, where a PTB7-Th: IEICO-4F system was prepared via inkjet printing [55]. More recently, Chen et al. [61] developed LBL inkjet printing with sequential deposition to balance the film aggregation and optimize the vertical phase separation. Through the in-depth investigation of the mechanism in film formation during the inkjet printing possess, they found that the distributions of the donor and acceptor were ununiform on the surface. The higher solubility of ITIC-4F tended to be redissolved when the adjacent droplets coalescent during the spraying process. On the other hand, the donor showed less redissolution, as shown in Figure 10c–e. Thus, the ITIC-4F is preferred to enrich at the center of the printed lines, and the PBDB-T-2F was nearly uniform in the direction of printing, which exhibited a periodical phase separation distribution on the film. Due to the optimization of the dropping temperature and morphology, the molecular aggregation was suppressed efficiently, and exciton dissociation was also enhanced in non-fullerene OSCs based on PBDB-T-2F: BTP-BO-4Cl. As a result, the best OSC delivered an average PCE of 13.09%, which is the highest value of lab-scale OSCs fabricated by a sequential inkjet printing method, as depicted in Figure 10f–g.

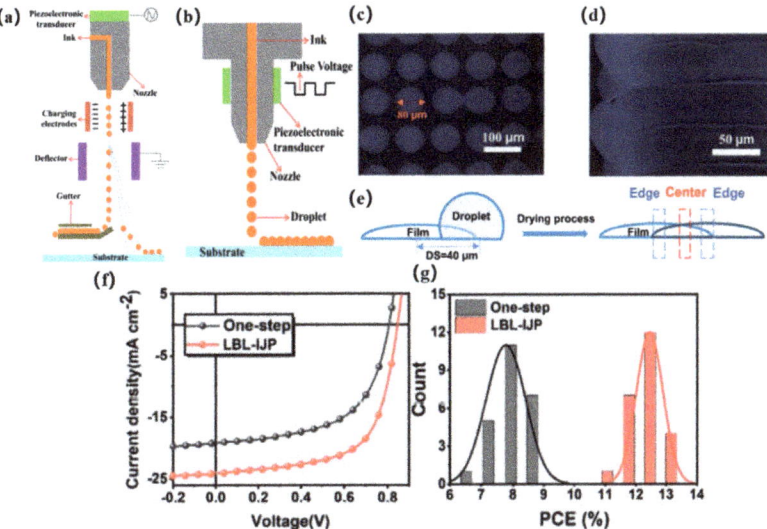

Figure 10. Schematic diagrams of (**a**) Continuous inkjet printing (CIJ), (**b**) Drop-on-demand (DOD) inkjet printing. Adapted with permission from Ref. [106]. Copyright 2019 The Royal Society of Chemistry; (**c**,**d**) Photographs of the ink drop and printed lines when printed at 50 °C. (**e**) The schematic diagram of droplet coalesces during inkjet printing with DS of 40 μm. (**f**) *J*–*V* curves of PBDB-T-2F: BTP–BO–4Cl devices from one-step and LBL–IJP. (**g**) The histogram of the one-step and LBL–IJP processed OSCs. Adapted with permission from Ref. [61]. Copyright 2022 Wiley–VCH.

3.4. Screen Printing

Screen printing is a versatile technique used to print patterns through the screen and supports full 2D printing films with high material utilization. Differing from other printing technologies, screen printing usually requires high-viscosity ink that enables the printing of nanoscale films, especially for some high-conductivity electrodes. Its reproducibility is affected by paste properties or screen tension [16,23,112–116]. Generally, screen printing can be divided into flatbed screen printing and rotary screen printing. Figure 11a depicts a schematic diagram of a flatbed screen printing method designed for printing on a flat substrate. Flatbed screen printing is a continuous process in which the ink is dispersed on the screen first, and the screen is contacted with the top of the substrate, then a squeegee is moved across the screen. Therefore, the ink is shear thin and transferred to the substrate to form a specific pattern. This printing process can be further repeated by raising the screen to move forward. Rotary screen printing, as shown in Figure 11b, is more suitable for high-throughput printing production owing to its higher printing speed and resolution than flatbed screen printing. In the process of rotary screen printing, the screen includes ink, squeegee, and patterns assembled as a cylinder, which rotates together with the substrate. The ink is pushed into the mesh through the stationary squeegee to reproduce the pattern. It allows the printing speeds to reach a high level (over 100 m min^{-1}). However, in terms of pattern operability, flatbed screen printing exhibits more advantages in printing flexibility and interval control than rotary. Krebs et al. [117], in 2009, first reported the preparation of OSCs by using a screen printing method. They discovered that the ink viscosity and solvent volatility should meet the requirements to screen print successfully. High viscosity is a prerequisite for ink through the mesh as smooth and low volatility can help the ink disperse on the substrate completely instead of drying at the screen. They developed thermos-cleavable solvents to solve the ink problem and prepared devices based on P3MHOCT. Screen printing is also used for fabricating conventional fullerene-based OSCs, especially for the donor polymer material of MEH-PPV [118]. At present, no records have been found for preparing the non-fullerene active layer by using the screen printing method. However, this technique exhibits advantages in printing electrode or transport layer materials [112,113,119]. Figure 11d describes the schematic of a fully roll-to-roll device. The transparent conductor of PEDOT:PSS was printed fully by the using flatbed screen printing method, while the silver electrode was covered on it by the same strategy, as shown in Figure 11c.

3.5. Gravure Printing

Gravure printing is a mature printing technique that produces high-resolution patterns, which is widely used in package and graphic publishing. Gravure printing is composed of two cylinders, called the gravure cylinder and the impression cylinder. During the printing process, the gravure cylinder with engraving patterns is embedded with an ink duct. The pattern of the cylinder determines the effective area for ink transfer. With the rotation of the gravure cylinder, the excess ink is scraped off by the blade. Meanwhile, the substrate completes the transfer of the ink by contacting the impression cylinder to form the pattern. The thickness of the films is determined by the depth of the mesh hole on the gravure cylinder. For gravure printing, the inks usually have relatively low viscosity to support printing with high speeds over 1000 m/min. Therefore, low boiling point solvents are widely used as ink solvents to enable the fast drying of the printed wet film [120–122]. However, gravure printing is still a challenging technique for OSCs. It tends to form a relatively thin layer. Until now, there have been few reports on the active layers deposited using this method. Similar to screen printing, the complex patterns in gravure printing will increase the cost of the fabrication. This fixed pattern mode is ideal for printing electrode materials or transport layers [123–126]. Figure 11e describes the gravure printing process of the silver-nanowire-based transparent electrodes by Wang et al. They prepared 1 cm^2 flexible solar cells based on PM6: Y6, as shown in Figure 11d, with an efficiency of 13.61%, demonstrating the high potential of gravure printing in the prepattern and mass fabrication of the devices.

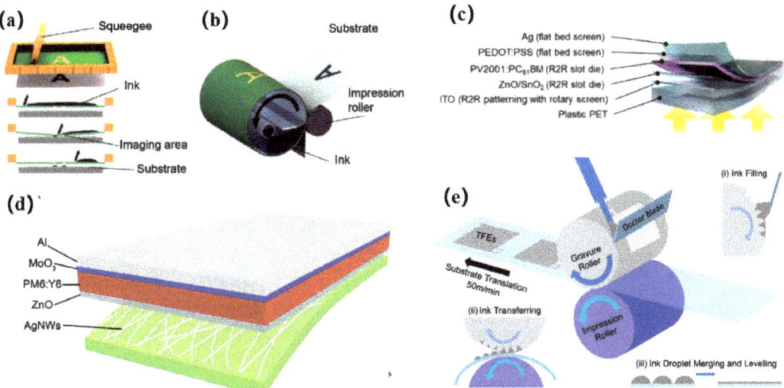

Figure 11. Schematic diagrams of (a) Flatbed screen printing, (b) Rotary screen printing. Adapted with permission from Ref. [112]. Copyright 2016 Wiley–VCH. (c) Fully roll-to-roll flexible device, (d) Device structure of the 1 cm^2 flexible OSCs. Adapted with permission from Ref. [119]. Copyright 2020 IOP Publishing Ltd. (e) Silver nanowire electrodes fabricated by high-speed gravure printing process. Adapted with permission from Ref. [126]. Copyright 2020 Wiley–VCH.

4. Conclusions and Prospects

With the rapid development of solvent-processed OSCs, it is necessary to replace the fabrication technique from the laboratory-scale spin coating method with efficient printing technologies, which are more compatible with the high-throughput mass production of devices in the future. In this review, we have summarized the recent progress of printing technologies, including blade coating, slot-die coating, inkjet printing, screen printing, and gravure printing, and made comparisons on their ink requirements, printing speed, and pattern dimensions as shown in Table 2. In view of all the printing technologies described in this review, blade coating and slot-die coating are the most desirable technologies for the scalable fabrication of large-area OSCs, showing high efficiency. The blade coating process is relatively simple and can print at high speed with low material waste. Similar to blade coating, slot-die coating can print one-dimensional patterns, which are favorable for sequential deposition processes. Inkjet printing has merits in changeable digital printing patterns and the contactless process. Although there are some achievements made in lab-scale OSCs, the relatively slow printing speed and complex ink preparation inhibit the further application of large-area OSCs. As for screen printing, this printing method is suitable for highly viscous ink, which contrasts gravure printing with a low viscosity of the ink. Due to the high requirement for ink, there are few reports on active layers fabricated by these two technologies. Instead, both printing methods are ideal for the fabrication of electrodes of large-area OSC modules

Table 2. The comparison of different fabrication technologies.

Technique	Ink Preparation	Ink Viscosity	Ink Usage Rate	Speed	Pattern [1]
Spin coating	Simple	Low	Low	-	0
Blade coating	Simple	Low	High	Low-High	0
Slot-die coating	Simple	Low-High	High	High	1
Screen printing	Moderate	High	High	Medium	2
Gravure printing	Difficult	Low	High	High	2
Inkjet printing	Moderate	Low	High	Medium	2

Pattern [1]: 0 (0-dimensional); 1 (1-dimensional); 2 (2-dimensional).

Although the efficiency of the printed OSCs fabricated via these technologies has been improved continuously, the mechanism of the film formation through these printing methods is still unclear, and there is a big gap in device efficiency between large and small-area devices. Thus, we propose several directions that deserve more attention. Firstly, we should recognize that most of the high-efficient materials used in OSCs require complex synthesis and treatment processes, which are unfavorable to the industrial fabrication of the devices. The design of organic materials with simple synthesis steps and easy post-processing is always crucial for the manufacture of printable OSCs. Secondly, the film morphology and quality determine the device performance directly. How to control fluid flow during the printing process is still a problem. It is of great significance to investigate the hydrodynamics and crystallization kinetics of active layer ink for obtaining high-quality printed films. Additive and solvent engineering have also been demonstrated as effective strategies to improve the film's quality. In addition, the application of non-toxic green solvents with a high boiling point can reduce the solution aggregation during the printing process and prevent the pollution caused by halogen solvents, which are important research topics in printable OSCs. Thirdly, the low stability of organic photovoltaic materials is a big challenge which needs to be addressed before the wide application and commercialization of OSCs. In order to improve the lifetime of devices and reduce their sensitivity to light, heat, or other complex conditions, the modification of active layer materials and transport layer materials are effective approaches to improve the device's stability. Additionally, for large-area ITO-free flexible OSCs, it is quite necessary to maintain their excellent mechanical stability by developing novel packaging techniques. These bottlenecks are expected to be solved to promote the application of printing technologies toward the commercialization of OSCs.

Author Contributions: Conceptualization, Z.X., T.C., X.L., Q.Z., J.S. and S.L., Writing—review & editing, Z.X. All authors have read and agreed to the published version of the manuscript.

Funding: This research was funded by the Shenzhen Science and Technology Program (Grant No. RCBS20200714114922263), and the Natural Science Foundation of Guangdong Province, China (Grant No. 2022A1515010622).

Data Availability Statement: All data generated or analyzed during this study are included in this published article.

Conflicts of Interest: The authors declare no conflict of interest.

References

1. Green, M.A.; Dunlop, E.D.; Hohl-Ebinger, J.; Yoshita, M.; Kopidakis, N.; Hao, X. Solar cell efficiency tables (Version 58). *Prog. Photovolt. Res. Appl.* **2021**, *29*, 657–667. [CrossRef]
2. Xin, J.; Feng, J.; Lin, B.; Naveed, H.B.; Xue, J.; Zheng, N.; Ma, W. The Importance of Nonequilibrium to Equilibrium Transition Pathways for the Efficiency and Stability of Organic Solar Cells. *Small* **2022**, *18*, e2200608. [CrossRef] [PubMed]
3. Xing, Z.; Meng, X.; Sun, R.; Hu, T.; Huang, Z.; Min, J.; Hu, X.; Chen, Y. An Effective Method for Recovering Nonradiative Recombination Loss in Scalable Organic Solar Cells. *Adv. Funct. Mater.* **2020**, *30*, 1800002. [CrossRef]
4. Li, G.; Chang, W.-H.; Yang, Y. Low-bandgap conjugated polymers enabling solution-processable tandem solar cells. *Nat. Rev. Mater.* **2017**, *2*, 1703080. [CrossRef]
5. Xu, X.; Yu, L.; Meng, H.; Dai, L.; Yan, H.; Li, R.; Peng, Q. Polymer Solar Cells with 18.74% Efficiency: From Bulk Heterojunction to Interdigitated Bulk Heterojunction. *Adv. Funct. Mater.* **2021**, *32*, 2108797. [CrossRef]
6. Cui, Y.; Yang, C.; Yao, H.; Zhu, J.; Wang, Y.; Jia, G.; Gao, F.; Hou, J. Efficient Semitransparent Organic Solar Cells with Tunable Color enabled by an Ultralow-Bandgap Nonfullerene Acceptor. *Adv. Mater.* **2017**, *29*, 1703080. [CrossRef]
7. Dai, S.; Zhan, X. Nonfullerene Acceptors for Semitransparent Organic Solar Cells. *Adv. Energy Mater.* **2018**, *8*, 1800002. [CrossRef]
8. Li, Y.; Xu, G.; Cui, C.; Li, Y. Flexible and Semitransparent Organic Solar Cells. *Adv. Energy Mater.* **2018**, *8*, 1701791. [CrossRef]
9. Lin, Y.; Wang, J.; Zhang, Z.G.; Bai, H.; Li, Y.; Zhu, D.; Zhan, X. An electron acceptor challenging fullerenes for efficient polymer solar cells. *Adv. Mater.* **2015**, *27*, 1170–1174. [CrossRef]
10. Li, D.; Deng, N.; Fu, Y.; Guo, C.; Zhou, B.; Wang, L.; Zhou, J.; Liu, D.; Li, W.; Wang, K.; et al. Fibrillization of Non-Fullerene Acceptors Enables 19% Efficiency Pseudo-Bulk Heterojunction Organic Solar Cells. *Adv. Mater.* **2022**, 2208211. [CrossRef]
11. Peng, W.; Lin, Y.; Jeong, S.Y.; Genene, Z.; Magomedov, A.; Woo, H.Y.; Chen, C.; Wahyudi, W.; Tao, Q.; Deng, J.; et al. Over 18% ternary polymer solar cells enabled by a terpolymer as the third component. *Nano Energy* **2022**, *92*, 106681. [CrossRef]

12. Cui, Y.; Xu, Y.; Yao, H.; Bi, P.; Hong, L.; Zhang, J.; Zu, Y.; Zhang, T.; Qin, J.; Ren, J.; et al. Single-Junction Organic Photovoltaic Cell with 19% Efficiency. *Adv. Mater.* **2021**, *33*, 2102420. [CrossRef]
13. Liu, F.; Zhou, L.; Liu, W.; Zhou, Z.; Yue, Q.; Zheng, W.; Sun, R.; Liu, W.; Xu, S.; Fan, H.; et al. Organic Solar Cells with 18% Efficiency Enabled by an Alloy Acceptor: A Two-in-One Strategy. *Adv. Mater.* **2021**, *33*, 2100830. [CrossRef]
14. Liu, Q.; Jiang, Y.; Jin, K.; Qin, J.; Xu, J.; Li, W.; Xiong, J.; Liu, J.; Xiao, Z.; Sun, K.; et al. 18% Efficiency organic solar cells. *Sci. Bull.* **2020**, *65*, 272–275. [CrossRef]
15. Lin, Y.; Firdaus, Y.; Isikgor, F.H.; Nugraha, M.I.; Yengel, E.; Harrison, G.T.; Hallani, R.; El-Labban, A.; Faber, H.; Ma, C.; et al. Self-Assembled Monolayer Enables Hole Transport Layer-Free Organic Solar Cells with 18% Efficiency and Improved Operational Stability. *ACS Energy Lett.* **2020**, *5*, 2935–2944. [CrossRef]
16. Jing, J.; Dong, S.; Zhang, K.; Xie, B.; Zhang, J.; Song, Y.; Huang, F. In-situ self-organized anode interlayer enables organic solar cells with simultaneously simplified processing and greatly improved efficiency to 17.8%. *Nano Energy* **2022**, *93*, 2210675. [CrossRef]
17. Xie, C.; Liu, Y.; Wei, W.; Zhou, Y. Large-Area Flexible Organic Solar Cells with a Robust Silver Nanowire-Polymer Composite as Transparent Top Electrode. *Adv. Funct. Mater.* **2022**, *33*, 2210675. [CrossRef]
18. Zhang, L.; Lin, B.; Hu, B.; Xu, X.; Ma, W. Blade-Cast Non-fullerene Organic Solar Cells in Air with Excellent Morphology, Efficiency, and Stability. *Adv. Mater.* **2018**, *30*, 1800343. [CrossRef]
19. Zhang, Z.; Yang, L.; Hu, Z.; Yu, J.; Liu, X.; Wang, H.; Cao, J.; Zhang, F.; Tang, W. Charge density modulation on asymmetric fused-ring acceptors for high-efficiency photovoltaic solar cells. *Mater. Chem. Front.* **2020**, *4*, 1747–1755. [CrossRef]
20. Cui, Y.; Yao, H.; Hong, L.; Zhang, T.; Tang, Y.; Lin, B.; Xian, K.; Gao, B.; An, C.; Bi, P.; et al. Organic photovoltaic cell with 17% efficiency and superior processability. *Natl. Sci. Rev.* **2020**, *7*, 1239–1246. [CrossRef]
21. Yu, H.; Ma, R.; Xiao, Y.; Zhang, J.; Liu, T.; Luo, Z.; Chen, Y.; Bai, F.; Lu, X.; Yan, H.; et al. Improved organic solar cell efficiency based on the regulation of an alkyl chain on chlorinated non-fullerene acceptors. *Mater. Chem. Front.* **2020**, *4*, 2428–2434. [CrossRef]
22. Cui, Y.; Yao, H.; Zhang, J.; Xian, K.; Zhang, T.; Hong, L.; Wang, Y.; Xu, Y.; Ma, K.; An, C.; et al. Single-Junction Organic Photovoltaic Cells with Approaching 18% Efficiency. *Adv. Mater.* **2020**, *32*, 1908205. [CrossRef] [PubMed]
23. Ng, L.W.T.; Lee, S.W.; Chang, D.W.; Hodgkiss, J.M.; Vak, D. Organic Photovoltaics' New Renaissance: Advances Toward Roll-to-Roll Manufacturing of Non-Fullerene Acceptor Organic Photovoltaics. *Adv. Mater. Technol.* **2022**, *7*, 2101556. [CrossRef]
24. Hamukwaya, S.L.; Hao, H.; Zhao, Z.; Dong, J.; Zhong, T.; Xing, J.; Hao, L.; Mashingaidze, M.M. A Review of Recent Developments in Preparation Methods for Large-Area Perovskite Solar Cells. *Coatings* **2022**, *12*, 252. [CrossRef]
25. Yu, R.; Wei, X.; Wu, G.; Tan, Z.A. Layer-by-layered organic solar cells: Morphology optimizing strategies and processing techniques. *Aggregate* **2021**, *3*, e107. [CrossRef]
26. Meng, X.; Zhang, L.; Xie, Y.; Hu, X.; Xing, Z.; Huang, Z.; Liu, C.; Tan, L.; Zhou, W.; Sun, Y.; et al. A General Approach for Lab-to-Manufacturing Translation on Flexible Organic Solar Cells. *Adv. Mater.* **2019**, *31*, 1903649. [CrossRef]
27. Wang, Y.; Wang, X.; Lin, B.; Bi, Z.; Zhou, X.; Naveed, H.B.; Zhou, K.; Yan, H.; Tang, Z.; Ma, W. Achieving Balanced Crystallization Kinetics of Donor and Acceptor by Sequential-Blade Coated Double Bulk Heterojunction Organic Solar Cells. *Adv. Energy Mater.* **2020**, *10*, 2000826. [CrossRef]
28. Wang, G.; Zhang, J.; Yang, C.; Wang, Y.; Xing, Y.; Adil, M.A.; Yang, Y.; Tian, L.; Su, M.; Shang, W.; et al. Synergistic Optimization Enables Large-Area Flexible Organic Solar Cells to Maintain over 98% PCE of the Small-Area Rigid Devices. *Adv. Mater.* **2020**, *32*, 2005153. [CrossRef]
29. Shen, Y.F.; Zhang, H.; Zhang, J.; Tian, C.; Shi, Y.; Qiu, D.; Zhang, Z.; Lu, K.; Wei, Z. In-situ Absorption Characterization Guided Slot-Die-Coated High-Performance Large-area Flexible Organic Solar Cells and Modules. *Adv. Mater.* **2022**, 2209030. [CrossRef]
30. Yu, A.G.; Gao, J.; Hummelen, J.C.; Wudl, F.; Heeger, A.J. Polymer Photovoltaic Cells: Enhanced Efficiencies via a Network of Internal Donor-Acceptor Heterojunctions. *Science* **1995**, *270*, 1789. [CrossRef]
31. Kim, J.Y.; Kim, S.H.; Lee, H.H.; Lee, K.; Ma, W.; Gong, X.; Heeger, A.J. New Architecture for High-Efficiency Polymer Photovoltaic Cells Using Solution-Based Titanium Oxide as an Optical Spacer. *Adv. Mater.* **2006**, *18*, 572–576. [CrossRef]
32. Choi, H.; Kim, H.B.; Ko, S.J.; Kim, J.Y.; Heeger, A.J. An organic surface modifier to produce a high work function transparent electrode for high performance polymer solar cells. *Adv. Mater.* **2015**, *27*, 892–896. [CrossRef]
33. Feng, S.; Zhang, C.; Liu, Y.; Bi, Z.; Zhang, Z.; Xu, X.; Ma, W.; Bo, Z. Fused-Ring Acceptors with Asymmetric Side Chains for High-Performance Thick-Film Organic Solar Cells. *Adv. Mater.* **2017**, *29*, 1703527. [CrossRef]
34. Li, C.; Fu, H.; Xia, T.; Sun, Y. Asymmetric Nonfullerene Small Molecule Acceptors for Organic Solar Cells. *Adv. Energy Mater.* **2019**, *9*, 1900999. [CrossRef]
35. Li, S.; Zhan, L.; Jin, Y.; Zhou, G.; Lau, T.K.; Qin, R.; Shi, M.; Li, C.Z.; Zhu, H.; Lu, X.; et al. Asymmetric Electron Acceptors for High-Efficiency and Low-Energy-Loss Organic Photovoltaics. *Adv. Mater.* **2020**, *32*, 2001160. [CrossRef]
36. Peng, W.; Zhang, G.; Zhu, M.; Xia, H.; Zhang, Y.; Tan, H.; Liu, Y.; Chi, W.; Peng, Q.; Zhu, W. Simple-Structured NIR-Absorbing Small-Molecule Acceptors with a Thiazolothiazole Core: Multiple Noncovalent Conformational Locks and D-A Effect for Efficient OSCs. *ACS Appl. Mater. Interfaces* **2019**, *11*, 48128–48133. [CrossRef]
37. Qin, R.; Wang, D.; Zhou, G.; Yu, Z.-P.; Li, S.; Li, Y.; Liu, Z.-X.; Zhu, H.; Shi, M.; Lu, X.; et al. Tuning terminal aromatics of electron acceptors to achieve high-efficiency organic solar cells. *J. Mater. Chem. A* **2019**, *7*, 27632–27639. [CrossRef]
38. Jee, M.H.; Ryu, H.S.; Lee, D.; Lee, W.; Woo, H.Y. Recent Advances in Nonfullerene Acceptor-Based Layer-by-Layer Organic Solar Cells Using a Solution Process. *Adv. Sci. (Weinh.)* **2022**, *9*, e2201876. [CrossRef]

39. Zhang, J.; Kan, B.; Pearson, A.J.; Parnell, A.J.; Cooper, J.F.K.; Liu, X.-K.; Conaghan, P.J.; Hopper, T.R.; Wu, Y.; Wan, X.; et al. Efficient non-fullerene organic solar cells employing sequentially deposited donor–acceptor layers. *J. Mater. Chem. A* **2018**, *6*, 18225–18233. [CrossRef]
40. Cui, Y.; Zhang, S.; Liang, N.; Kong, J.; Yang, C.; Yao, H.; Ma, L.; Hou, J. Toward Efficient Polymer Solar Cells Processed by a Solution-Processed Layer-By-Layer Approach. *Adv. Mater.* **2018**, *30*, 1802499. [CrossRef]
41. Hu, M.; Zhang, Y.; Liu, X.; Zhao, X.; Hu, Y.; Yang, Z.; Yang, C.; Yuan, Z.; Chen, Y. Layer-by-Layer Solution-Processed Organic Solar Cells with Perylene Diimides as Acceptors. *ACS Appl. Mater. Interfaces* **2021**, *13*, 29876–29884. [CrossRef] [PubMed]
42. Chen, Z.; Cai, P.; Chen, J.; Liu, X.; Zhang, L.; Lan, L.; Peng, J.; Ma, Y.; Cao, Y. Low band-gap conjugated polymers with strong interchain aggregation and very high hole mobility towards highly efficient thick-film polymer solar cells. *Adv. Mater.* **2014**, *26*, 2586–2591. [CrossRef] [PubMed]
43. Hu, X.; Yi, C.; Wang, M.; Hsu, C.-H.; Liu, S.; Zhang, K.; Zhong, C.; Huang, F.; Gong, X.; Cao, Y. High-Performance Inverted Organic Photovoltaics with Over 1-μm Thick Active Layers. *Adv. Energy Mater.* **2014**, *4*, 1400378. [CrossRef]
44. Liu, Y.; Zhao, J.; Li, Z.; Mu, C.; Ma, W.; Hu, H.; Jiang, K.; Lin, H.; Ade, H.; Yan, H. Aggregation and morphology control enables multiple cases of high-efficiency polymer solar cells. *Nat. Commun.* **2014**, *5*, 5293. [CrossRef] [PubMed]
45. Nian, L.; Chen, Z.; Herbst, S.; Li, Q.; Yu, C.; Jiang, X.; Dong, H.; Li, F.; Liu, L.; Wurthner, F.; et al. Aqueous Solution Processed Photoconductive Cathode Interlayer for High Performance Polymer Solar Cells with Thick Interlayer and Thick Active Layer. *Adv. Mater.* **2016**, *28*, 7521–7526. [CrossRef]
46. Ji, G.; Zhao, W.; Wei, J.; Yan, L.; Han, Y.; Luo, Q.; Yang, S.; Hou, J.; Ma, C.-Q. 12.88% efficiency in doctor-blade coated organic solar cells through optimizing the surface morphology of a ZnO cathode buffer layer. *J. Mater. Chem. A* **2019**, *7*, 212–220. [CrossRef]
47. Wu, Q.; Guo, J.; Sun, R.; Guo, J.; Jia, S.; Li, Y.; Wang, Y.; Min, J. Slot-die printed non-fullerene organic solar cells with the highest efficiency of 12.9% for low-cost PV-driven water splitting. *Nano Energy* **2019**, *61*, 559–566. [CrossRef]
48. Zhang, L.; Zhao, H.; Lin, B.; Yuan, J.; Xu, X.; Wu, J.; Zhou, K.; Guo, X.; Zhang, M.; Ma, W. A blade-coated highly efficient thick active layer for non-fullerene organic solar cells. *J. Mater. Chem. A* **2019**, *7*, 22265–22273. [CrossRef]
49. Zhao, H.; Zhang, L.; Naveed, H.B.; Lin, B.; Zhao, B.; Zhou, K.; Gao, C.; Zhang, C.; Wang, C.; Ma, W. Processing-Friendly Slot-Die-Cast Nonfullerene Organic Solar Cells with Optimized Morphology. *ACS Appl. Mater. Interfaces* **2019**, *11*, 42392–42402. [CrossRef]
50. Sun, R.; Guo, J.; Wu, Q.; Zhang, Z.; Yang, W.; Guo, J.; Shi, M.; Zhang, Y.; Kahmann, S.; Ye, L.; et al. A multi-objective optimization-based layer-by-layer blade-coating approach for organic solar cells: Rational control of vertical stratification for high performance. *Energy Environ. Sci.* **2019**, *12*, 3118–3132. [CrossRef]
51. Sun, R.; Wu, Q.; Guo, J.; Wang, T.; Wu, Y.; Qiu, B.; Luo, Z.; Yang, W.; Hu, Z.; Guo, J.; et al. A Layer-by-Layer Architecture for Printable Organic Solar Cells Overcoming the Scaling Lag of Module Efficiency. *Joule* **2020**, *4*, 407–419. [CrossRef]
52. Lin, B.; Zhou, X.; Zhao, H.; Yuan, J.; Zhou, K.; Chen, K.; Wu, H.; Guo, R.; Scheel, M.A.; Chumakov, A.; et al. Balancing the pre-aggregation and crystallization kinetics enables high efficiency slot-die coated organic solar cells with reduced non-radiative recombination losses. *Energy Environ. Sci.* **2020**, *13*, 2467–2479. [CrossRef]
53. Jeong, S.; Park, B.; Hong, S.; Kim, S.; Kim, J.; Kwon, S.; Lee, J.H.; Lee, M.S.; Park, J.C.; Kang, H.; et al. Large-Area Nonfullerene Organic Solar Cell Modules Fabricated by a Temperature-Independent Printing Method. *ACS Appl. Mater. Interfaces* **2020**, *12*, 41877–41885. [CrossRef]
54. Zhao, H.; Naveed, H.B.; Lin, B.; Zhou, X.; Yuan, J.; Zhou, K.; Wu, H.; Guo, R.; Scheel, M.A.; Chumakov, A.; et al. Hot Hydrocarbon-Solvent Slot-Die Coating Enables High-Efficiency Organic Solar Cells with Temperature-Dependent Aggregation Behavior. *Adv. Mater.* **2020**, *32*, e2002302. [CrossRef]
55. Corzo, D.; Bihar, E.; Alexandre, E.B.; Rosas-Villalva, D.; Baran, D. Ink Engineering of Transport Layers for 9.5% Efficient All-Printed Semitransparent Nonfullerene Solar Cells. *Adv. Funct. Mater.* **2020**, *31*, 2005763. [CrossRef]
56. Yuan, J.; Liu, D.; Zhao, H.; Lin, B.; Zhou, X.; Naveed, H.B.; Zhao, C.; Zhou, K.; Tang, Z.; Chen, F.; et al. Patterned Blade Coating Strategy Enables the Enhanced Device Reproducibility and Optimized Morphology of Organic Solar Cells. *Adv. Energy Mater.* **2021**, *11*, 2100098. [CrossRef]
57. Wu, Q.; Wang, W.; Wu, Y.; Chen, Z.; Guo, J.; Sun, R.; Guo, J.; Yang, Y.; Min, J. High-Performance All-Polymer Solar Cells with a Pseudo-Bilayer Configuration Enabled by a Stepwise Optimization Strategy. *Adv. Funct. Mater.* **2021**, *31*, 2010411. [CrossRef]
58. Zhang, Y.; Liu, K.; Huang, J.; Xia, X.; Cao, J.; Zhao, G.; Fong, P.W.K.; Zhu, Y.; Yan, F.; Yang, Y.; et al. Graded bulk-heterojunction enables 17% binary organic solar cells via nonhalogenated open air coating. *Nat. Commun.* **2021**, *12*, 4815. [CrossRef]
59. Zhang, J.; Zhang, L.; Wang, X.; Xie, Z.; Hu, L.; Mao, H.; Xu, G.; Tan, L.; Chen, Y. Reducing Photovoltaic Property Loss of Organic Solar Cells in Blade-Coating by Optimizing Micro-Nanomorphology via Nonhalogenated Solvent. *Adv. Energy Mater.* **2022**, *12*, 2200165. [CrossRef]
60. Xue, J.; Naveed, H.B.; Zhao, H.; Lin, B.; Wang, Y.; Zhu, Q.; Wu, B.; Bi, Z.; Zhou, X.; Zhao, C.; et al. Kinetic processes of phase separation and aggregation behaviors in slot-die processed high efficiency Y6-based organic solar cells. *J. Mater. Chem. A* **2022**, *10*, 13439–13447. [CrossRef]
61. Chen, X.; Huang, R.; Han, Y.; Zha, W.; Fang, J.; Lin, J.; Luo, Q.; Chen, Z.; Ma, C.Q. Balancing the Molecular Aggregation and Vertical Phase Separation in the Polymer: Nonfullerene Blend Films Enables 13.09% Efficiency of Organic Solar Cells with Inkjet-Printed Active Layer. *Adv. Energy Mater.* **2022**, *12*, 2200044. [CrossRef]

62. Zhao, H.; Lin, B.; Xue, J.; Naveed, H.B.; Zhao, C.; Zhou, X.; Zhou, K.; Wu, H.; Cai, Y.; Yun, D.; et al. Kinetics Manipulation Enables High-Performance Thick Ternary Organic Solar Cells via R2R-Compatible Slot-Die Coating. *Adv. Mater.* **2022**, *34*, e2105114. [CrossRef] [PubMed]
63. Zhang, B.; Yang, F.; Chen, S.; Chen, H.; Zeng, G.; Shen, Y.; Li, Y.; Li, Y. Fluid Mechanics Inspired Sequential Blade-Coating for High-Performance Large-Area Organic Solar Modules. *Adv. Funct. Mater.* **2022**, *32*, 2202011. [CrossRef]
64. Gao, L.; Zhang, Z.G.; Bin, H.; Xue, L.; Yang, Y.; Wang, C.; Liu, F.; Russell, T.P.; Li, Y. High-Efficiency Nonfullerene Polymer Solar Cells with Medium Bandgap Polymer Donor and Narrow Bandgap Organic Semiconductor Acceptor. *Adv. Mater.* **2016**, *28*, 8288–8295. [CrossRef]
65. Son, H.J.; Wang, W.; Xu, T.; Liang, Y.; Wu, Y.; Li, G.; Yu, L. Synthesis of fluorinated polythienothiophene-co-benzodithiophenes and effect of fluorination on the photovoltaic properties. *J. Am. Chem. Soc.* **2011**, *133*, 1885–1894. [CrossRef]
66. Zhang, H.; Yao, H.; Hou, J.; Zhu, J.; Zhang, J.; Li, W.; Yu, R.; Gao, B.; Zhang, S.; Hou, J. Over 14% Efficiency in Organic Solar Cells Enabled by Chlorinated Nonfullerene Small-Molecule Acceptors. *Adv. Mater.* **2018**, *30*, 1800613. [CrossRef]
67. Yuan, J.; Zhang, Y.; Zhou, L.; Zhang, G.; Yip, H.-L.; Lau, T.-K.; Lu, X.; Zhu, C.; Peng, H.; Johnson, P.A.; et al. Single-Junction Organic Solar Cell with over 15% Efficiency Using Fused-Ring Acceptor with Electron-Deficient Core. *Joule* **2019**, *3*, 1140–1151. [CrossRef]
68. Jiang, K.; Wei, Q.; Lai, J.Y.L.; Peng, Z.; Kim, H.K.; Yuan, J.; Ye, L.; Ade, H.; Zou, Y.; Yan, H. Alkyl Chain Tuning of Small Molecule Acceptors for Efficient Organic Solar Cells. *Joule* **2019**, *3*, 3020–3033. [CrossRef]
69. Azzopardi, B.; Emmott, C.J.M.; Urbina, A.; Krebs, F.C.; Mutale, J.; Nelson, J. Economic assessment of solar electricity production from organic-based photovoltaic modules in a domestic environment. *Energy Environ. Sci.* **2011**, *4*, 3741–3753. [CrossRef]
70. Emmott, C.J.M.; Urbina, A.; Nelson, J. Environmental and economic assessment of ITO-free electrodes for organic solar cells. *Sol. Energy Mater. Sol. Cells* **2012**, *97*, 14–21. [CrossRef]
71. Irimia-Vladu, M. "Green" electronics: Biodegradable and biocompatible materials and devices for sustainable future. *Chem. Soc. Rev.* **2014**, *43*, 588–610. [CrossRef]
72. Sun, Y.; Liu, T.; Kan, Y.; Gao, K.; Tang, B.; Li, Y. Flexible Organic Solar Cells: Progress and Challenges. *Small Sci.* **2021**, *1*, 2100001. [CrossRef]
73. Howells, C.T.; Saylan, S.; Kim, H.; Marbou, K.; Aoyama, T.; Nakao, A.; Uchiyama, M.; Samuel, I.D.W.; Kim, D.-W.; Dahlem, M.S.; et al. Influence of perfluorinated ionomer in PEDOT:PSS on the rectification and degradation of organic photovoltaic cells. *J. Mater. Chem. A* **2018**, *6*, 16012–16028. [CrossRef]
74. Xu, B.; Gopalan, S.A.; Gopalan, A.I.; Muthuchamy, N.; Lee, K.P.; Lee, J.S.; Jiang, Y.; Lee, S.W.; Kim, S.W.; Kim, J.S.; et al. Functional solid additive modified PEDOT:PSS as an anode buffer layer for enhanced photovoltaic performance and stability in polymer solar cells. *Sci. Rep.* **2017**, *7*, 45079. [CrossRef]
75. Kee, S.; Kim, N.; Kim, B.S.; Park, S.; Jang, Y.H.; Lee, S.H.; Kim, J.; Kim, J.; Kwon, S.; Lee, K. Controlling Molecular Ordering in Aqueous Conducting Polymers Using Ionic Liquids. *Adv. Mater.* **2016**, *28*, 8625–8631. [CrossRef]
76. Zheng, Z.; Hu, Q.; Zhang, S.; Zhang, D.; Wang, J.; Xie, S.; Wang, R.; Qin, Y.; Li, W.; Hong, L.; et al. A Highly Efficient Non-Fullerene Organic Solar Cell with a Fill Factor over 0.80 Enabled by a Fine-Tuned Hole-Transporting Layer. *Adv. Mater.* **2018**, *30*, 1801801. [CrossRef]
77. Alharbi, N.S.; Wang, C.; Alsaadi, F.E.; Rabah, S.O.; Tan, Z.A. A General Approach of Adjusting the Surface-Free Energy of the Interfacial Layer for High-Performance Organic Solar Cells. *Adv. Sustain. Syst.* **2020**, *4*, 2000054. [CrossRef]
78. Peng, R.; Wan, Z.; Song, W.; Yan, T.; Qiao, Q.; Yang, S.; Ge, Z.; Wang, M. Improving Performance of Nonfullerene Organic Solar Cells over 13% by Employing Silver Nanowires-Doped PEDOT:PSS Composite Interface. *ACS Appl. Mater. Interfaces* **2019**, *11*, 42447–42454. [CrossRef]
79. Xu, H.; Yuan, F.; Zhou, D.; Liao, X.; Chen, L.; Chen, Y. Hole transport layers for organic solar cells: Recent progress and prospects. *J. Mater. Chem. A* **2020**, *8*, 11478–11492. [CrossRef]
80. Yang, Q.; Yu, S.; Fu, P.; Yu, W.; Liu, Y.; Liu, X.; Feng, Z.; Guo, X.; Li, C. Boosting Performance of Non-Fullerene Organic Solar Cells by 2D g-C_3N_4 Doped PEDOT:PSS. *Adv. Funct. Mater.* **2020**, *30*, 1910205. [CrossRef]
81. Fan, P.; Zhang, D.; Wu, Y.; Yu, J.; Russell, T.P. Polymer-Modified ZnO Nanoparticles as Electron Transport Layer for Polymer-Based Solar Cells. *Adv. Funct. Mater.* **2020**, *30*, 2002932. [CrossRef]
82. Wang, Y.; Zheng, Z.; Wang, J.; Liu, X.; Ren, J.; An, C.; Zhang, S.; Hou, J. New Method for Preparing ZnO Layer for Efficient and Stable Organic Solar Cells. *Adv. Mater.* **2022**, 2208305. [CrossRef] [PubMed]
83. Xu, J.; Chen, Z.; Zapien, J.A.; Lee, C.S.; Zhang, W. Surface engineering of ZnO nanostructures for semiconductor-sensitized solar cells. *Adv. Mater.* **2014**, *26*, 5337–5367. [CrossRef] [PubMed]
84. Sun, B.; Sirringhaus, H. Surface Tension and Fluid Flow Driven Self-Assembly of Ordered ZnO Nanorod Films for High-Performance Field Effect Transistors. *J. Am. Chem. Soc.* **2006**, *128*, 16231–16237. [CrossRef] [PubMed]
85. Zhu, G.; Yang, R.; Wang, S.; Wang, Z.L. Flexible high-output nanogenerator based on lateral ZnO nanowire array. *Nano Lett.* **2010**, *10*, 3151–3155. [CrossRef]
86. Liu, X.; Zheng, Z.; Wang, J.; Wang, Y.; Xu, B.; Zhang, S.; Hou, J. Fluidic Manipulating of Printable Zinc Oxide for Flexible Organic Solar Cells. *Adv. Mater.* **2022**, *34*, e2106453. [CrossRef]

87. Yao, J.; Qiu, B.; Zhang, Z.G.; Xue, L.; Wang, R.; Zhang, C.; Chen, S.; Zhou, Q.; Sun, C.; Yang, C.; et al. Cathode engineering with perylene-diimide interlayer enabling over 17% efficiency single-junction organic solar cells. *Nat. Commun.* **2020**, *11*, 2726. [CrossRef]
88. Liu, M.; Jiang, Y.; Liu, D.; Wang, J.; Ren, Z.; Russell, T.P.; Liu, Y. Imidazole-Functionalized Imide Interlayers for High Performance Organic Solar Cells. *ACS Energy Lett.* **2021**, *6*, 3228–3235. [CrossRef]
89. Bi, Z.; Naveed, H.B.; Mao, Y.; Yan, H.; Ma, W. Importance of Nucleation during Morphology Evolution of the Blade-Cast PffBT4T-2OD-Based Organic Solar Cells. *Macromolecules* **2018**, *51*, 6682–6691. [CrossRef]
90. Krebs, F.C. Polymer solar cell modules prepared using roll-to-roll methods: Knife-over-edge coating, slot-die coating and screen printing. *Sol. Energy Mater. Sol. Cells* **2009**, *93*, 465–475. [CrossRef]
91. Søndergaard, R.; Hösel, M.; Angmo, D.; Larsen-Olsen, T.T.; Krebs, F.C. Roll-to-roll fabrication of polymer solar cells. *Mater. Today* **2012**, *15*, 36–49. [CrossRef]
92. Wengeler, L.; Schmidt-Hansberg, B.; Peters, K.; Scharfer, P.; Schabel, W. Investigations on knife and slot die coating and processing of polymer nanoparticle films for hybrid polymer solar cells. *Chem. Eng. Process. Process Intensif.* **2011**, *50*, 478–482. [CrossRef]
93. Sun, H.; Wang, Q.; Qian, J.; Yin, Y.; Shi, Y.; Li, Y. Unidirectional coating technology for organic field-effect transistors: Materials and methods. *Semicond. Sci. Technol.* **2015**, *30*, 054001. [CrossRef]
94. Gu, X.; Shaw, L.; Gu, K.; Toney, M.F.; Bao, Z. The meniscus-guided deposition of semiconducting polymers. *Nat. Commun.* **2018**, *9*, 534.
95. Sun, R.; Guo, J.; Sun, C.; Wang, T.; Luo, Z.; Zhang, Z.; Jiao, X.; Tang, W.; Yang, C.; Li, Y.; et al. A universal layer-by-layer solution-processing approach for efficient non-fullerene organic solar cells. *Energy Environ. Sci.* **2019**, *12*, 384–395.
96. Mens, R.; Adriaensens, P.; Lutsen, L.; Swinnen, A.; Bertho, S.; Ruttens, B.; D'Haen, J.; Manca, J.; Cleij, T.; Vanderzande, D.; et al. NMR study of the nanomorphology in thin films of polymer blends used in organic PV devices: MDMO-PPV/PCBM. *J. Polym. Sci. Part A Polym. Chem.* **2008**, *46*, 138–145. [CrossRef]
97. Zhao, W.; Zhang, Y.; Zhang, S.; Li, S.; He, C.; Hou, J. Vacuum-assisted annealing method for high efficiency printable large-area polymer solar cell modules. *J. Mater. Chem. C* **2019**, *7*, 3206–3211.
98. Li, Y.; Liu, H.; Wu, J.; Tang, H.; Wang, H.; Yang, Q.; Fu, Y.; Xie, Z. Additive and High-Temperature Processing Boost the Photovoltaic Performance of Nonfullerene Organic Solar Cells Fabricated with Blade Coating and Nonhalogenated Solvents. *ACS Appl. Mater. Interfaces* **2021**, *13*, 10239–10248. [CrossRef] [PubMed]
99. Zhao, W.; Zhang, S.; Zhang, Y.; Li, S.; Liu, X.; He, C.; Zheng, Z.; Hou, J. Environmentally Friendly Solvent-Processed Organic Solar Cells that are Highly Efficient and Adaptable for the Blade-Coating Method. *Adv. Mater.* **2018**, *30*, 1704837. [CrossRef]
100. Ma, Q.; Jia, Z.; Meng, L.; Zhang, J.; Zhang, H.; Huang, W.; Yuan, J.; Gao, F.; Wan, Y.; Zhang, Z.; et al. Promoting charge separation resulting in ternary organic solar cells efficiency over 17.5%. *Nano Energy* **2020**, *78*, 105272. [CrossRef]
101. Na, S.I.; Seo, Y.H.; Nah, Y.C.; Kim, S.S.; Heo, H.; Kim, J.E.; Rolston, N.; Dauskardt, R.H.; Gao, M.; Lee, Y.; et al. High Performance Roll-to-Roll Produced Fullerene-Free Organic Photovoltaic Devices via Temperature-Controlled Slot Die Coating. *Adv. Funct. Mater.* **2018**, *29*, 1805825. [CrossRef]
102. Strohm, S.; Machui, F.; Langner, S.; Kubis, P.; Gasparini, N.; Salvador, M.; McCulloch, I.; Egelhaaf, H.J.; Brabec, C.J. P3HT: Non-fullerene acceptor based large area, semi-transparent PV modules with power conversion efficiencies of 5%, processed by industrially scalable methods. *Energy Environ. Sci.* **2018**, *11*, 2225–2234. [CrossRef]
103. Yang, Y.; Feng, E.; Li, H.; Shen, Z.; Liu, W.; Guo, J.; Luo, Q.; Zhang, J.; Lu, G.; Ma, C.; et al. Layer-by-layer slot-die coated high-efficiency organic solar cells processed using twin boiling point solvents under ambient condition. *Nano Res.* **2021**, *14*, 4236–4242. [CrossRef]
104. Yang, F.; Huang, Y.; Li, Y.; Li, Y. Large-area flexible organic solar cells. *npj Flex. Electron.* **2021**, *5*, 30. [CrossRef]
105. Basaran, O.A.; Gao, H.; Bhat, P.P. Nonstandard Inkjets. *Annu. Rev. Fluid Mech.* **2013**, *45*, 85–113. [CrossRef]
106. Karunakaran, S.K.; Arumugam, G.M.; Yang, W.; Ge, S.; Khan, S.N.; Lin, X.; Yang, G. Recent progress in inkjet-printed solar cells. *J. Mater. Chem. A* **2019**, *7*, 13873–13902. [CrossRef]
107. Bihar, E.; Corzo, D.; Hidalgo, T.C.; Rosas-Villalva, D.; Salama, K.N.; Inal, S.; Baran, D. Fully Inkjet-Printed, Ultrathin and Conformable Organic Photovoltaics as Power Source Based on Cross-Linked PEDOT:PSS Electrodes. *Adv. Mater. Technol.* **2020**, *5*, 2000226. [CrossRef]
108. Derby, B. Inkjet Printing of Functional and Structural Materials: Fluid Property Requirements, Feature Stability, and Resolution. *Annu. Rev. Mater. Res.* **2010**, *40*, 395–414. [CrossRef]
109. Eggenhuisen, T.M.; Galagan, Y.; Biezemans, A.F.K.V.; Slaats, T.M.W.L.; Voorthuijzen, W.P.; Kommeren, S.; Shanmugam, S.; Teunissen, J.P.; Hadipour, A.; Verhees, W.J.H.; et al. High efficiency, fully inkjet printed organic solar cells with freedom of design. *J. Mater. Chem. A* **2015**, *3*, 7255–7262. [CrossRef]
110. Maisch, P.; Tam, K.C.; Lucera, L.; Egelhaaf, H.-J.; Scheiber, H.; Maier, E.; Brabec, C.J. Inkjet printed silver nanowire percolation networks as electrodes for highly efficient semitransparent organic solar cells. *Org. Electron.* **2016**, *38*, 139–143. [CrossRef]
111. Corzo, D.; Almasabi, K.; Bihar, E.; Macphee, S.; Rosas-Villalva, D.; Gasparini, N.; Inal, S.; Baran, D. Digital Inkjet Printing of High-Efficiency Large-Area Nonfullerene Organic Solar Cells. *Adv. Mater. Technol.* **2019**, *4*, 1900040. [CrossRef]
112. Hu, G.; Kang, J.; Ng, L.W.T.; Zhu, X.; Howe, R.C.T.; Jones, C.G.; Hersam, M.C.; Hasan, T. Functional inks and printing of two-dimensional materials. *Chem. Soc. Rev.* **2018**, *47*, 3265–3300. [CrossRef]

113. Kang, B.; Lee, W.H.; Cho, K. Recent advances in organic transistor printing processes. *ACS Appl. Mater. Interfaces* **2013**, *5*, 2302–2315. [CrossRef]
114. Parida, B.; Singh, A.; Kalathil Soopy, A.K.; Sangaraju, S.; Sundaray, M.; Mishra, S.; Liu, S.F.; Najar, A. Recent Developments in Upscalable Printing Techniques for Perovskite Solar Cells. *Adv. Sci. (Weinh.)* **2022**, *9*, e2200308. [CrossRef]
115. Wang, G.; Adil, M.A.; Zhang, J.; Wei, Z. Large-Area Organic Solar Cells: Material Requirements, Modular Designs, and Printing Methods. *Adv. Mater.* **2019**, *31*, e1805089. [CrossRef]
116. Xue, P.; Cheng, P.; Han, R.P.S.; Zhan, X. Printing fabrication of large-area non-fullerene organic solar cells. *Mater. Horiz.* **2022**, *9*, 194–219. [CrossRef]
117. Jørgensen, M.; Hagemann, O.; Alstrup, J.; Krebs, F.C. Thermo-cleavable solvents for printing conjugated polymers: Application in polymer solar cells. *Sol. Energy Mater. Sol. Cells* **2009**, *93*, 413–421. [CrossRef]
118. Krebs, F.C.; Jørgensen, M.; Norrman, K.; Hagemann, O.; Alstrup, J.; Nielsen, T.D.; Fyenbo, J.; Larsen, K.; Kristensen, J. A complete process for production of flexible large area polymer solar cells entirely using screen printing—First public demonstration. *Sol. Energy Mater. Sol. Cells* **2009**, *93*, 422–441. [CrossRef]
119. Ylikunnari, M.; Välimäki, M.; Väisänen, K.-L.; Kraft, T.M.; Sliz, R.; Corso, G.; Po, R.; Barbieri, R.; Carbonera, C.; Gorni, G.; et al. Flexible OPV modules for highly efficient indoor applications. *Flex. Print. Electron.* **2020**, *5*, 014008. [CrossRef]
120. Kapnopoulos, C.; Mekeridis, E.D.; Tzounis, L.; Polyzoidis, C.; Zachariadis, A.; Tsimikli, S.; Gravalidis, C.; Laskarakis, A.; Vouroutzis, N.; Logothetidis, S. Fully gravure printed organic photovoltaic modules: A straightforward process with a high potential for large scale production. *Sol. Energy Mater. Sol. Cells* **2016**, *144*, 724–731. [CrossRef]
121. Nguyen, H.A.D.; Lee, C.; Shin, K.-H.; Lee, D. An Investigation of the Ink-Transfer Mechanism During the Printing Phase of High-Resolution Roll-to-Roll Gravure Printing. *IEEE Trans. Compon. Packag. Manuf. Technol.* **2015**, *5*, 1516–1524. [CrossRef]
122. Schneider, A.; Traut, N.; Hamburger, M. Analysis and optimization of relevant parameters of blade coating and gravure printing processes for the fabrication of highly efficient organic solar cells. *Sol. Energy Mater. Sol. Cells* **2014**, *126*, 149–154. [CrossRef]
123. Park, J.D.; Lim, S.; Kim, H. Patterned silver nanowires using the gravure printing process for flexible applications. *Thin Solid Films* **2015**, *586*, 70–75. [CrossRef]
124. Peng, Y.; Du, B.; Xu, X.; Yang, J.; Lin, J.; Ma, C. Transparent triboelectric sensor arrays using gravure printed silver nanowire electrodes. *Appl. Phys. Express* **2019**, *12*, 066503. [CrossRef]
125. Scheideler, W.J.; Smith, J.; Deckman, I.; Chung, S.; Arias, A.C.; Subramanian, V. A robust, gravure-printed, silver nanowire/metal oxide hybrid electrode for high-throughput patterned transparent conductors. *J. Mater. Chem. C* **2016**, *4*, 3248–3255. [CrossRef]
126. Wang, Z.; Han, Y.; Yan, L.; Gong, C.; Kang, J.; Zhang, H.; Sun, X.; Zhang, L.; Lin, J.; Luo, Q.; et al. High Power Conversion Efficiency of 13.61% for 1 cm^2 Flexible Polymer Solar Cells Based on Patternable and Mass-Producible Gravure-Printed Silver Nanowire Electrodes. *Adv. Funct. Mater.* **2020**, *31*, 2007276. [CrossRef]

Disclaimer/Publisher's Note: The statements, opinions and data contained in all publications are solely those of the individual author(s) and contributor(s) and not of MDPI and/or the editor(s). MDPI and/or the editor(s) disclaim responsibility for any injury to people or property resulting from any ideas, methods, instructions or products referred to in the content.

Article

BiVO$_4$ Photoanodes Modified with Synergetic Effects between Heterojunction Functionalized FeCoO$_x$ and Plasma Au Nanoparticles

Huangzhaoxiang Chen [1], Qian Zhang [1,*], Aumber Abbas [2], Wenran Zhang [1], Shuzhou Huang [1], Xiangguo Li [1], Shenghua Liu [1,*] and Jing Shuai [1]

[1] School of Materials, Shenzhen Campus, Sun Yat-Sen University, No. 66, Gongchang Road, Guangming District, Shenzhen 518107, China; chenhzhx@mail2.sysu.edu.cn (H.C.); zhangwr28@mail2.sysu.edu.cn (W.Z.); huangshzh6@mail2.sysu.edu.cn (S.H.); lixguo@mail.sysu.edu.cn (X.L.); shuaij3@mail.sysu.edu.cn (J.S.)
[2] Songshan Lake Materials Laboratory, Room 425, C1 Building, University Innovation City, Songshan Lake, Dongguan 523000, China; aumber.abbas@sslab.org.cn
* Correspondence: zhangqian6@mail2.sysu.edu.cn (Q.Z.); liushengh@mail.sysu.edu.cn (S.L.)

Citation: Chen, H.; Zhang, Q.; Abbas, A.; Zhang, W.; Huang, S.; Li, X.; Liu, S.; Shuai, J. BiVO$_4$ Photoanodes Modified with Synergetic Effects between Heterojunction Functionalized FeCoO$_x$ and Plasma Au Nanoparticles. Catalysts 2023, 13, 1063. https://doi.org/10.3390/catal13071063

Academic Editor: Bruno Fabre

Received: 22 May 2023
Revised: 24 June 2023
Accepted: 26 June 2023
Published: 1 July 2023

Copyright: © 2023 by the authors. Licensee MDPI, Basel, Switzerland. This article is an open access article distributed under the terms and conditions of the Creative Commons Attribution (CC BY) license (https://creativecommons.org/licenses/by/4.0/).

Abstract: The design and development of high-performance photoanodes are the key to efficient photoelectrochemical (PEC) water splitting. Based on the carrier transfer characteristics and localized surface plasmon resonance effect of noble metals, gold nanoparticles (AuNPs) have been used to improve the performance of photoanodes. In this study, a novel efficient composite BiVO$_4$/Au/FeCoO$_x$ photoanode is constructed, and the quantitative analysis of its performance is systematically conducted. The results reveal that the co-modification of AuNPs and FeCoO$_x$ plays a synergetic role in enhancing the absorption of ultraviolet and visible light of BiVO$_4$, which is mainly attributed to the localized surface plasmon resonance effect induced by AuNPs and the extended light absorption edge position induced by the BiVO$_4$/FeCoO$_x$ heterojunction. The BiVO$_4$/Au/FeCoO$_x$ photoanode exhibits a high photocurrent density of 4.11 mA cm^{-2} at 1.23 V versus RHE at room temperature under AM 1.5 G illumination, which corresponds to a 299% increase compared to a pristine BiVO$_4$ photoanode. These results provide practical support for the design and preparation of PEC photoanodes decorated with AuNPs and FeCoO$_x$.

Keywords: solar water splitting; plasmon resonance effect; nanomaterials; heterojunction; photoelectrochemical

1. Introduction

Since the discovery of photoelectrochemical (PEC) water splitting based on semiconductor electrodes, scientists have devoted substantial efforts to transforming solar energy into clean and carbon-neutral hydrogen (H$_2$) energy in a stable, cost-effective, and efficient way [1]. The overall PEC water splitting consists of three main stages: (I) the absorption of light and the generation of electron–hole pairs in the semiconductor; (II) the separation of the electron and hole and their transfer to the surface of the semiconductor; (III) the occurrence of an oxygen evolution reaction (OER) on the photoanode and a hydrogen evolution reaction (HER) on the photocathode [2]. Due to the fracture of an O–H bond and the generation of an O–O bond, the oxygen evolution reaction on the surface of the photoanode is dynamically sluggish [3]. Thus, discovering and developing efficient photoanodes has always been regarded as the key to photoelectrochemical water splitting. In the past few decades, a wide variety of semiconductors, such as TiO$_2$ [4], Fe$_2$O$_3$ [5], and BiVO$_4$ [6], have been studied and employed as photoanodes.

Among the available photoanodes, BiVO$_4$ is undoubtedly an appealing photocatalyst material. It is an n-type semiconductor with a relatively narrow band gap of about 2.4 eV, which enables it to absorb ultraviolet light and a wide range of visible light [7,8]. In addition, it has a suitable conduction band edge position which is close to the thermodynamic precipitation potential of O_2 [9,10]. Moreover, it is nontoxic and has impressive PEC performance in solution [11,12]. However, the slow carrier dynamics result in the recombination of bulk and interface charges of the unmodified $BiVO_4$, which limits its solar-to-hydrogen conversion efficiency. It has been widely verified that the actual maximum photocurrent density under AM 1.5 G illumination (100 mW cm^{-2}) of unmodified $BiVO_4$ is far below the theoretical value (7.5 mA cm^{-2}) [13].

In order to boost the OER of photoanodes, it is an effective strategy to modify $BiVO_4$ with oxygen evolution cocatalysts (OECs), which can reduce electron–hole recombination and enhance the photochemical performance. OECs like $FeCoO_x$ [14], Cu_2O [15], NiOOH [16], and FeOOH [17] have been utilized and experimented with in modifying $BiVO_4$ photoanodes. Among them, a new type of $FeCoO_x$ has attracted researchers' attention due to its effective suppression of electron–hole recombination on the surface of $BiVO_4$ photoanode [18,19]. According to theoretical and experimental studies, it is clear that the substitution of cations in CoO_x with iron ions results in abundant oxygen vacancies and forms a heterojunction with $BiVO_4$ [20,21]. $FeCoO_x$ also decreases the carrier transfer resistance to a certain extent, which enables the transfer of more photogenerated holes in the valence band of $BiVO_4$ to the $FeCoO_x$ layer. This leads to the generation of more reactive hydroxyl radicals to participate in water oxidation reactions efficiently [22,23]. However, the recombination center of the electron–hole may emerge on the interface of the heterojunction, leading to a decrease in PEC performance.

To promote bulk charge separation and carrier transfer, a practical method is to insert an electron transport layer (such as Pt, Au, Ni, etc.) into the middle of the heterojunction. As a classic plasmon nanostructure, the gold nanoparticles (AuNPs) are able to confine light in the vicinity of the surface and generate hot electron flow, which has a great effect on the chemical and PEC performance of the metal–oxide interface [24]. Since the work function of $BiVO_4$ is greater than gold, a Schottky junction is naturally formed when there is direct contact between $BiVO_4$ and Au. The Schottky junction may reduce the charge recombination on the metal–semiconductor interface, promote carrier transfer, and adjust the electronic band structure. Thus, the lifetime of photogenerated electrons and holes ought to increase, resulting in more electrons and holes involved in the PEC process [25].

On the other hand, the impact of geometric factors such as the size and shape of AuNPs on photoelectrochemical water splitting has been experimentally verified. AuNPs present particular photoelectronic and electrochemical properties when the particle size is adequately small (quantum size effect), such as localized surface plasmon resonance (LSPR) characteristics and electrochemical catalytic properties with great dependence on particle size [26,27]. Under the illumination of visible light or infrared light, the energy of oscillating electrons and the localized electromagnetic field generated by LSPR will be transferred to the conduction band of $BiVO_4$ via direct electron transfer (DET) or plasmon resonant energy transfer (PRET) [10,28,29]. In the case of $BiVO_4$/Au, numerous experiments indicate that energy transfers via DET, implying that the excited hot electron of Au may overcome the metal–semiconductor barrier and transfer to the conduction band of $BiVO_4$, contributing to improved photocurrent density.

Herein, in order to progressively enhance the PEC properties of $BiVO_4$ photoanodes and verify the synergism of AuNPs and $FeCoO_x$, the $BiVO_4$/Au/$FeCoO_x$ photoanode is constructed through a template route, which is followed by electrochemical deposition of AuNPs and $FeCoO_x$, respectively. The optimized $BiVO_4$/Au/$FeCoO_x$ photoanode exhibits a high photocurrent density of 4.11 mA cm^{-2} at 1.23 V versus RHE under AM 1.5 G (100 mW cm^{-2}) illumination, which is over three-fold that of the pristine $BiVO_4$ photoanode and a 29.7% increase in that of the $BiVO_4$/$FeCoO_x$ counterpart.

2. Results and Discussion

Figure 1a displays a schematic diagram of the preparation process for the $BiVO_4/Au/FeCoO_x$ photoanode, and the experimental details are presented in the experimental section. The $BiVO_4/Au/FeCoO_x$ photoanode X-ray diffraction (XRD) patterns are shown in Figure 1b. The characteristic peaks of crystal planes such as (020), (011), and (121) of $BiVO_4$ have been labeled. The peaks corresponding to the (111) and (200) planes of Au have also been marked, but due to the small content and size, more diffraction peaks of AuNPs and $FeCoO_x$ are too difficult to detect. Figure 1c displays the scanning electron microscopy (SEM) image of the $BiVO_4/Au/FeCoO_x$ photoanode, which implies that the co-modification of AuNPs and $FeCoO_x$ makes the originally striped $BiVO_4$ thicker and more compact. The surface of the $BiVO_4/Au/FeCoO_x$ photoanode material is uniformly filled with a similar cluster structure. Figure S1a shows the SEM image of a pristine $BiVO_4$ photoanode. It can be seen that $BiVO_4$ is a thin film formed by particles, strips, or clusters, with particle sizes ranging from 100 nm to 300 nm, while the length of the strip formed by the particles is approximately 300 nm to 1 μm. Additionally, due to issues with the preparation methods, there are unfilled holes in $BiVO_4$. Figure S1b shows the SEM image of the $BiVO_4/Au$ photoanode, in which the particles with significantly brighter brightness than $BiVO_4$ are gold nanoparticles, with a particle size of approximately 80~150 nm. Figure S1c shows the SEM image of the $BiVO_4/FeCoO_x$ photoanode. It can be seen that $FeCoO_x$ adheres to the surface of $BiVO_4$ in a paste-like form, making the original strip or cluster structure of $BiVO_4$ more flexible.

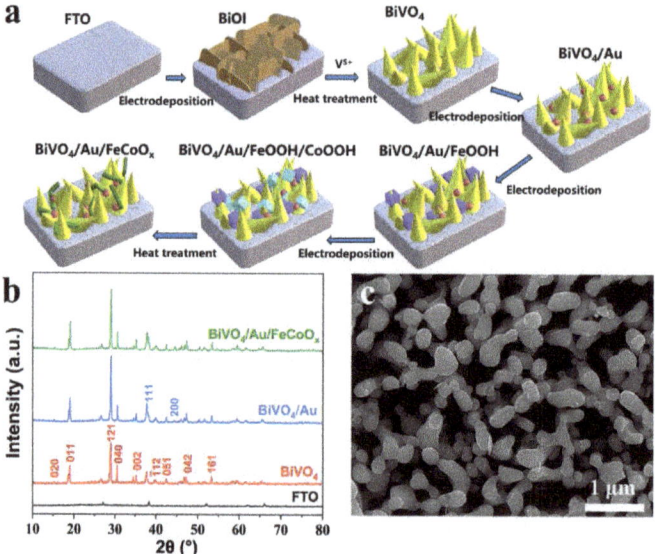

Figure 1. (a) Diagram for the preparation procedure of the $BiVO_4/Au/FeCoO_x$ photoanode. (b) XRD patterns of photoanodes. (c) SEM image of $BiVO_4/Au/FeCoO_x$.

Figure 2a–g display the energy dispersive spectroscopy (EDS) spectrum of the $BiVO_4/Au/FeCoO_x$ composite photoanode. The distribution of six elements, Bi, V, O, Au, Fe, and Co, can be seen from the EDS elemental mapping, which can also prove the existence of these elements. From the mapping of Au in Figure 2e, it can be seen that there are three local positions with dense Au distribution in the upper left, upper right, and lower middle directions, which may be caused by large-size AuNPs. As shown in Figure 2h and Table S1, the content information of each element in the selected area indicates that the contents of Au, Fe, and Co are very low among the six kinds of elements and appear to have atomic ratios of 1.11%, 1.28%, and 0.58%, respectively, indicating that Au and FeCoO$_x$ mainly play a modifying role in the photoanodes. Moreover, it can also be calculated that the atomic ratio of Fe and Co in the selected area is Fe:Co = 2.2:1. In the relevant publications on the preparation of Au and FeCoO$_x$, the content of Au, Fe, and Co is generally below 5% [30], and the most significant improvement in the PEC performance of FeCoO$_x$ at present belongs to the work of Wang et al., where the relative atomic ratio of Fe:Co is Fe:Co = 1.1:1 [14].

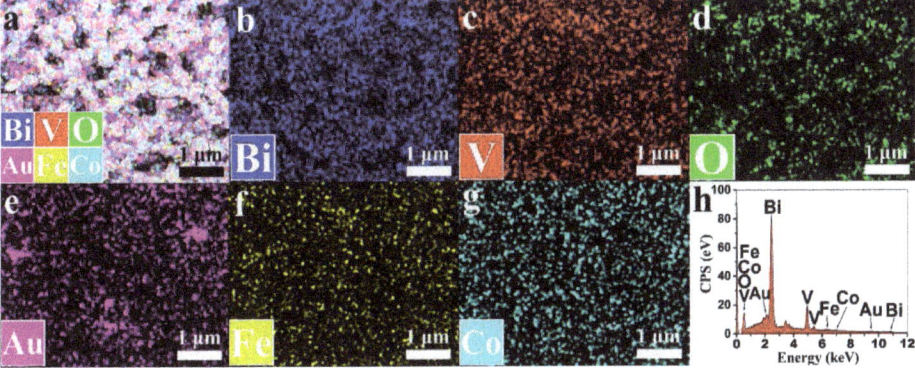

Figure 2. Elemental mapping of BiVO$_4$/Au/FeCoO$_x$. (**a**) Overall image. (**b**) Bi. (**c**) V. (**d**) O. (**e**) Au. (**f**) Fe. (**g**) Co. (**h**) EDS energy distribution.

Figure 3 displays the high-resolution X-ray photoelectron spectroscopy (XPS) elemental spectrum of the BiVO$_4$/Au/FeCoO$_x$ photoanode, which contains information on the composition and chemical states of elements in the composite photoanode. The high-resolution Bi 4f spectrum can be fitted by two main peaks, located at 159.11 eV and 164.41 eV, respectively, corresponding to the binding energy of the Bi 4f$_{7/2}$ spin state and the Bi 4f$_{5/2}$ spin state. The high-resolution V 2p spectrum displays two main peaks at 516.70 eV and 524.16 eV, respectively, and they correspond to the binding energy of the V 2p$_{3/2}$ spin state and the V 2p$_{1/2}$ spin state. Figure 3c displays the high-resolution O 1s spectrum, fitted with three main peaks located at 530.45 eV, 531.55 eV, and 532.35 eV, respectively. These three peaks correspond to O^{2-} in the lattice (O$_L$), hydroxyl bound to metal cations (O$_V$), and oxygen in the chemically adsorbed or dissociated state (such as dissociated CO$_3^{2-}$, absorbed H$_2$O, absorbed O$_2$, etc., O$_C$). From Figure 3c, it can be seen that the O$_L$ peak is quite prominent compared with the O$_V$ and O$_C$ peaks, reflecting that the oxygen element is mainly present in the form of O^{2-} in the lattice. The existence of the O$_V$ peak proves the presence of a large number of oxygen defects or vacancies on the surface of the BiVO$_4$/Au/FeCoO$_x$ photoanode, while the existence of the O$_C$ peak proves the presence of surface hydroxyl groups on the surface of the BiVO$_4$/Au/FeCoO$_x$ photoanode. The presence of surface hydroxyl groups can promote the transport and capture of photogenerated electrons and photogenerated holes, thereby reducing the recombination of photogenerated carriers and improving the photoelectric conversion efficiency of BiVO$_4$/Au/FeCoO$_x$. The high-resolution Au 4f spectrum displays two main peaks, 83.82 eV and 87.51 eV, respectively, corresponding to the binding energy of the Au 4f$_{7/2}$ spin state and the Au 4f$_{5/2}$ spin state. The high-resolution Fe 2p spectrum is fitted with six peaks, including three main peaks and three satellite peaks. The three main peaks are located at 710.31 eV, 712.42 eV, and 723.97 eV. Among them, the peak located at 723.97 eV corresponds to the Fe 2p$_{1/2}$ peak, and the first two of the three main peaks represent Fe^{2+} and Fe^{3+}, respectively, which together form the Fe 2p$_{3/2}$ peak. The atomic ratio of these two peaks can also be calculated from the peak position table as Fe^{3+}:Fe^{2+} = 1.89:1. The high-resolution Co 2p spectrum consists of six fitted peaks, including three main peaks and three satellite peaks. The three main peaks are located at 779.60 eV, 780.60 eV, and 795.74 eV, where the peak at 795.74 eV corresponds to the Co $2p_{1/2}$ peak. The first two of the three main peaks represent Co^{3+} and Co^{2+}, respectively, and together, they form the Co $2p_{3/2}$ peak. The atomic ratio of these two ions can be calculated from the peak position table shown in Table S2, as $Co^{2+}:Co^{3+}$ = 3.59:1. The atomic ratio of Fe and Co can also be calculated as Fe:Co = 1.37:1.

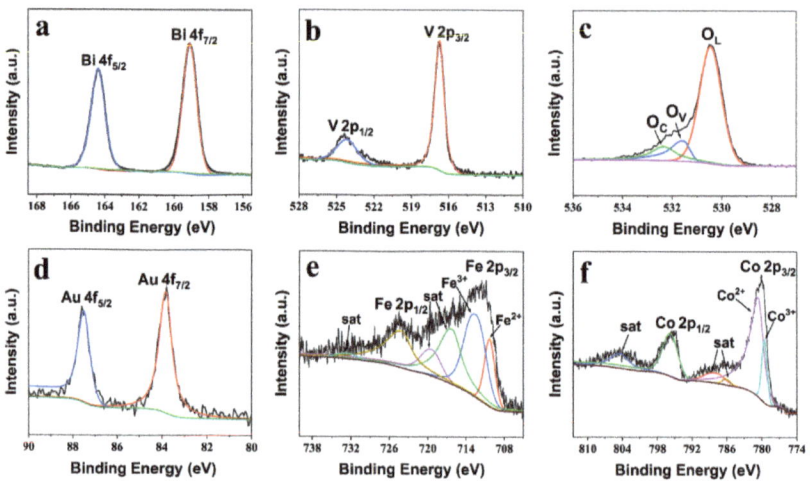

Figure 3. High-resolution XPS spectrum of the $BiVO_4/Au/FeCoO_x$. (**a**) Bi 4f, (**b**) V 2p, (**c**) O 1s, (**d**) Fe 2p, (**e**) Co 2p, and (**f**) Au 4f.

Figure 4a displays the ultraviolet–visible (UV-vis) absorption spectra of $BiVO_4$, $BiVO_4/Au$, and $BiVO_4/Au/FeCoO_x$ photoanodes. The FTO substrate has been used as the background to eliminate the influence of FTO on the absorbance. It can be seen that the light absorption edge of $BiVO_4$ is around 510 nm, which is consistent with its band gap of 2.3~2.5 eV. The light absorption edge of $BiVO_4/Au$ remains almost unchanged, but its absorption wavelength range from 300 nm to 900 nm is higher than that of pristine $BiVO_4$. The light absorption edge of $BiVO_4/Au/FeCoO_x$ photoanodes is about 660 nm, and the light absorption at wavelengths of 300~900 nm is stronger than that of $BiVO_4$ and $BiVO_4/Au$. The loading of AuNPs and $FeCoO_x$ has played a synergic role in enhancing the absorption of UV-vis light of $BiVO_4$, which is mainly due to the LSPR effect induced by AuNPs and the extended light absorption edge position induced by the $BiVO_4/FeCoO_x$ heterojunction.

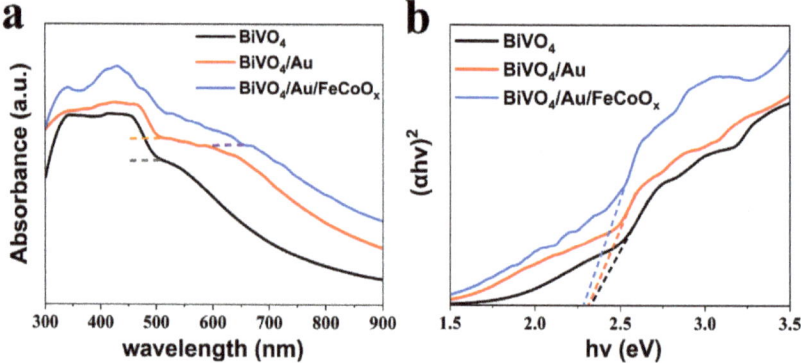

Figure 4. Light absorbance of photoanodes. (**a**) UV-vis diffusion absorption spectra. (**b**) Tauc plots of $BiVO_4$ films modified with cocatalysts.

The band gap can also be obtained by Tauc plots. Since most of the valence band electrons and conduction band electrons of the semiconductor are distributed near the band gap, when the energy of a photon is close to the band gap, a large number of electrons can be excited from the top of the valence band to the bottom of the conduction band by absorbing the photon energy. At this moment, the absorption coefficient of the semiconductor will increase with the increase in the number of photons. The relationship between the band gap and the absorption coefficient of the semiconductor materials can be expressed as follows:

$$(\alpha h\nu)^n = A * (h\nu - E_g) \qquad (1)$$

where α is the absorption coefficient, $h \approx 4.13567 \times 10^{-15}$ eV s, as Planck's coefficient, ν is the frequency of the incident photon, A is the proportional coefficient, and E_g is the band gap of the semiconductor. The value of n is related to the type of semiconductor. When the semiconductor is a direct band gap semiconductor, that is, the conduction band bottom and valence band top of the semiconductor correspond to the same wave vector of electronic states, a direct band edge transition can occur, where n = 2. When the band gap of a semiconductor is an indirect band gap, that is, the conduction band bottom and valence band top of the semiconductor are not at the same wave vector of electronic states, the band edge transition requires the participation of phonons, where n = 1/2 at this time. Since $BiVO_4$ is a direct band gap semiconductor, n is set as 2. Substitute the absorption coefficient and others into Equation (1) and create an $(\alpha h\nu)^2 - h\nu$ diagram. As shown in Figure 4b, the band gap of $BiVO_4$ and $BiVO_4/Au$ is about 2.32 eV, while the band gap of $BiVO_4/Au/FeCoO_x$ is about 2.28 eV. These results indicate that the modification of AuNPs has little effect on the band gap of $BiVO_4$, while the band gap narrowing of $BiVO_4/Au/FeCoO_x$ photoanodes can be attributed to the narrow band gap of $FeCoO_x$ (2.04 eV) compared to $BiVO_4$ (2.3~2.5 eV).

Figure 5a displays the linear sweep voltammetry (LSV) curve of the photoanodes. It can be seen that the photocurrent density under dark conditions is only 7×10^{-7} mA cm^{-2} at 1.23 V vs. RHE, and the magnitude of this photocurrent density can be regarded as almost no photocurrent generation. Under AM 1.5 G illumination at 1.23 V vs. RHE, the photocurrent density of $BiVO_4/Au/FeCoO_x$ reached 4.11 mA cm^{-2}, which is nearly four times than that of pristine $BiVO_4$ (1.03 mA cm^{-2}). This photocurrent density is increased by 71.8% compared to $BiVO_4/Au$ (1.77 mA cm^{-2}) and increased by 29.7% compared to $BiVO_4/FeCoO_x$ (3.17 mA cm^{-2}). Additionally, the $BiVO_4/Au/FeCoO_x$ photoanode exhibits an onset potential of 0.245 V vs. RHE, which is the lowest among the photoanodes and cathodically shifted by about 27 mV compared with that of pristine $BiVO_4$. There are multiple reasons for this increase in the photocurrent density. For example, according to the discussions of Figure 4a, the modification of $FeCoO_x$ has expanded the original light absorption edge of $BiVO_4$ from only about 510 nm to about 660 nm. On the other hand, the modification of AuNPs and $FeCoO_x$ has played a positive role in improving the absorbance of $BiVO_4$ for ultraviolet and visible light, which is consistent with the results of UV-vis spectra. It is not difficult to understand that under the assumption of the same photoelectric conversion efficiency, the stronger the absorption of light by the photoanode, the greater the photocurrent density generated. Moreover, the synergism effect of AuNPs and $FeCoO_x$ may contribute to the enhancement of the photocurrent density. For example, the photogenerated holes in the valence band of $BiVO_4$ will be captured in the $FeCoO_x$ surface and participate in the PEC water oxidation, while the excited hot electron of AuNPs may overcome the metal–semiconductor barrier and transfer to the conduction band of $BiVO_4$. Namely, the quality of $FeCoO_x$ as an oxygen evolution cocatalyst is of vital significance in promoting PEC water oxidation, which may cause differences in the photocurrent between photoanode counterparts modified with $FeCoO_x$ [14].

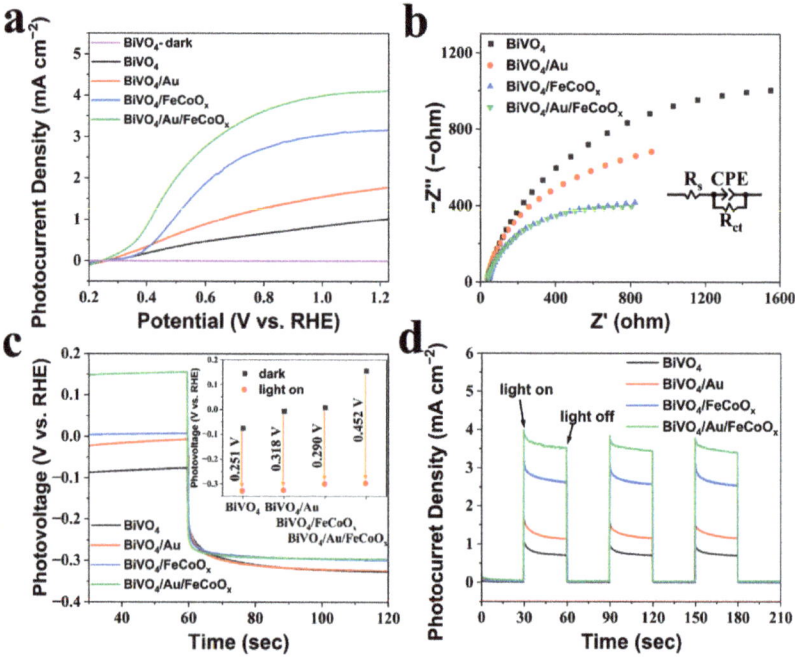

Figure 5. PEC performance of photoanodes in 0.5 M potassium phosphate buffer (pH = 7.0) AM 1.5 G (100 mW cm^{-2}). (**a**) LSV curves. (**b**) EIS curves. (**c**) Photovoltage curves. (**d**) j-t curve of the photoanodes at 1.23 V vs. RHE under intermittent illumination.

The increase in photocurrent density can also be attributed to a decrease in surface charge recombination. Figure 5b represents the electrochemical impedance spectroscopy (EIS) of the prepared BiVO$_4$, BiVO$_4$/Au, BiVO$_4$/FeCoO$_x$, and BiVO$_4$/Au/FeCoO$_x$ photoanodes. The EIS spectra can be explained with the Randles equivalent circuit model, which contains two basic elements, R$_s$ and R$_{ct}$ [31]. The element R$_s$ is the resistance relating to carrier transfer, containing the resistance of FTO substrates, the electrolyte, and wire connections in the circuit. The diameter of the circle corresponding to the semicircle or arc in the middle-frequency region of EIS reflects the interface resistance at the photoelectrode–electrolyte interface, also known as the elements of R$_{ct}$ or charge transfer resistance. Figure 5b mainly captures the arced part of the middle-frequency region. It can be easily seen from Figure 5b that the modification of AuNPs and FeCoO$_x$, respectively, reduced the diameters of the BiVO$_4$ photoanodes to some extent, while the diameter of the Nyquist plot of BiVO$_4$/Au/FeCoO$_x$ is the lowest among the photoanodes. These results indicate that the co-modification of photoanode with AuNPs and FeCoO$_x$ plays a positive role in reducing interface resistance and the recombination of photogenerated electrons and holes on the photoanode-electrolyte interface synergically, thereby promoting the participation of photogenerated holes in the water oxidation reaction and increasing the photocurrent density. Figure 5c displays the photovoltage changes of the photoanodes with or without illumination. Theoretically, the photovoltage is equal to the difference between the open circuit voltage under light and dark conditions. Compared to the pristine BiVO$_4$ with a photovoltage of 0.251 V, the photovoltage of BiVO$_4$/Au increases to 0.318 V. The photovoltage of BiVO$_4$/FeCoO$_x$ increases to 0.290 V, and the photovoltage of BiVO$_4$/Au/FeCoO$_x$ increases to 0.452 V, which is 80.1% higher than the pristine BiVO$_4$. The photovoltage results are in agreement with the onset potential of photoanodes shown in LSV curves in Figure 5a. It can be concluded that the loading of AuNPs and FeCoO$_x$, to some extent, increased the photovoltage of the photoanodes. This results in the cathodic onset potential shifting, indicating that the synergism of AuNPs and FeCoO$_x$ not only boosts OER kinetics on the photoanodes but also contributes to thermodynamics, providing a greater driving force on PEC water splitting [32].

Figure 5d displays the j–t curve of the photoanodes at 1.23 V vs. RHE and intermittent AM 1.5 G illumination (100 mW cm^{-2}), with periods of 30~60 s, 90~120 s, and 150~180 s. It can be seen that when the condition first enters the illumination from the dark, each photoanode exhibits a peak of decreased photocurrent density, and the sharpness of this peak reflects the degree of surface charge recombination. Therefore, it can be said that the surface charge recombination on each photoanode is still quite severe.

Figure S2 displays the applied bias photon-to-current efficiency (ABPE) diagram of the photoanodes by substituting the LSV curve of the photoanode material into Equation (3) with a Faraday efficiency of 100% and a P$_{in}$ of 100 mW cm^{-2}. From Figure S2, it can be seen that BiVO$_4$ reached a peak ABPE of 0.31% at a potential of 0.69 V vs. RHE, while BiVO$_4$/Au reached a peak of 0.57% at about 0.68 V vs. RHE. BiVO$_4$/FeCoO$_x$ reached a peak of 1.32% at 0.70 V vs. RHE, and the target photoanode BiVO$_4$/Au/FeCoO$_x$ reached a peak ABPE of 1.75% at about 0.64 V vs. RHE, which is more than five times the peak value of pristine BiVO$_4$. These data are sufficient to demonstrate the significant impact of AuNP modification on improving photoelectric conversion efficiency and reducing surface charge recombination, among which, the most attractive aspect for researchers is the significant increase in photocurrent density.

Figure S3 displays the durability test of the photoanodes at 1.23 V vs. RHE under AM 1.5 G. It can be seen that within about 30 min after the start of durability testing, the photocurrent density of each photoelectrochemical system decreased rapidly. In this period, the photocurrent density of BiVO$_4$/Au/FeCoO$_x$ decreased by 56.3% from the initial value of 3.98 mA cm^{-2} to 1.75 mA cm^{-2}. The photocurrent density of BiVO$_4$/FeCoO$_x$ decreased from the initial 3.07 mA cm^{-2} to 1.25 mA cm^{-2}, a relative decrease of 59.3%. The photocurrent density of BiVO$_4$ decreased from the initial 1.08 mA cm^{-2} to 0.64 mA cm^{-2}, a relative decrease of 40.7%. These results indicate that there is still room for improvement in the quality of FeCoO$_x$ and AuNPs, while they may quickly dissolve in the electrolyte over time, which is consistent with the results of the photocurrent density compared with works in other publications. Namely, the stability of BiVO$_4$/Au/FeCoO$_x$ has decreased to a certain extent, which is one of the effects of the modification of AuNPs and FeCoO$_x$. However, compared to the weakened stability of BiVO$_4$/Au/FeCoO$_x$, the beneficial effect of increasing photocurrent density is more significant.

The schematic diagram of a photoelectrochemical system using BiVO$_4$/Au/FeCoO$_x$ as a composite photoanode is shown in Figure 6. The following is the proposed working mechanism of the BiVO$_4$/Au/FeCoO$_x$ composite photoanode under illumination in phosphoric acid buffer, which consists of five processes: (I) electrons on AuNPs are excited by LSPR under incident light irradiation, (II) photogenerated hot electrons of AuNPs are transferred to the conduction band of BiVO$_4$ and collected by external circuits, (III) photogenerated holes on the valence band of BiVO$_4$ can be extracted and stored in the FeCoO$_x$ layer, and photogenerated electrons are extracted and transferred to the AuNPs, (IV) water is oxidized by holes on FeCoO$_x$, leading to the formation of oxygen and the elimination of holes, and (V) electron-deficient AuNPs are reduced by photogenerated electrons from BiVO$_4$, returning to their original metal state.

Due to the fact that the work function of BiVO$_4$ is different from Au, this difference in work function causes the energy bands of BiVO$_4$ to bend, which naturally forms a Schottky junction or Schottky barrier. Under illumination, the photogenerated hot electrons of Au can be amplified by LSPR, therefore becoming more likely to flow across the Schottky barrier into the conduction band of BiVO$_4$ and be collected by external circuits [24]. Thus, the electrons transmitted from AuNPs to BiVO$_4$ in process II are reasonable. In addition, the introduction of AuNPs and FeCoO$_x$ plays a synergic role in the photoelectrocatalysis kinetics and thermodynamics. The promoted charge transfer process will reduce the recombination of photogenerated electron–hole pairs on the electrode–electrolyte interface [28,33].

Therefore, compared to the pristine BiVO$_4$, the lifetime of photogenerated electrons and photogenerated holes is expected to improve. More electrons and holes participate in the PEC water splitting, indicating that the charge utilization rate will also be improved. Meanwhile, shown in Figure 6 as a hole storage layer, FeCoO$_x$ also forms a functionalized p-n junction with BiVO$_4$ and boosts the photogenerated holes in the valence band of BiVO$_4$ to be captured and stored in the FeCoO$_x$ surface. The stored holes can further participate in the effective water oxidation reaction to form reactive hydroxyl radicals. Thus, FeCoO$_x$ cocatalysts promote the catalytic kinetics in the electrode–electrolyte interface. Based on the above analysis, it is believed that AuNPs and FeCoO$_x$ can exert a synergistic effect on the PEC performance improvement of the BiVO$_4$.

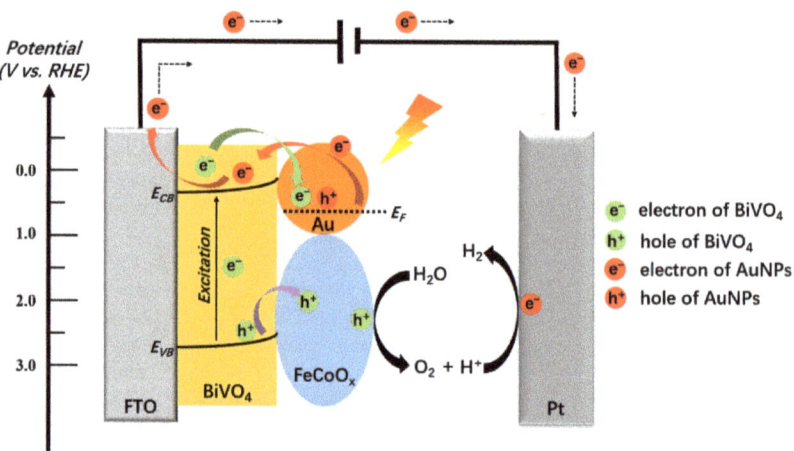

Figure 6. Proposed mechanism of the BiVO$_4$/Au/FeCoO$_x$ photoanode for PEC water splitting.

3. Materials and Methods

3.1. Preparation of BiVO$_4$ Films

Ethanol solution with 0.23 M 1,4-benzoquinone was mixed with water solution with 0.4 M Bi (NO$_3$)$_3$ and 0.4 M KI in a ratio of 7:17 to prepare the electrolyte, and the pH was adjusted to 1.70 with nitric acid. Ag/AgCl was used as the reference electrode, platinum electrode as the counter electrode, and FTO glass as the working electrode. The bismuth precursor film BiOI was deposited on FTO glass for 300 s at −0.1 V. Dimethyl sulfoxide (DMSO) dissolved in 0.2 M acetylacetone vanadium oxide was dropped onto BiOI and kept in air at 450 °C for 2 h at a heating rate of approximately 2 °C/min. Finally, BiOI was soaked in 1 M NaOH for 30 min to remove V$_2$O$_5$ and was rinsed with pure water.

3.2. Preparation of AuNPs

Briefly, 0.005 M chloroauric acid solution was used as the electrolyte, Ag/AgCl was the reference electrode, and a platinum electrode was the counter electrode. AuNPs were deposited at 0.15 V for 40 s. Then, they were rinsed with pure water.

3.3. Preparation of FeCoO$_x$ Cocatalysts

The FeCoO$_x$ cocatalyst was deposited through two steps under AM 1.5 G. Briefly, 0.1 M FeSO$_4$ solution was used as the electrolyte, while Ag/AgCl was used as the reference electrode, and platinum electrode as the counter electrode. The FeOOH layer was deposited at 0.25 V for 300 s. Then, 0.025 M (CH$_3$COO)$_2$Co solution was used as the electrolyte, platinum electrode as the counter electrode, and Ag/AgCl as the reference electrode, and CoO$_x$ was deposited at 0.25 V for 40 s. Finally, the FeCoO$_x$ cocatalyst was obtained after keeping it in air at 500 °C for 2 h

3.4. Characterization

The morphologies and structures of the photoanodes were characterized using SEM (JSM-IT500, JEOL) coupled with EDS, and XRD (D8 Advance, Bruker) with Cu Kα radiation, respectively. The composition and chemical states of photoanodes were characterized by XPS (ESCALAB xi+). UV-vis absorption spectra were recorded by an Evolution 220 spectrophotometer.

3.5. PEC measurements

The PEC measurements were completed through electrochemical workstations and xenon lamps. The test circuit was constructed under AM 1.5 G illumination using a classic three-electrode system (platinum as the counter electrode, Ag/AgCl as the reference electrode) and 0.5 M phosphoric acid buffer (pH = 7.00). Photocurrent density potential curves were obtained using linear sweep voltammetry ranging from approximately 0.1 to 1.4 V vs. RHE with a scanning rate of 0.01 V s^{-1}. Electrochemical impedance spectra were obtained at open circuit potential with a frequency ranging from roughly 100 kHz to 0.01 Hz and a signal amplitude of 0.01 V. Photovoltage was measured at open circuit potential under AM 1.5 G for 60 s and then continued for 60 s in the dark condition.

All the potentials vs. RHE can be converted from the potentials vs. Ag/AgCl following

$$E_{RHE} = E_{Ag/AgCl} + E^0_{Ag/AgCl} + 0.059 * pH \qquad (2)$$

where E_{RHE} is converted potential vs. RHE, $E^0_{Ag/AgCl}$ is 0.197 V at room temperature (25 °C), and $E_{Ag/AgCl}$ is the potential vs. Ag/AgCl.

ABPE can be calculated according to

$$ABPE = \frac{j * (1.23 - V_b)}{P_{in}} * 100\% \qquad (3)$$

where j is the photocurrent density (mA cm^{-2}), V_b is the corresponding potential vs. RHE, and P_{in} is the illumination intensity (100 mW cm^{-2}).

4. Conclusions

In summary, to progressively enhance the PEC properties of BiVO$_4$ photoanodes, a novel BiVO$_4$/Au/FeCoO$_x$ photoanode is designed and the synergetic role of AuNPs and FeCoO$_x$ is verified. The characterization, optical properties, and PEC performance of optimized photoanodes are thoroughly investigated. The results reveal that the co-modification of AuNPs and FeCoO$_x$ plays a synergic role in enhancing the absorption of ultraviolet and visible light of BiVO$_4$. This is mainly attributed to the LSPR effect induced by AuNPs and the extended light absorption edge position induced by the BiVO$_4$/FeCoO$_x$ heterojunction. Moreover, the appropriately sized Au nanoparticles and FeCoO$_x$ cocatalysts played a positive role in increasing the photovoltage of BiVO$_4$ and in reducing the charge transfer resistance on the electrode–electrolyte interface, consequently leading to the promotion of the separation and migration of photogenerated carriers and the prominent enhancement of the photocurrent density. This work provides practical experimental support and a theoretical explanation for the design of effective PEC photoanodes.

Supplementary Materials: The following supporting information can be downloaded at: https://www.mdpi.com/article/10.3390/catal13071063/s1, Figure S1: SEM images of (a) BiVO$_4$, (b) BiVO$_4$/Au, and (c) BiVO$_4$/FeCoO$_x$.; Figure S2: ABPE curves of photoanodes in 0.5 M potassium phosphate buffer (pH=7.0) under AM 1.5G (100 mW cm−2); Figure S3: Durability tests of photoanodes in 0.5 M potassium phosphate buffer (pH=7.0) under AM 1.5G (100 mW cm−2) at 1.23 V vs. RHE. Table S1: Content Information of Each Element in Mapping of BiVO$_4$/Au/FeCoO$_x$. Table S2: Peak position and atomic content in XPS results.

Author Contributions: Methodology, H.C. and Q.Z.; Project administration, J.S.; Resources, Q.Z. and S.L.; Writing original draft, H.C. and Q.Z.; Writing review and editing, H.C., Q.Z., A.A., S.H. and S.L. Data curation, X.L. and W.Z.; Formal analysis, S.L.; Investigation, J.S. All authors have read and agreed to the published version of the manuscript.

Funding: This research was funded by the Shenzhen Science and Techonlogy Program (Grant No. 202206193000001), Guangdong Basic and Applied Basic Research Foundation (Grant No. 2023A151501207, 22022A1515110619), the Open Project Fund from Guangdong Provincial Key Laboratory of Materials and Technology for Energy Conversion, Guangdong Technion-Israel Institute of Technology, Grant No.MATEC2023KF003.

Data Availability Statement: All data generated or analyzed during this study are included in this published article.

Conflicts of Interest: The authors declare no conflict of interest.

References

1. Zou, S.; Burke, M.S.; Kast, M.G.; Fan, J.; Danilovic, N.; Boettcher, S.W. Fe (Oxy)hydroxide Oxygen Evolution Reaction Electrocatalysis: Intrinsic Activity and the Roles of Electrical Conductivity, Substrate, and Dissolution. *Chem. Mater.* **2015**, *27*, 8011–8020. [CrossRef]
2. Corby, S.; Rao, R.R.; Steier, L.; Durrant, J.R. The kinetics of metal oxide photoanodes from charge generation to catalysis. *Nat. Rev. Mater.* **2021**, *6*, 1136–1155. [CrossRef]
3. Ran, J.; Zhang, J.; Yu, J.; Jaroniec, M.; Qiao, S. Earth-Abundant Cocatalysts for Semiconductor-Based Photocatalytic Water Splitting. *Chem. Soc. Rev.* **2014**, *43*, 7787–7812. [CrossRef] [PubMed]
4. Li, G.; Lian, Z.; Wang, W.; Zhang, D.; Li, H. Nanotube-confinement induced size-controllable g-C_3N_4 quantum dots modified single-crystalline TiO_2 nanotube arrays for stable synergetic photoelectrocatalysis. *Nano Energy* **2016**, *19*, 446–454. [CrossRef]
5. Cesar, I.; Kay, A.; Gonzalez Martinez, J.A.; Grätzel, M. Translucent Thin Film Fe_2O_3 Photoanodes for Efficient Water Splitting by Sunlight: Nanostructure-Directing Effect of Si-Doping. *J. Am. Chem. Soc.* **2006**, *128*, 4582–4583. [CrossRef]
6. Cui, J.; Zhu, S.; Zou, Y.; Zhang, Y.; Yuan, S.; Li, T.; Guo, S.; Liu, H.; Wang, J. Improved Photoelectrochemical Performance of $BiVO_4$ for Water Oxidation Enabled by the Integration of the Ni@NiO Core–Shell Structure. *Catalysts* **2022**, *12*, 1456. [CrossRef]
7. Li, T.; He, J.; Peña, B.; Berlinguette, C.P. Curing $BiVO_4$ Photoanodes with Ultraviolet Light Enhances Photoelectrocatalysis. *Angew. Chem. Int. Ed.* **2016**, *55*, 1769–1772. [CrossRef]
8. Zhou, B.; Zhao, X.; Liu, H.; Qu, J.; Huang, C.P. Visible-light sensitive cobalt-doped $BiVO_4$ (Co-$BiVO_4$) photocatalytic composites for the degradation of methylene blue dye in dilute aqueous solutions. *Appl. Catal. B* **2010**, *99*, 214–221. [CrossRef]
9. Park, Y.; McDonald, K.J.; Choi, K.S. Progress in bismuth vanadate photoanodes for use in solar water oxidation. *Chem. Soc. Rev.* **2013**, *42*, 2321–2337. [CrossRef]
10. Jason, K.C.; Sheraz, G.; Francesca, M.T.; Le, C.; Glans, P.A.; Guo, J.; Joel, W.A.; Yano, J.; Ian, D. Sharp. Electronic Structure of Monoclinic $BiVO_4$. *Chem. Mater.* **2014**, *26*, 5365–5373. [CrossRef]
11. Xu, X.; Xu, Y.; Xu, F.; Jiang, G.; Jian, J.; Yu, H.; Zhang, E.; Shchukin, D.; Kaskel, S.; Wang, H. Black $BiVO_4$: Size tailored synthesis, rich oxygen vacancies, and sodium storage performance. *J. Mater. Chem. A* **2020**, *8*, 1636–1645. [CrossRef]
12. Wang, S.; Wang, X.; Liu, B.; Guo, Z.; Ostrikov, K.; Wang, L.; Huang, W. Vacancy defect engineering of $BiVO_4$ photoanodes for photoelectrochemical water splitting. *Nanoscale* **2021**, *13*, 17989–18009. [CrossRef] [PubMed]
13. Kalanoor, B.S.; Seo, H.; Kalanur, S.S. Recent developments in photoelectrochemical water-splitting using $WO_3/BiVO_4$ heterojunction photoanode: A review. *Mater. Sci. Energy Technol.* **2018**, *1*, 49–62. [CrossRef]
14. Wang, S.; Yun, J.; Hu, Y.; Xiao, M.; Wang, L.; He, T.; Du, A. New Iron-Cobalt Oxide Catalysts Promoting $BiVO_4$ Films for Photoelectrochemical Water Splitting. *Adv. Funct. Mater.* **2018**, *28*, 1802685. [CrossRef]
15. Wang, W.; Huang, X.; Wu, S.; Zhou, Y.; Wang, L.; Shi, H.; Liang, Y.; Zou, B. Preparation of p–n junction $Cu_2O/BiVO_4$ heterogeneous nanostructures with enhanced visible-light photocatalytic activity. *Appl. Catal. B Environ.* **2013**, *134–135*, 293–301. [CrossRef]
16. Dionigi, F.; Strasser, P. NiFe-Based (Oxy)hydroxide Catalysts for Oxygen Evolution Reaction in Non-Acidic Electrolytes. *Adv. Energy Mater.* **2016**, *6*, 1600621. [CrossRef]
17. Zhang, B.; Wang, L.; Zhang, Y.; Bi, Y.; Ding, Y. Ultrathin FeOOH Nanolayers with Abundant Oxygen Vacancies on $BiVO_4$ Photoanodes for Efficient Water Oxidation. *Angew. Chem. Int. Ed.* **2018**, *57*, 2248–2252. [CrossRef]
18. Kang, Z.; Sun, Z.; Zang, Y.; Wan, S.; Zheng, Y.; Tao, X. Dual functions of heterometallic FeCo oxyhydroxides in borate-treated $BiVO_4$ photoanodes toward boosted activity and photostability in photoelectrochemical water oxidation. *Chem. Eng. J.* **2022**, *431*, 133379. [CrossRef]
19. Rosa, W.S.; Rabelo, L.G.; Tiveron Zampaulo, L.G.; Gonçalves, R.V. Ternary Oxide $CuWO_4/BiVO_4/FeCoO_x$ Films for Photoelectrochemical Water Oxidation: Insights into the Electronic Structure and Interfacial Band Alignment. *ACS Appl. Mater. Interfaces* **2022**, *14*, 22858–22869. [CrossRef]
20. Trotochaud, L.; Young, S.L.; Ranney, J.K.; Boettcher, S.W. Nickel-Iron oxyhydroxide oxygen-evolution electrocatalysts: The role of intentional and incidental iron incorporation. *J. Am. Chem. Soc.* **2014**, *136*, 6744–6753. [CrossRef]

21. Wang, Q.; Lei, Y.; Wang, Y.; Xiong, X.; Xue, X.; Feng, Y.; Wang, D.; Li, Y. Engineering of Electronic States on Co_3O_4 Ultrathin Nanosheets by Cation Substitution and Anion Vacancies for Oxygen Evolution Reaction. *Small* **2020**, *16*, 2001571. [CrossRef] [PubMed]
22. Kang, Q.; Xi, C.; Zhang, Y.; Zhang, R.; Li, Z.; Sheng, G.R.; Liu, H.; Dong, C.K.; Chen, Y.J.; Du, X.W. Laser-induced oxygen vacancies in $FeCo_2O_4$ nanoparticles for boosting oxygen evolution and reduction. *Chem. Commun.* **2019**, *55*, 8579–8582. [CrossRef] [PubMed]
23. Chang, X.; Wang, T.; Zhang, P.; Zhang, J.; Li, A.; Gong, J. Enhanced Surface Reaction Kinetics and Charge Separation of p-n Heterojunction $Co_3O_4/BiVO_4$ Photoanodes. *Am. Chem. Soc.* **2015**, *137*, 8356–8359. [CrossRef]
24. Park, J.Y.; Kim, S.M.; Lee, H.; Nedrygailov, I.I. Hot-Electron-Mediated Surface Chemistry: Toward Electronic Control of Catalytic Activity. *Acc. Chem. Res.* **2015**, *48*, 2475–2483. [CrossRef] [PubMed]
25. Daniel, M.C.; Astruc, D. Gold Nanoparticles: Assembly, Supramolecular Chemistry, Quantum-Size-Related Properties, and Applications toward Biology, Catalysis, and Nanotechnology. *Chem. Rev.* **2004**, *104*, 293–346. [CrossRef] [PubMed]
26. Agrawal, A.; Cho, S.H.; Zandi, O.; Ghosh, S.; Johns, R.W.; Milliron, D.J. Localized Surface Plasmon Resonance in Semiconductor Nanocrystals. *Chem. Rev.* **2018**, *118*, 3121–3207. [CrossRef] [PubMed]
27. Sau, T.K.; Murphy, C.J. Room Temperature, High-Yield Synthesis of Multiple Shapes of Gold Nanoparticles in Aqueous Solution. *J. Am. Chem. Soc.* **2004**, *126*, 8648–8649. [CrossRef]
28. Lee, M.G.; Moon, C.W.; Park, H.; Sohn, W.; Kang, S.B.; Lee, S.; Choi, K.J.; Jang, H.W. Dominance of Plasmonic Resonant Energy Transfer over Direct Electron Transfer in Substantially Enhanced Water Oxidation Activity of $BiVO_4$ by Shape-Controlled Au Nanoparticles. *Small* **2017**, *13*, 1701644. [CrossRef]
29. Tang, G.; Li, H.; Cheng, C. Au nanoparticles embedded in $BiVO_4$ films photoanode with enhanced photoelectrochemical performance. *Nanotechnology* **2019**, *30*, 445401. [CrossRef]
30. Hazarika, K.K.; Chutia, B.; Changmai, R.R.; Bharali, P.; Boruah, P.K.; Das, M.R. $Fe_xCo_{3-x}O_4$ Nanohybrids Anchored on a Carbon Matrix for High-Performance Oxygen Electrocatalysis in Alkaline Media. *ChemElectroChem* **2022**, *9*, e202200867. [CrossRef]
31. Geng, H.; Ying, P.; Zhao, Y.; Gu, X. Cactus shaped $FeOOH/Au/BiVO_4$ photoanodes for efficient photoelectrochemical water splitting. *Int. J. Hydrog. Energy* **2021**, *46*, 35280–35289. [CrossRef]
32. Tong, H.; Jiang, Y.; Zhang, Q.; Jiang, W.; Wang, K.; Luo, X.; Lin, Z.; Xia, L. Boosting Photoelectrochemical Water Oxidation with Cobalt Phosphide Nanosheets on Porous $BiVO_4$. *ACS Sustain. Chem. Eng.* **2019**, *7*, 769–778. [CrossRef]
33. Huang, C.C.; Song, Z.Y.; Li, H.Q.; Yu, X.Y.; Cui, Y.M.; Yang, M.; Huang, X.J. Enhanced As(III) detection under near-neutral conditions: Synergistic effect of boosted adsorption by oxygen vacancies and valence cycle over activated Au NPs loaded on $FeCoO_x$ nanosheets. *Sens. Actuators B Chem.* **2023**, *382*, 133489. [CrossRef]

Disclaimer/Publisher's Note: The statements, opinions and data contained in all publications are solely those of the individual author(s) and contributor(s) and not of MDPI and/or the editor(s). MDPI and/or the editor(s) disclaim responsibility for any injury to people or property resulting from any ideas, methods, instructions or products referred to in the content.

Article

Exceptional Photocatalytic Performance of the LaFeO$_3$/g-C$_3$N$_4$ Z-Scheme Heterojunction for Water Splitting and Organic Dyes Degradation

Muhammad Humayun [1,*], Ayesha Bahadur [2], Abbas Khan [1,3] and Mohamed Bououdina [1,*]

1. Department of Mathematics and Sciences, College of Humanities and Sciences, Energy, Water, and Environment Lab, Prince Sultan University, Riyadh 11586, Saudi Arabia; abbas053@gmail.com
2. Department of Chemistry, Bacha Khan University Charsadda, Charsadda 24420, Pakistan; aliali95141@gmail.com
3. Department of Chemistry, Abdul Wali Khan University Mardan, Mardan 23200, Pakistan
* Correspondence: mhumayun@psu.edu.sa (M.H.); mbououdina@psu.edu.sa (M.B.)

Abstract: To simulate natural photosynthesis, scientists have developed an artificial Z-scheme system that splits water into hydrogen and oxygen using two different semiconductors. Researchers are striving to improve the performance of Z-scheme systems by improving light absorption, developing redox couples with high stability, and finding new cocatalysts. Here, we report the synthesis and utilization of LaFeO$_3$/g-C$_3$N$_4$ as a Z-scheme system for water reduction to produce hydrogen and organic dye degradation under visible light irradiation. The as-fabricated photocatalyst revealed exceptional activity for H$_2$ production (i.e., 351 µmol h^{-1}g^{-1}), which is 14.6 times higher compared to that of the single-component g-C$_3$N$_4$ (i.e., 24 µmol h^{-1}g^{-1}). In addition, the composite photocatalyst degraded 87% of Methylene Blue (MB) and 94% of Rhodamine B (RhB) in 2 h. Various experimental analyses confirmed that the exceptional performance of the LaFeO$_3$/g-C$_3$N$_4$ Z-scheme catalyst is due to remarkably enhanced charge carrier separation and improved light absorption. The development of this highly effective Z-scheme heterostructure photocatalyst will pave the way for the sustainable development of newly designed Z-scheme scheme systems that will tackle energy and environmental crises.

Keywords: photocatalysis; hydrogen evolution; visible light; organic dyes; Z-scheme

Citation: Humayun, M.; Bahadur, A.; Khan, A.; Bououdina, M. Exceptional Photocatalytic Performance of the LaFeO$_3$/g-C$_3$N$_4$ Z-Scheme Heterojunction for Water Splitting and Organic Dyes Degradation. *Catalysts* 2023, 13, 907. https://doi.org/10.3390/catal13050907

Academic Editors: Yongming Fu and Qian Zhang

Received: 23 March 2023
Revised: 14 May 2023
Accepted: 18 May 2023
Published: 20 May 2023

Copyright: © 2023 by the authors. Licensee MDPI, Basel, Switzerland. This article is an open access article distributed under the terms and conditions of the Creative Commons Attribution (CC BY) license (https://creativecommons.org/licenses/by/4.0/).

1. Introduction

Energy and environmental issues are intricately linked and have emerged as two of the world's most important concerns [1]. The ongoing use of fossil fuels for energy generation has caused substantial environmental damage, such as climate change, air pollution, and water contamination [2,3]. In consequence, there has been an increasing trend toward adopting cleaner renewable energy sources such as wind, solar, and hydropower to alleviate the impact of these challenges [4]. Energy efficiency and conservation techniques are also being adopted in order to reduce energy usage and greenhouse gas emissions [5]. The transition to a more sustainable energy system is critical to ensure a healthier and more sustainable environment for future generations [6–8].

Recently, heterostructure semiconductor photocatalytic technology has gained increasing interest due to its potential to directly utilize solar energy and transform it into chemical energy, which can then be employed in a variety of applications such as the generation of solar fuels such as hydrogen and hydrocarbon fuels, air cleaning, as well as the degradation of various contaminants [9–14]. The method employs semiconductors, which absorb light energy and generate electron–hole pairs, which subsequently react with other molecules to form new compounds [15–17]. Photocatalytic processes take place on the semiconductors' surfaces, which serve as catalysts. One of the major advantages of semiconductor photocatalysis is its ability to perform under ambient conditions without the need for additional energy sources. As a result, it is a potential technology for long-term and cost-effective environmental remediation and energy generation. However, increasing the process's efficiency and selectivity remains a difficulty, and further research is needed to increase the practicality and scalability of semiconductor photocatalysis [18–21].

Recently, the 2D layered material g-C_3N_4 received promising attention in photocatalysis due to its unique features compared to the other semiconductors. g-C_3N_4 is composed of C and N atoms that are stacked together. Since g-C_3N_4 possesses a narrow bandgap (i.e., 2.7 eV), making it an effective catalyst for visible-light-driven reactions. It has exceptional chemical stability, biocompatibility, and an environmentally friendly nature. Various photocatalytic reactions including CO_2 reduction, organic pollutant degradation, and hydrogen production have all been performed by g-C_3N_4 photocatalysts. Nevertheless, enhancing the photocatalytic activities and understanding the fundamental mechanism of g-C_3N_4 photocatalysis still remain challenging [22–24].

Several strategies including structural modification, heterojunction formation, and co-catalyst loading have been developed with the aim to optimize the catalytic activities of g-C_3N_4. Among these, the heterojunction formation involves the coupling of other semiconductors with g-C_3N_4, to enhance charge separation and transfer. This strategy has shown significant improvements in the catalytic performance of g-C_3N_4 for a wide range of purposes [25]. Among various sorts of heterojunctions, g-C_3N_4-based Z-scheme systems have a great potential for energy and environmental applications due to their high efficiency and flexibility.

The $LaFeO_3$ photocatalyst, on the other hand, is a promising semiconductor photocatalytic utilized for various applications owing to its exceptional electronic and structural features. It is a perovskite-type oxide composed of lanthanum, iron, and oxygen atoms organized in a crystalline lattice. $LaFeO_3$ exhibits a narrow bandgap (i.e., 1.8–2.1 eV), which renders it a highly effective photocatalyst for catalytic reactions triggered by visible light. It also possesses excellent chemical and thermal stability, as well as outstanding biocompatibility. However, due to its inadequate conduction band potential for the water reduction reaction, the practical application of the $LaFeO_3$ photocatalyst remains challenging. However, it has excellent performance for pollutant degradation [26].

Thus, the coupling of $LaFeO_3$ with g-C_3N_4 is highly crucial to improve its catalytic performance. When the two materials are coupled, the absorbed light can be efficiently utilized over a wider spectral range, resulting in enhanced photocatalytic performance. Additionally, the $LaFeO_3$/g-C_3N_4 heterojunction can facilitate charge carrier separation and transfer, thereby enhancing photocatalytic efficiency. It has been shown that the coupling of $LaFeO_3$ with g-C_3N_4 remarkably improves its catalytic activity for various applications [27–29].

Herein, we report the synthesis of $LaFeO_3$/g-C_3N_4 as a Z-scheme heterostructure photocatalyst for H_2 production and degradation of organic dyes. The heterostructure photocatalyst revealed enhanced performance by producing 351 μmol $h^{-1}g^{-1}$ of H_2 through water splitting and degraded 87 and 94% of MB and RhB dyes, respectively. The superior activities of the $LaFeO_3$/g-C_3N_4 catalyst are accredited to the promoted charge transfer and separation and the extended light absorption. This work provides a promising route for the design and implementation of efficient and sustainable catalysts for environmental remediation and energy production.

2. Results and Discussion

2.1. Structural Characterization and Chemical Composition

The XRD patterns of the synthesized photocatalysts are displayed in Figure 1a. The (100) and (002) crystal planes in g-C_3N_4 are represented by diffraction peaks at 12.9° and 27.7°, respectively (JCPDS No. 87-1526). In-plane trigonal nitrogen-linked tri-s-triazine connected-layer units correlate to the weaker peak at the 2-theta value of 12.9°. The stronger peak at a 2-theta value of 27.4° is associated with the stacking of heptazine rings, which has a 0.32 nm interlayer-distance (inter-planar stacking of conjugated aromatic segments) [30,31]. Diffraction peaks at 22.6°, 32.2°, 39.6°, 46.1°, 57.4°, and 67.3° in LaFeO$_3$ photocatalyst are indexed to the orthorhombic phase and reflect the (002), (112), (022), (004), (204), and (040) crystal planes, respectively (JCPDS No. 37-1493) [32,33]. The distinctive peak of LaFeO$_3$ in the LaFeO$_3$/g-C$_3$N$_4$ catalyst corresponds to the (202) plane, which indicates that the LaFeO$_3$ and g-C$_3$N$_4$ surfaces coexist without any observable impurities, indicating that the nanocomposite was successfully formed.

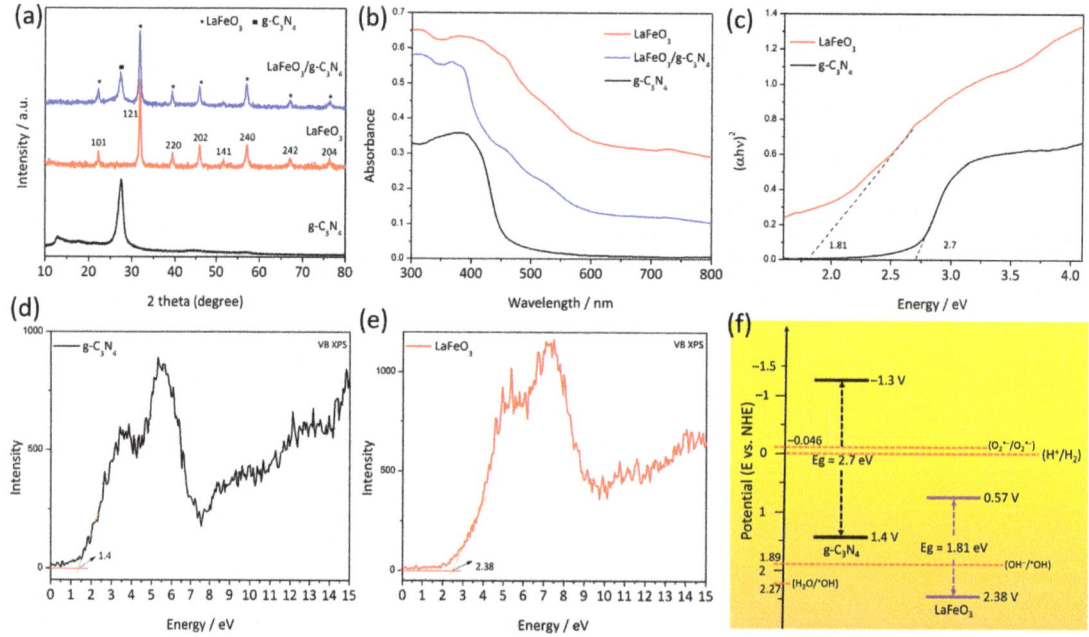

Figure 1. (a) XRD patterns and (b) UV-visible absorption spectra of g-C$_3$N$_4$, LaFeO$_3$, and LaFeO$_3$/g-C$_3$N$_4$ photocatalysts. (c) Estimated energy band gaps, (d,e) valence band XPS spectra, and (f) band edge positions versus the reduction potential of NHE of g-C$_3$N$_4$ and LaFeO$_3$ photocatalysts.

As shown in Figure 1b, the UV-visible absorption spectra of the as-synthesized LaFeO$_3$, g-C$_3$N$_4$, and LaFeO$_3$/g-C$_3$N$_4$ photocatalysts were measured using a UV-visible absorption spectrometer. A characteristic absorption edge is seen at 458 nm for pure g-C$_3$N$_4$, which is in line with the previously reported optical energy band gap value of 2.7 eV. LaFeO$_3$ has an absorption edge at 685 nm, which means it absorbs light in the visible spectrum (E_g = 1.81 eV). In perovskite-type oxide materials, the transition of electrons between the valence band (O 2p) and the conduction band (Fe 3d) is principally responsible for the strong absorption edges [34,35]. The absorption characteristic peak of LaFeO$_3$/g-C$_3$N$_4$ revealed extended visible light absorption, demonstrating that the synergistic interaction of the nanomaterials can alter the optical features of the base materials. From Tauc's plots shown in Figure 1c, we may infer the photocatalysts' band gaps. The predicted values for the energy band gaps (E_g) of the g-C$_3$N$_4$ and LaFeO$_3$ catalysts are 2.7 and 1.81 eV, respectively. After being exposed to visible light, semiconductors typically generate electron–hole pairs in the valence band and then the excited electrons are transferred to their conduction band. As shown in Figure 1d,e, we have explored the valence band XPS spectra of bare g-C$_3$N$_4$ and LaFeO$_3$ photocatalysts to gain insight into their precise valence and conduction bands. The predicted valence band potential values of pristine g-C$_3$N$_4$ and LaFeO$_3$ photocatalysts are 1.4 and 2.38 V, respectively. The conduction band potential values of the catalysts were calculated according to the equation mentioned in our previous report [36]. Thus, the predicted conduction band potential values of pristine g-C$_3$N$_4$ and LaFeO$_3$ are −1.3 and 0.57 V vs. the NHE, respectively. Figure 1f demonstrates the energy band gaps and band edge potentials of g-C$_3$N$_4$ and LaFeO$_3$ photocatalysts.

To verify the microstructure of g-C$_3$N$_4$ and LaFeO$_3$/g-C$_3$N$_4$ catalysts, TEM and high-resolution TEM (HR-TEM) micrographs were recorded (Figure 2a–d). Figure 2a reveals the TEM micrograph of g-C$_3$N$_4$, which demonstrates that the material is composed of stacked layers of ultra-thin flat surface nanosheets with thicknesses of 80–100 nm. An HRTEM image displaying the distinctive structure of g-C$_3$N$_4$ is shown in Figure 2b. The aggregation of LaFeO$_3$ particles on the surface of g-C$_3$N$_4$ nanosheets could be obviously seen in the TEM micrograph of the LaFeO$_3$/g-C$_3$N$_4$ heterostructure (Figure 2c). The distinct lattice fringes of g-C$_3$N$_4$ and LaFeO$_3$ catalysts can be clearly seen in the HRTEM micrograph of the LaFeO$_3$/g-C$_3$N$_4$ catalyst (Figure 2d) [37]. This confirms the successful fabrication of LaFeO$_3$/g-C$_3$N$_4$ heterojunction.

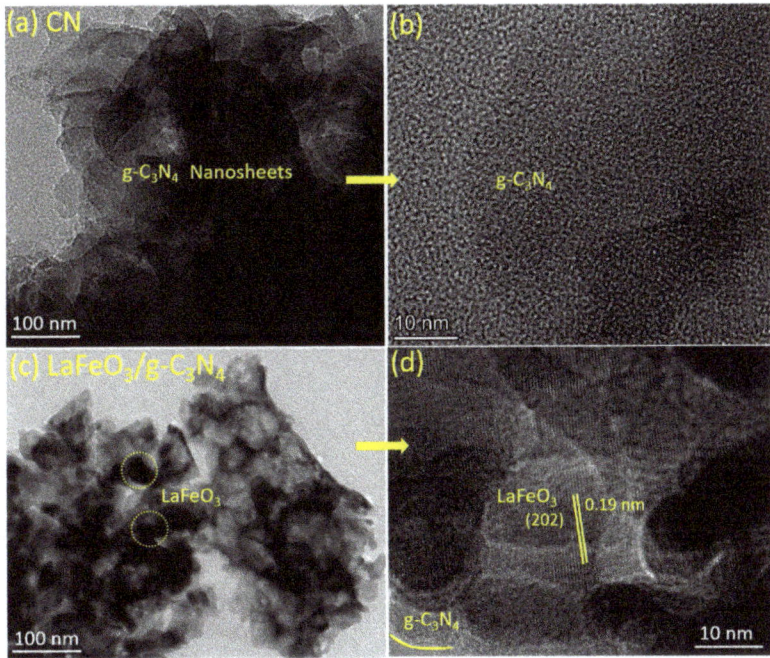

Figure 2. (**a**) TEM, and (**b**) HRTEM micrographs of g-C$_3$N$_4$ photocatalyst. (**c**) TEM and (**d**) HRTEM micrographs of LaFeO$_3$/g-C$_3$N$_4$ photocatalyst.

The elemental compositions of the g-C$_3$N$_4$ and LaFeO$_3$/g-C$_3$N$_4$ catalysts were studied by means of XPS analysis. The existence of relevant elements is revealed by the XPS survey spectra of the photocatalysts (Figure 3a). The deconvoluted C1s spectra (Figure 3b) demonstrate the existence of peaks at 284.75 and 288.15 eV, which are accredited to the Sp2 hybridized C atoms (i.e., N−C=N) and graphitic-carbon (i.e., C−N), respectively. Noticeably, after LaFeO$_3$ coupling, the binding energy peaks of g-C$_3$N$_4$ are somewhat shifted toward the higher binding energies side, possibly due to the charge transfer at the interface of the as-fabricated LaFeO$_3$/g-C$_3$N$_4$ heterojunction. The N 1s spectrum of bare g-C$_3$N$_4$ (Figure 3c) reveals two peaks at 398.6 and 401.2 eV, respectively, accredited to the Sp2-hybridized N atoms (i.e., C-N=C) by heptazine rings and the tertiary N atoms (i.e., N-(C)$_3$). The electron delocalization effect causes a comparable shift in the N 1s peak of the LaFeO$_3$/g-C$_3$N$_4$ catalyst [38].

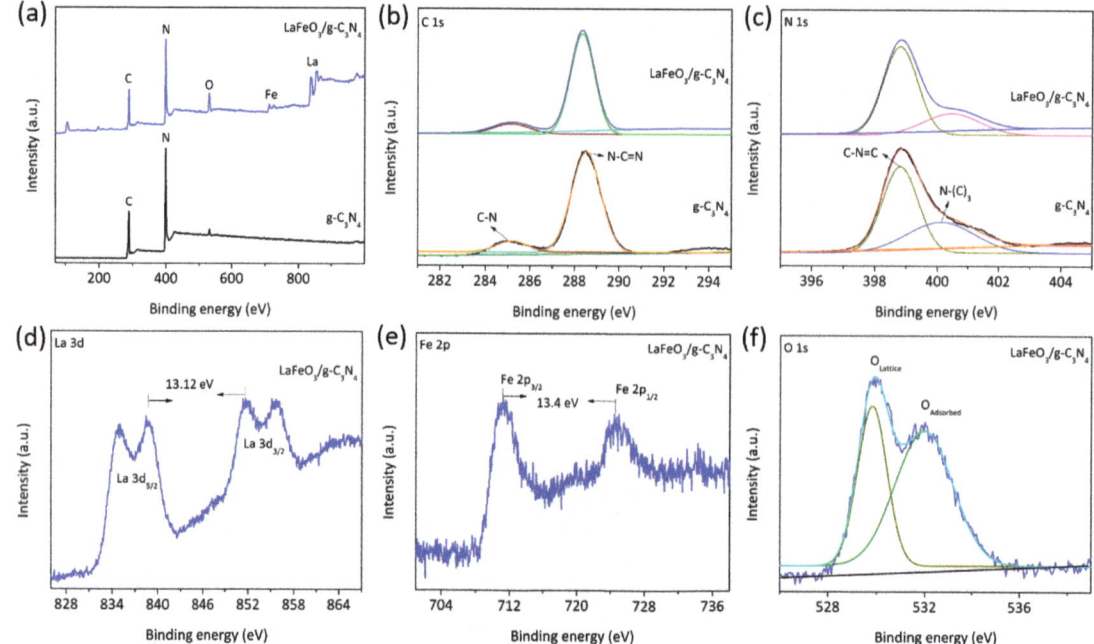

Figure 3. (**a**) XPS survey spectra, (**b**) deconvoluted high resolution C 1s spectra, and (**c**) deconvoluted high resolution N 1s spectra of g-C$_3$N$_4$ and LaFeO$_3$/g-C$_3$N$_4$ photocatalysts. (**d**) High-resolution La 3d spectrum, (**e**) high-resolution Fe 2p spectrum, and (**f**) deconvoluted high-resolution O 1s spectrum of LaFeO$_3$/g-C$_3$N$_4$ photocatalyst.

In addition, two distinct peaks can be seen in the La 3d XPS spectrum of the LaFeO$_3$/g-C$_3$N$_4$ catalyst (Figure 3d) at 834.65 and 851.55 eV. In addition, two small peaks originated at 838.6 and 855.2 eV corresponding to the satellite peaks of the La 3d$_{5/2}$ and La 3d$_{3/2}$ orbitals. These peaks are produced by the transferring of an electron from a 2p to an empty 4f orbital in the O$_2$ ligands. The spin-orbital splitting of the La 3d$_{5/2}$ and La 3d$_{3/2}$ orbitals is 16.6 eV, demonstrating the +3-oxidation state of La. The peaks originating at 710.1 (i.e., 2p$_{3/2}$) and 723.65 eV (i.e., 2p$_{1/2}$) in the Fe 2p spectrum of the LaFeO$_3$/g-C$_3$N$_4$ composite (Figure 3e) indicate the +3-oxidation state of the Fe in the composite. Lattice oxygen (i.e., La-O) and surface-adsorbed hydroxyl groups (•OH) are responsible for the two typical peaks at 530.15 and 532.55 eV in the deconvoluted O 1s XPS spectra of the LaFeO$_3$/g-C$_3$N$_4$ catalyst (Figure 3f) [27,39]. The XPS analysis shows that LaFeO$_3$ and g-C$_3$N$_4$ have strong interactions with each other.

2.2. Photogenerated Charge Separation

Photoluminescence (PL) spectroscopy was utilized to explore the photo-physics of photogenerated charges in the g-C$_3$N$_4$ and LaFeO$_3$/g-C$_3$N$_4$. The PL method can provide us with useful information about the presence of surface defects, vacancies, as well as charge recombination at the catalyst surface [40,41]. As is evident from Figure 4a, the g-C$_3$N$_4$ photocatalyst has a robust PL signal, which favors efficient recombination of charge carriers. However, the PL peak intensity is significantly suppressed for the LaFeO$_3$/g-C$_3$N$_4$ catalyst. Based on these findings, we may conclude that the LaFeO$_3$/g-C$_3$N$_4$ composite exhibits much lower charge recombination compared to that of the pristine g-C$_3$N$_4$. In addition, surface photo-voltage (SPV) spectroscopy indicates enhanced charge transfer and separation in the LaFeO$_3$/g-C$_3$N$_4$ catalyst. In semiconductors, SPV spectroscopy is typically employed to learn more about electron–hole pairs' excitation, separation, as well as transfer at the surface. SPV typically detects a signal representing the charge separation. Increased charge carrier separation results in a robust SPV signal [42]. Figure 4b shows that the SPV signal from g-C_3N_4 is quite low. The LaFeO$_3$/g-C_3N_4 composite, on the other hand, shows a significantly higher SPV signal, indicating better charge separation and transfer in the fabricated heterojunction. Electrochemical impedance spectra (EIS) measurement is employed to confirm the findings of PL and SPV. In most cases, the arc radius of the resulting EIS Nyquist plots might be used to assess the charge transfer resistance of the catalysts. The huge arc radius is indicative of weak charge separation in semiconducting nanomaterials, as has been widely stated [43]. Figure 4c shows that the LaFeO$_3$/g-C_3N_4 catalyst has highly effective charge transfer and separation in comparison to that of the bare g-C_3N_4 catalyst, as proven by the short arc radius of the EIS Nyquist plot. The equivalent circuit model is provided as the inset of Figure 4c.

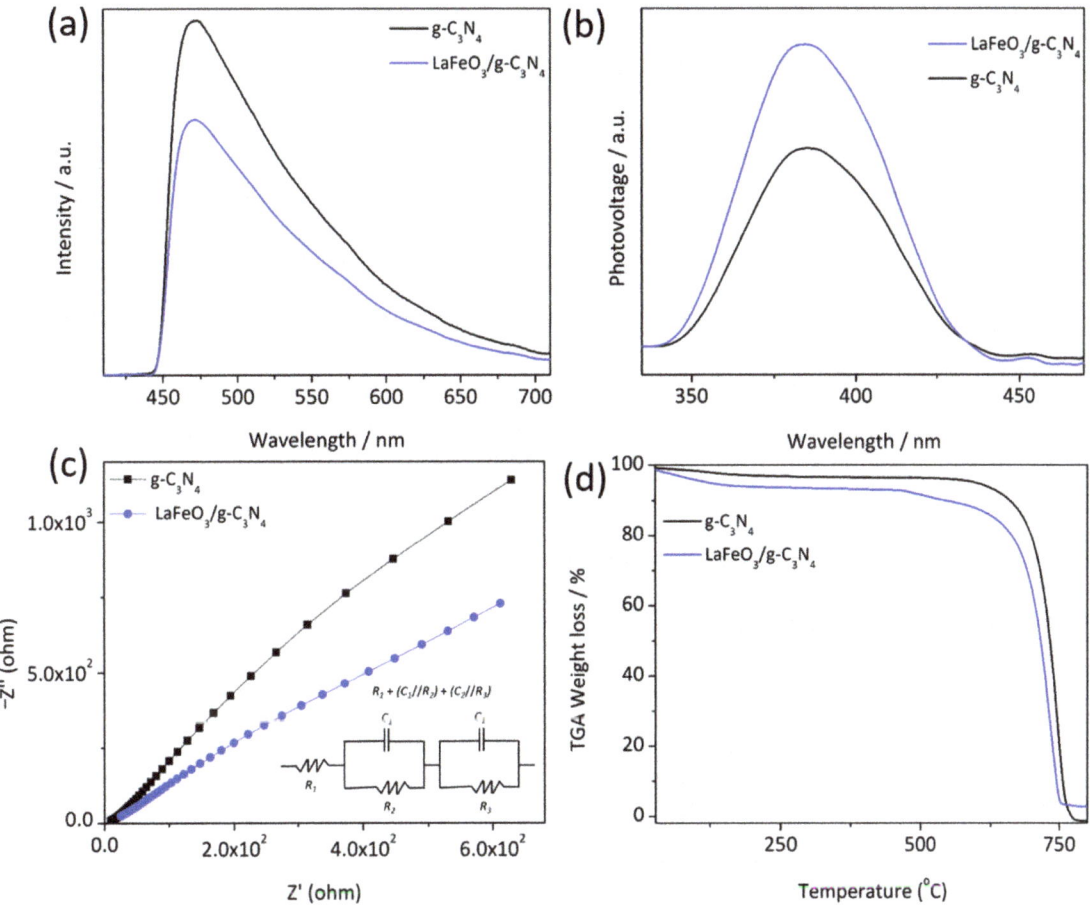

Figure 4. (**a**) PL spectra; (**b**) SPV spectra; (**c**) EIS Nyquist plots with inset the equivalent circuit model in which R_1 is the FTO glass resistance, C_1/R_2 is the impedance of the composite/electrolyte interface, and C_2/R_3 is the impedance of counter electrode/electrolyte interface; (**d**) TGA spectra of g-C_3N_4 and LaFeO$_3$/g-C_3N_4 photocatalysts.

2.3. Thermogravimetric Analysis

Figure 4d displays the thermogravimetric analysis (TGA) spectra for g-C_3N_4 and $LaFeO_3/g$-C_3N_4 catalysts. It is worth noting that between 750 and 770 °C, pure g-C_3N_4 loses a substantial amount of weight, likely due to disintegration or combustion. However, between 700 and 750 °C, the $LaFeO_3/g$-C_3N_4 composite shows a dramatic loss of weight. Like other g-C_3N_4-based photocatalysts, the lower weight is due to the coupling of $LaFeO_3$, which reduces the thermal stability of pure g-C_3N_4. After heating the photocatalyst to temperatures above 700 °C, the g-C_3N_4 percentage composition in the $LaFeO_3/g$-C_3N_4 catalyst is estimated to be around 92%.

2.4. Photocatalytic Activities

The visible light catalytic activity of g-C_3N_4 and $LaFeO_3/g$-C_3N_4 catalysts for hydrogen generation was investigated. The photocatalyst's performance for H_2 production was evaluated at 1 h intervals throughout the 4 h photocatalytic experiment. Upon exposure to visible light irradiations ($\lambda > 420$ nm), pristine g-C_3N_4 produced a tiny amount of H_2 (i.e., 24.0 μmol h^{-1} g^{-1}) with the assistance of co-catalyst Pt, as shown in Figure 5a. It is vital to highlight that the $LaFeO_3$ did not produce H_2 because of its unsuitable conduction band potential (i.e., 0.47 V) for water reduction. The amount of H_2 produced over the composite of $LaFeO_3/g$-C_3N_4 is 351 μmol h^{-1} g^{-1}, remarkably higher in comparison to that of the pristine g-C_3N_4 component. The enhanced separation and transfer of electron–hole pairs via the Z-scheme transfer system led to the exceptional activity of the $LaFeO_3/g$-C_3N_4 catalyst. A stability test of the $LaFeO_3/g$-C_3N_4 catalyst for H_2 evolution under visible light irradiation was conducted. Recyclability testing (Figure 5b) shows that after four consecutive photocatalytic re-cycles, the photocatalyst activity did not decline significantly. These findings identify the superior recycling performance and stability of $LaFeO_3/g$-C_3N_4 photocatalyst.

To validate the H_2 evolution results, the photocatalytic performance of the catalysts for the degradation of RhB and MB dyes was evaluated under visible-light irradiation for 2 h. Figure 5c shows that the RhB degradation over pristine g-C_3N_4 and $LaFeO_3$ photocatalysts is approximately 33 and 44%, respectively. Interestingly, the RhB degradation over the $LaFeO_3/g$-C_3N_4 photocatalyst is much more significant (i.e., 94%). Similarly, the pure g-C_3N_4 and $LaFeO_3$ photocatalysts degraded by about 27 and 39% of the MB, respectively. After 2 h of visible light irradiation, the $LaFeO_3/g$-C_3N_4 catalyst decomposed 87% of MB (Figure 5d). Meanwhile, the photocatalytic recyclability tests for the degradation of RhB and MB dyes were measured in order to confirm the stability of the $LaFeO_3/g$-C_3N_4 catalyst. As is obvious from Figure 5e,f, after 8 h of catalytic cycles (each lasting 2 h), the composite catalyst does not show any apparent decrease in catalytic performance for the degradation of RhB and MB dyes. Based on the above experiments, it is confirmed that coupling $LaFeO_3$ with g-C_3N_4 can greatly increase its surface redox ability due to the improved photogenerated charge separation in a Z-scheme direction.

In order to confirm which reactive intermediate species are involved in the degradation of RhB and MB dyes over the as-prepared $LaFeO_3/g$-C_3N_4 photocatalyst, radical trapping experiments were carried out to quench the reactive species such as holes (h^+), hydroxyl radicals ($^{\bullet}OH$), and superoxide radicals ($O_2^{\bullet-}$). Scavenging species such as ethylenediaminetetraacetic acid disodium salt (EDTA-2Na), isopropyl alcohol (IPA), and p-Benzoquinone (BQ) were employed to assess the contributions of h^+, $^{\bullet}OH$, and $O_2^{\bullet-}$ to the oxidation of RhB and MB dyes [40]. As revealed in Figure 6a,b, after 2 h of irradiation in the absence of any scavenger, approximately 94 and 87% of RhB and MB dyes were degraded over the $LaFeO_3/g$-C_3N_4 photocatalyst, respectively. However, after the addition of scavenger solutions (1 mmol) into the suspension containing RhB and MB dyes, only the IPA and BQ scavengers significantly reduced the degradation of both dyes. This suggests that the $^{\bullet}OH$, and $O_2^{\bullet-}$ radicals contributed considerably to the total degradation of RhB and MB dyes, while the role of h^+ was less significant. EPR spectroscopic investigation of the $^{\bullet}O_2$ and $^{\bullet}OH$ production during photocatalysis was carried out under dark and light (10 min) conditions at room temperature, as shown in Figure 6c,d, respectively. The trapping reagents DMPO-•OH and DMPO-•O$_2$ were added to the solution containing the LaFeO$_3$/g-C$_3$N$_4$ photocatalyst under continuous stirring prior to EPR analysis. As expected, no peaks of DMPO-•OH and DMPO-•O$_2$ can be seen in the dark. Surprisingly, distinct peaks of DMPO-•OH and DMPO-•O$_2$ were identified following a 10 min visible light catalytic reaction. The significant redox power of the LaFeO$_3$/g-C$_3$N$_4$ photocatalyst is clarified, and the Z-scheme charge transfer mechanism is confirmed.

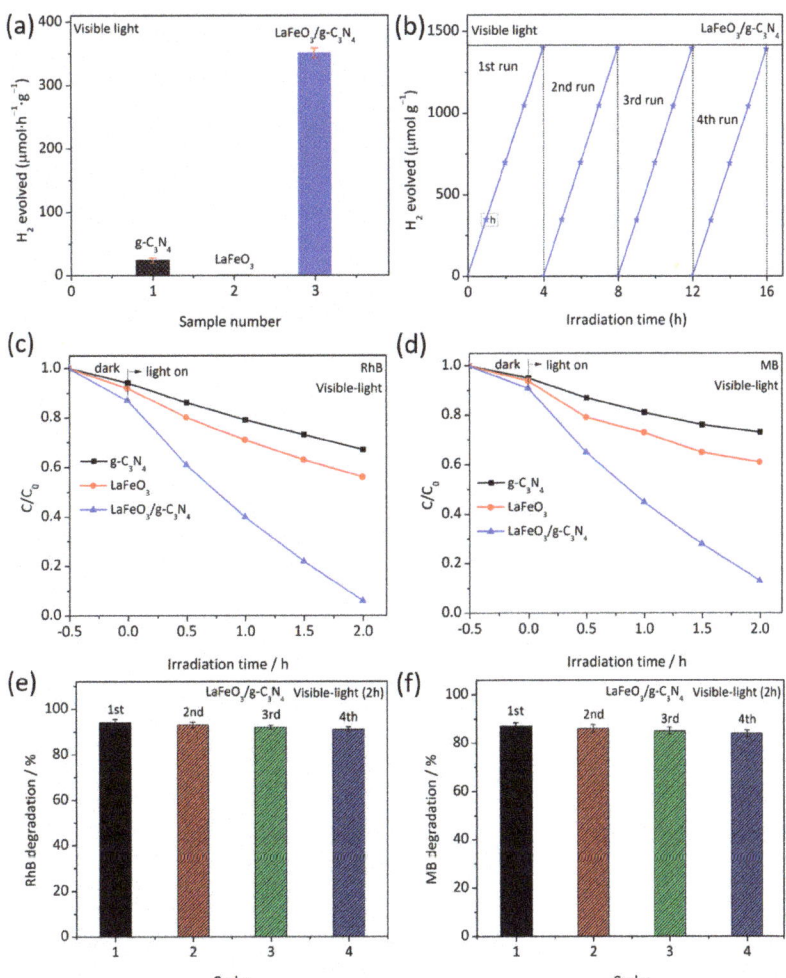

Figure 5. (**a**) Photocatalytic H$_2$ evolution over g-C$_3$N$_4$, LaFeO$_3$, and LaFeO$_3$/g-C$_3$N$_4$ photocatalysts. (**b**) Stability test for H$_2$ evolution LaFeO$_3$/g-C$_3$N$_4$ photocatalyst. (**c**) Photocatalytic degradation of RhB and (**d**) photocatalytic degradation of MB over the g-C$_3$N$_4$, LaFeO$_3$, and LaFeO$_3$/g-C$_3$N$_4$ photocatalysts. (**e**) Photocatalytic recyclability test for RhB degradation and (**f**) photocatalytic recyclability test for MB degradation over LaFeO$_3$/g-C$_3$N$_4$ photocatalyst.

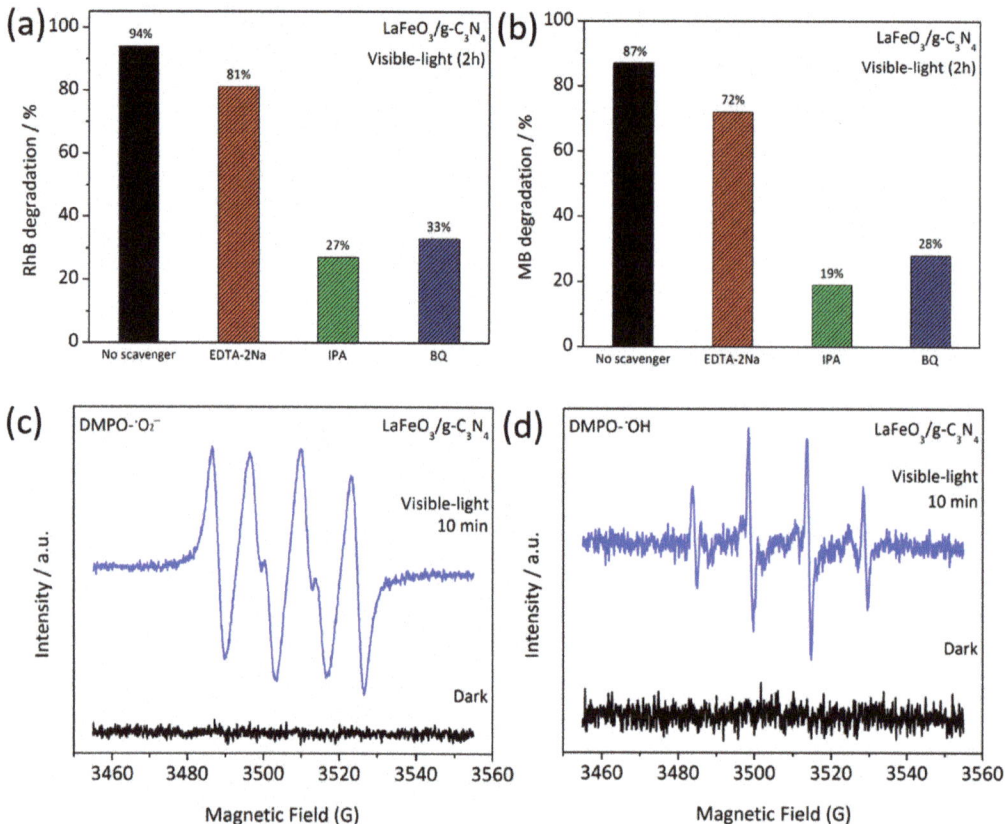

Figure 6. (**a**) Scavenging experiments for RhB degradation and (**b**) scavenging experiments for MB degradation. (**c**) EPR spectra of •O_2^- and (**d**) EPR spectra of •OH for LaFeO$_3$/g-C$_3$N$_4$ photocatalyst.

2.5. Mechanism

Based on the obtained results, a schematic for charge carrier generation, separation, transfer, and photocatalytic activities over the LaFeO$_3$/g-C$_3$N$_4$ Z-scheme system is depicted in Figure 7. The predicted band gap values of LaFeO$_3$ and g-C$_3$N$_4$ catalysts are 1.81 and 2.7 eV, respectively. Notably, the valence band potentials of g-C$_3$N$_4$ and LaFeO$_3$ photocatalysts are 1.4 and 2.38 V, respectively. In addition, the conduction band potentials of g-C$_3$N$_4$ and LaFeO$_3$ photocatalysts are calculated to be −1.3 and 0.57 V, respectively. When LaFeO$_3$ is coupled with g-C$_3$N$_4$, a heterojunction of LaFeO$_3$/g-C$_3$N$_4$ is formed. Charge carriers (electron–hole pairs) are generated in both components of the LaFeO$_3$/g-C$_3$N$_4$ heterojunction when exposed to visible light. Consequently, the electrons in both components are stimulated to their respective conduction bands. Meanwhile, holes continue to appear in their corresponding valence bands. The recombination between electrons of LaFeO$_3$ and holes of g-C$_3$N$_4$ occurs due to the proximity of the conduction band of LaFeO$_3$ and the valence band of g-C$_3$N$_4$. Thus, the transfer of charges occurs in a Z-scheme direction in the as-fabricated LaFeO$_3$/g-C$_3$N$_4$ catalyst. This improves charge carriers' separation in the as-fabricated LaFeO$_3$/g-C$_3$N$_4$ catalyst. It is important to note that the standard potentials for water reduction and superoxide radical (•O_2^-) generation are 0 and −0.046 V, respectively, versus the reduction potential of a Normal Hydrogen Electrode (NHE) [44]. The standard potential for hydroxyl radical (•OH) production is 2.27 V vs. the NHE [45]. Consequently, photogenerated electrons in the g-C$_3$N$_4$ conduction band reduce water to H$_2$ and O$_2$ to •O_2^-. Moreover, the valence band holes of LaFeO$_3$ will react with water and surface-adsorbed hydroxyl groups to produce •OH. The EPR study confirms the existence of both the •OH and •O$_2$ radicals. Therefore, the RhB and MB dyes are oxidized by a combination of •OH and •O$_2$. Furthermore, the photogeneration could also contribute to the total degradation of organic dyes. The above results reveal that the charge separation in the fabricated LaFeO$_3$/g-C$_3$N$_4$ Z-scheme heterojunction is significantly promoted which led to improved H$_2$ evolution and dyes degradation performance. According to the charge transfer phenomenon in the LaFeO$_3$/g-C$_3$N$_4$ Z-scheme photocatalyst, the catalytic reactions would proceed in a fashion as mentioned in Equations (1)–(7):

$$\text{LaFeO}_3/\text{g-C}_3\text{N}_4 + h\nu \rightarrow \text{LaFeO}_3\ (e^- + h^+)/\text{g-C}_3\text{N}_4\ (e^- + h^+) \tag{1}$$

$$\text{g-C}_3\text{N}_4\ (e^-) + 2H^+ \rightarrow H_2 \tag{2}$$

$$\text{g-C}_3\text{N}_4\ (e^-) + O_2 \rightarrow {}^\bullet O_2^- \tag{3}$$

$$\text{LaFeO}_3\ (h^+) + OH^-/H_2O \rightarrow {}^\bullet OH \tag{4}$$

$$\text{LaFeO}_3\ (h^+) + \text{Dyes} \rightarrow \text{Degradation Products} \tag{5}$$

$${}^\bullet O_2^- + \text{Dyes} \rightarrow \text{Degradation Products} \tag{6}$$

$${}^\bullet OH + \text{Dyes} \rightarrow \text{Degradation Products} \tag{7}$$

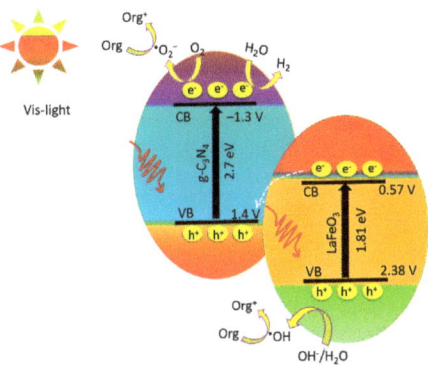

Figure 7. The energy band gaps, valence band, conduction band potentials, charge carriers' separation and transfer, and surface redox reactions over the Z-scheme LaFeO$_3$/g-C$_3$N$_4$ photocatalyst.

3. Materials and Methods

Analytical-grade solvents/reagents were used in this study without being further purified.

3.1. Fabrication of g-C$_3$N$_4$

An amount of 5 g of melamine was directly calcined in an air environment at 550 °C (5 °C min^{-1}) for 2 h in a furnace. After reaching room temperature, the product was collected and milled into a fine powder.

3.2. Fabrication of LaFeO$_3$ Nanoparticles

A sol–gel process was used to fabricate LaFeO$_3$ nanoparticles. In a typical procedure, an equimolar (0.1 M) precursor solution of La(NO$_3$)$_3$·6H$_2$O and Fe(NO$_3$)$_3$·9H$_2$O was prepared in methanol/ethylene glycol (1:1) mixture and stirred under room temperature for half an hour. After that, both solutions were mixed under continuous stirring and then kept under ultrasonic treatment for half an hour. The mixed solution was then stirred for 16 h and then dried at 85 °C. Consequently, the dried powder was calcined in air at 650 °C for 2 h to obtain LaFeO$_3$ nanoparticles.

3.3. Fabrication of LaFeO$_3$/g-C$_3$N$_4$ Composite

A wet chemical approach was used to fabricate the LaFeO$_3$/g-C$_3$N$_4$ heterostructure composite. A 5wt% of LaFeO$_3$ to that of the corresponding g-C$_3$N$_4$ (total weight of the sample was 1 g) was dispersed in a beaker containing 25 mL of water/ethanol mixture (1:1) and continuously stirred for 12 h. Then, the powder dispersed in the solvent mixture was centrifuged and washed with de-ionized water and ethanol. Finally, it was dried in an oven at 85 °C, milled into a fine powder, and then calcined at 450 °C (5 °C min^{-1}) for 2 h.

3.4. Characterization

A Bruker-D8 powder diffractometer (Chiba, Japan) was employed to acquire the X-ray diffraction patterns of the catalysts using a CuKα radiation source. A UV-2550 Shimadzu-Kyoto, Japan Spectrophotometer was used to obtain the UV-vis absorption spectra. The TEM images were taken via a transmission electron microscope (TEM) model (JEOL Ltd-JEM-2100, Tokyo, Japan) set at 200 kV. An Ultra-DLD-Kratos-Axis X-ray photoelectron spectroscope (XPS) (Kyoto, Japan) with Al (mono) X-ray source was employed to detect the chemical composition and elemental states of the catalysts. An FP-6500 (Tokyo, Japan) fluorescence spectrometer was utilized to detect the PL spectra of the catalysts. Surface photovoltage (SPV) spectra were measured using equipment linked with a lock-in amplifier (model SR830, Sunnyvale, CA, USA) and a light chopper (model SR540, Sunnyvale, CA, USA). A Perkin Elmer TGA-8000 (USA) was used for a thermogravimetric study in the temperature range of 30–780 °C under air conditions. Electrochemical impedance spectroscopy (EIS) spectra were obtained with Shanghai Chenhua CHI-760E equipment (Shanghai, China) while employing a Ag/AgCl reference electrode. For electron paramagnetic resonance (EPR) spectra measurement, a Bruker-A300 (Beijing, China) apparatus was used. The EPR measurement was performed at room temperature, and the trapping reagent 5,5-Dimethyl-1-pyrroline-N-oxide (DMPO) was utilized.

3.5. Photocatalytic Experiments

The photocatalytic H$_2$ production experiments were carried out in a sealed 250 mL quartz reactor. During the test, methanol was added as a sacrificial agent. About 50 mg of the photocatalyst powder was diffused in a mixture of water (80 mL)/methanol (20 mL) under continuous stirring. To remove bubbles, the system was extensively evacuated for half an hour. The system was then irradiated under a Perfect-light 300 W Xenon lamp (Beijing Perfect light Technology Co., Ltd) with a cut-off filter (λ > 420 nm). The hydrogen produced during the photocatalytic reaction was measured at a fixed time interval (1 h) and detected via an online gas chromatograph (CEAULIGH, GC-7920 with carrier gas N$_2$) linked to a TCD detector. The photocatalytic recyclability test for H$_2$ production over the LaFeO$_3$/g-C$_3$N$_4$ photocatalyst was evaluated for 16 h (4 cycles, each lasting 4 h) under the same experimental conditions. The photocatalytic experiments for Rhodamine B (RhB) and Methylene Blue (MB) were performed in a 100 mL volume quartz reactor with the assistance of a 300 W Xe lamp (cut-off filter λ > 420 nm). The duration of each experiment was 2 h. Before photocatalytic reaction, the dye solutions containing catalysts (50 mg) were stirred in the dark for 30 min to attain adsorption equilibrium. Similarly, the photocatalytic recyclable tests for the dyes' degradation over the LaFeO$_3$/g-C$_3$N$_4$ photocatalyst were evaluated for 8 h (4 cycles, each lasting 2 h) under the same experimental conditions. The scavenger-trapping experiments were performed under the same experimental conditions with the assistance of ethylenediaminetetraacetic acid disodium salt (EDTA-2Na), isopropyl alcohol (IPA), and p-Benzoquinone (BQ) as the h^+, $^\bullet OH$, and $O_2^{\bullet-}$ trapping agents, respectively.

4. Conclusions

In summary, the fabrication of the Z-scheme $LaFeO_3/g-C_3N_4$ system for water reduction to evolve hydrogen and organic dye degradation is described in this study. Notably, the as-fabricated $LaFeO_3/g-C_3N_4$ composite catalyst revealed outstanding water reduction performance to generate H_2 (i.e., 351 mol $h^{-1}g^{-1}$) under visible light irradiation, which is significantly higher in comparison to that of the $g-C_3N_4$. Additionally, the photocatalyst oxidized 87% of the MB dye and 94% of the RhB dye after two hours of visible light irradiation. The results indicate that the significantly increased photo-activities of the $LaFeO_3/g-C_3N_4$ Z-scheme heterostructure are due to extended light absorption and dramatically accelerated charge carrier separation. The development of this highly efficient Z-scheme heterostructure photocatalyst offers a promising strategy for the design and implementation of efficient Z-scheme systems to address energy and environmental concerns.

Author Contributions: Conceptualization, M.H. and M.B.; methodology, A.K.; validation, M.H. and M.B.; formal analysis, M.H. and A.B.; investigation, M.H.; resources, M.B.; data curation, M.H.; writing—original draft preparation, M.H. and A.K.; writing—review and editing, M.B.; supervision, M.B.; project administration, M.H.; funding acquisition, M.H. and M.B. All authors have read and agreed to the published version of the manuscript.

Funding: This research received no external funding.

Data Availability Statement: Data will be available from the corresponding author upon reasonable request.

Acknowledgments: Authors thank Prince Sultan University for the financial support.

Conflicts of Interest: The authors declare no conflict of interest.

Sample Availability: Samples of the compounds are not available from the authors.

References

1. Fang, M.; Tan, X.; Liu, Z.; Hu, B.; Wang, X. Recent Progress on Metal-Enhanced Photocatalysis: A Review on the Mechanism. *Research* **2021**, *2021*, 9794329. [CrossRef] [PubMed]
2. Hu, J.; Guo, X.W.; Zhang, Y.C.; Zhang, F. Review on the Advancement of SnS_2 in the Photocatalysis. *J. Mater. Chem. A* **2023**, *11*, 7331–7343.
3. Nnabuife, S.G.; Ugbeh-Johnson, J.; Okeke, N.E.; Ogbonnaya, C. Present and Projected Developments in Hydrogen Production: A Technological Review. *Carbon Capture Sci. Technol.* **2022**, *3*, 100042. [CrossRef]
4. Qazi, A.; Hussain, F.; Rahim, N.A.; Hardaker, G.; Alghazzawi, D.; Shaban, K.; Haruna, K. Towards Sustainable Energy: A Systematic Review of Renewable Energy Sources, Technologies, and Public Opinions. *IEEE Access* **2019**, *7*, 63837–63851. [CrossRef]
5. Pan, X.; Shao, T.; Zheng, X.; Zhang, Y.; Ma, X.; Zhang, Q. Energy and Sustainable Development Nexus: A Review. *Energy Strategy Rev.* **2023**, *47*, 101078. [CrossRef]
6. Kabeyi, M.J.B.; Olanrewaju, O.A. Sustainable Energy Transition for Renewable and Low Carbon Grid Electricity Generation and Supply. *Front. Energy Res.* **2022**, *9*, 743114. [CrossRef]
7. Noureen, L.; Wang, Q.; Humayun, M.; Shah, W.A.; Xu, Q.; Wang, X. Recent Advances in Structural Engineering of Photocatalysts for Environmental Remediation. *Environ. Res.* **2023**, *219*, 115004. [CrossRef]
8. Khan, N.A.; Humayun, M.; Usman, M.; Ghazi, Z.A.; Naeem, A.; Khan, A.; Khan, A.L.; Tahir, A.A.; Ullah, H. Structural Characteristics and Environmental Applications of Covalent Organic Frameworks. *Energies* **2021**, *14*, 2267. [CrossRef]
9. Li, J.; Yuan, H.; Zhang, W.; Jin, B.; Feng, Q.; Huang, J.; Jiao, Z. Advances in Z-scheme Semiconductor Photocatalysts for the Photoelectrochemical Applications: A Review. *Carbon Energy* **2022**, *4*, 294–331. [CrossRef]
10. Wang, H.; Zhang, L.; Chen, Z.; Hu, J.; Li, S.; Wang, Z.; Liu, J.; Wang, X. Semiconductor Heterojunction Photocatalysts: Design, Construction, and Photocatalytic Performances. *Chem. Soc. Rev.* **2014**, *43*, 5234–5244. [CrossRef]
11. Liu, Z.; Yu, Y.; Zhu, X.; Fang, J.; Xu, W.; Hu, X.; Li, R.; Yao, L.; Qin, J.; Fang, Z. Semiconductor Heterojunctions for Photocatalytic Hydrogen Production and Cr(VI) Reduction: A review. *Mater. Res. Bull.* **2022**, *147*, 111636. [CrossRef]
12. Low, J.; Yu, J.; Jaroniec, M.; Wageh, S.; Al-Ghamdi, A.A. Heterojunction Photocatalysts. *Adv. Mater.* **2017**, *29*, 1601694. [CrossRef] [PubMed]

13. Tahir, M.; Tasleem, S.; Tahir, B. Recent Development in Band Engineering of Binary Semiconductor Materials for Solar Driven Photocatalytic Hydrogen Production. *Int. J. Hydrog. Energy* **2020**, *45*, 15985–16038. [CrossRef]
14. Cao, S.; Yu, J. g-C_3N_4-Based Photocatalysts for Hydrogen Generation. *J. Phys. Chem. Lett.* **2014**, *5*, 2101–2107. [CrossRef]
15. Alhebshi, A.; Sharaf Aldeen, E.; Mim, R.S.; Tahir, B.; Tahir, M. Recent Advances in Constructing Heterojunctions of Binary Semiconductor Photocatalysts for Visible Light Responsive CO_2 Reduction to Energy Efficient Fuels: A Review. *Int. J. Energy Res.* **2022**, *46*, 5523–5584. [CrossRef]
16. Lee, J.-T.; Lee, S.-W.; Wey, M.-Y. S-scheme g-C_3N_4/ZnO Heterojunction Photocatalyst with Enhanced Photodegradation of Azo Dye. *J. Taiwan Inst. Chem. Eng.* **2022**, *134*, 104357. [CrossRef]
17. Kadi, M.W.; El-Hout, S.I.; Shawky, A.; Mohamed, R.M. Enhanced Mercuric Ions Reduction over Mesoporous S-scheme $LaFeO_3$/ZnO p-n Heterojunction Photocatalysts. *J. Taiwan Inst. Chem. Eng.* **2022**, *138*, 104476. [CrossRef]
18. Irshad, M.; tul Ain, Q.; Zaman, M.; Aslam, M.Z.; Kousar, N.; Asim, M.; Rafique, M.; Siraj, K.; Tabish, A.N.; Usman, M. Photocatalysis and Perovskite Oxide-Based Materials: A Remedy for a Clean and Sustainable Future. *RSC Adv.* **2022**, *12*, 7009–7039. [CrossRef]
19. Huang, C.-W.; Hsu, S.-Y.; Lin, J.-H.; Jhou, Y.; Chen, W.-Y.; Lin, K.-Y.A.; Lin, Y.-T.; Nguyen, V.-H. Solar-Light-Driven $LaFe_xNi_{1-x}O_3$ Perovskite Oxides for Photocatalytic Fenton-like Reaction to Degrade Organic Pollutants. *Beilstein J. Nanotechnol.* **2022**, *13*, 882–895. [CrossRef]
20. Yaseen, M.; Humayun, M.; Khan, A.; Idrees, M.; Shah, N.; Bibi, S. Photo-Assisted Removal of Rhodamine B and Nile Blue Dyes from Water Using CuO-SiO_2 Composite. *Molecules* **2022**, *27*, 5343. [CrossRef]
21. Yaseen, M.; Khan, A.; Humayun, M.; Farooq, S.; Shah, N.; Bibi, S.; Khattak, Z.A.K.; Rehman, A.U.; Ahmad, S.; Ahmad, S.M.; et al. Facile Synthesis of Fe_3O_4–SiO_2 Nanocomposites for Wastewater Treatment. *Macromol. Mater. Eng.* **2023**, 2200695. [CrossRef]
22. Wen, J.; Xie, J.; Chen, X.; Li, X. A Review on g-C_3N_4-Based Photocatalysts. *Appl. Surf. Sci.* **2017**, *391*, 72–123. [CrossRef]
23. Liu, R.; Chen, Z.; Yao, Y.; Li, Y.; Cheema, W.A.; Wang, D.; Zhu, S. Recent Advancements in gC_3N_4-Based Photocatalysts for Photocatalytic CO_2 Reduction: A Mini Review. *RSC Adv.* **2020**, *10*, 29408–29418. [CrossRef] [PubMed]
24. Catherine, H.N.; Chiu, W.-L.; Chang, L.-L.; Tung, K.-L.; Hu, C. Gel-like Ag-Dicyandiamide Metal–Organic Supramolecular Network-Derived g-C_3N_4 for Photocatalytic Hydrogen Generation. *ACS Sustain. Chem. Eng.* **2022**, *10*, 8360–8369. [CrossRef]
25. Huang, H.; Jiang, L.; Yang, J.; Zhou, S.; Yuan, X.; Liang, J.; Wang, H.; Wang, H.; Bu, Y.; Li, H. Synthesis and Modification of Ultrathin g-C_3N_4 for Photocatalytic Energy and Environmental Applications. *Renew. Sustain. Energy Rev.* **2023**, *173*, 113110. [CrossRef]
26. Humayun, M.; Ullah, H.; Usman, M.; Habibi-Yangjeh, A.; Tahir, A.A.; Wang, C.; Luo, W. Perovskite-type Lanthanum Ferrite Based Photocatalysts: Preparation, Properties, and Applications. *J. Energy Chem.* **2022**, *66*, 314–338. [CrossRef]
27. Khan, I.; Luo, M.; Guo, L.; Khan, S.; Shah, S.A.; Khan, I.; Khan, A.; Wang, C.; Ai, B.; Zaman, S. Synthesis of Phosphate-Bridged g-C_3N_4/$LaFeO_3$ Nanosheets Z-scheme Nanocomposites as Efficient Visible Photocatalysts for CO_2 Reduction and Malachite Green Degradation. *Appl. Catal. A* **2022**, *629*, 118418. [CrossRef]
28. Hu, C.; Yu, B.; Zhu, Z.; Zheng, J.; Wang, W.; Liu, B. Construction of Novel S-scheme $LaFeO_3$/g-C_3N_4 Composite with Efficient Photocatalytic Capacity for Dye Degradation and Cr (VI) Reduction. *Colloids Surf. A Physicochem. Eng. Asp.* **2023**, *664*, 131189. [CrossRef]
29. Ismael, M.; Wu, Y. A Facile Synthesis Method for Fabrication of $LaFeO_3$/gC_3N_4 Nanocomposite as Efficient Visible-Light-Driven Photocatalyst for Photodegradation of RhB and 4-CP. *New J. Chem.* **2019**, *43*, 13783–13793. [CrossRef]
30. Raziq, F.; Hayat, A.; Humayun, M.; Mane, S.K.B.; Faheem, M.B.; Ali, A.; Zhao, Y.; Han, S.; Cai, C.; Li, W. Photocatalytic Solar Fuel Production and Environmental Remediation through Experimental and DFT Based Research on CdSe-QDs-coupled P-doped-g-C_3N_4 Composites. *Appl. Catal. B* **2020**, *270*, 118867. [CrossRef]
31. Raziq, F.; He, J.; Gan, J.; Humayun, M.; Faheem, M.B.; Iqbal, A.; Hayat, A.; Fazal, S.; Yi, J.; Zhao, Y. Promoting Visible-Light Photocatalytic Activities for Carbon Nitride Based 0D/2D/2D Hybrid System: Beyond the Conventional 4-electron Mechanism. *Appl. Catal. B* **2020**, *270*, 118870. [CrossRef]
32. Humayun, M.; Li, Z.; Sun, L.; Zhang, X.; Raziq, F.; Zada, A.; Qu, Y.; Jing, L. Coupling of Nanocrystalline Anatase TiO_2 to Porous Nanosized $LaFeO_3$ for Efficient Visible-Light Photocatalytic Degradation of Pollutants. *Nanomaterials* **2016**, *6*, 22. [CrossRef] [PubMed]
33. Aizat, A.; Aziz, F.; Mohd Sokri, M.N.; Sahimi, M.S.; Yahya, N.; Jaafar, J.; Wan Salleh, W.N.; Yusof, N.; Ismail, A.F. Photocatalytic Degradation of Phenol by $LaFeO_3$ Nanocrystalline Synthesized by Gel Combustion Method via Citric Acid Route. *SN Appl. Sci.* **2019**, *1*, 1–10. [CrossRef]
34. Zhu, J.; Li, H.; Zhong, L.; Xiao, P.; Xu, X.; Yang, X.; Zhao, Z.; Li, J. Perovskite Oxides: Preparation, Characterizations, and Applications in Heterogeneous Catalysis. *ACS Catal.* **2014**, *4*, 2917–2940. [CrossRef]
35. Badro, J.; Rueff, J.-P.; Vanko, G.; Monaco, G.; Fiquet, G.; Guyot, F. Electronic Transitions in Perovskite: Possible Nonconvecting Layers in the Lower Mantle. *Science* **2004**, *305*, 383–386. [CrossRef]
36. Neena, D.; Humayun, M.; Bhattacharyya, D.; Fu, D. Hierarchical Sr-ZnO/g-C_3N_4 Heterojunction with Enhanced Photocatalytic Activities. *J. Photochem. Photobiol. A* **2020**, *396*, 112515.
37. Xu, K.; Feng, J. Superior Photocatalytic Performance of $LaFeO_3$/gC_3N_4 Heterojunction Nanocomposites under Visible Light Irradiation. *RSC Adv.* **2017**, *7*, 45369–45376. [CrossRef]

38. Acharya, S.; Mansingh, S.; Parida, K. The Enhanced Photocatalytic Activity of gC$_3$N$_4$-LaFeO$_3$ for the Water Reduction Reaction through a Mediator free Z-scheme Mechanism. *Inorg. Chem. Front.* **2017**, *4*, 1022–1032. [CrossRef]
39. Zhang, J.; Zhu, Z.; Jiang, J.; Li, H. Fabrication of a Novel AgI/LaFeO$_3$/g-C$_3$N$_4$ dual Z-scheme Photocatalyst with Enhanced Photocatalytic Performance. *Mater. Lett.* **2020**, *262*, 127029. [CrossRef]
40. Humayun, M.; Sun, N.; Raziq, F.; Zhang, X.; Yan, R.; Li, Z.; Qu, Y.; Jing, L. Synthesis of ZnO/Bi-doped Porous LaFeO$_3$ Nanocomposites as Highly Efficient Nano-photocatalysts Dependent on the Enhanced Utilization of Visible-Light-Excited Electrons. *Appl. Catal. B* **2018**, *231*, 23–33. [CrossRef]
41. Humayun, M.; Zada, A.; Li, Z.; Xie, M.; Zhang, X.; Qu, Y.; Raziq, F.; Jing, L. Enhanced Visible-Light Activities of Porous BiFeO$_3$ by Coupling with Nanocrystalline TiO$_2$ and Mechanism. *Appl.Catal. B* **2016**, *180*, 219–226. [CrossRef]
42. Raziq, F.; Qu, Y.; Humayun, M.; Zada, A.; Yu, H.; Jing, L. Synthesis of SnO$_2$/BP Codoped g-C$_3$N$_4$ Nanocomposites as Efficient Cocatalyst-Free Visible-Light Photocatalysts for CO$_2$ Conversion and Pollutant Degradation. *Appl. Catal. B* **2017**, *201*, 486–494. [CrossRef]
43. Zhao, R.; Sun, X.; Jin, Y.; Han, J.; Wang, L.; Liu, F. Au/Pd/gC$_3$N$_4$ Nanocomposites for Photocatalytic Degradation of Tetracycline Hydrochloride. *J. Mater. Sci.* **2019**, *54*, 5445–5456. [CrossRef]
44. Usman, M.; Zeb, Z.; Ullah, H.; Suliman, M.H.; Humayun, M.; Ullah, L.; Shah, S.N.A.; Ahmed, U.; Saeed, M. A Review of Metal-Organic Frameworks/Graphitic Carbon Nitride Composites for Solar-Driven Green H$_2$ Production, CO$_2$ Reduction, and Water Purification. *J. Environ. Chem. Eng.* **2022**, *10*, 107548. [CrossRef]
45. Chen, D.; Cheng, Y.; Zhou, N.; Chen, P.; Wang, Y.; Li, K.; Huo, S.; Cheng, P.; Peng, P.; Zhang, R. Photocatalytic Degradation of Organic Pollutants using TiO$_2$-based Photocatalysts: A Review. *J. Clean. Prod.* **2020**, *268*, 121725. [CrossRef]

Disclaimer/Publisher's Note: The statements, opinions and data contained in all publications are solely those of the individual author(s) and contributor(s) and not of MDPI and/or the editor(s). MDPI and/or the editor(s) disclaim responsibility for any injury to people or property resulting from any ideas, methods, instructions or products referred to in the content.

Article

Vertical Growth of WO$_3$ Nanosheets on TiO$_2$ Nanoribbons as 2D/1D Heterojunction Photocatalysts with Improved Photocatalytic Performance under Visible Light

Ling Wang [1], Keyi Xu [1], Hongwang Tang [2] and Lianwen Zhu [2,*]

[1] College of Chemistry and Life Science, Zhejiang Normal University, Jinhua 321004, China
[2] School of Biology and Chemical Engineering, Jiaxing University, Jiaxing 314001, China
* Correspondence: lwzhu@mail.zjxu.edu.cn

Abstract: We report the construction of 2D/1D heterojunction photocatalysts through the hydrothermal growth of WO$_3$ nanosheets on TiO$_2$ nanoribbons for the first time. Two-dimensional WO$_3$ nanosheets were vertically arrayed on the surface of TiO$_2$ nanoribbons, and the growth density could be simply controlled by adjusting the concentration of the precursors. The construction of WO$_3$/TiO$_2$ heterojunctions not only decreases the band gap energy of TiO$_2$ from 3.12 to 2.30 eV and broadens the photoresponse range from the UV region to the visible light region but also significantly reduces electron–hole pair recombination and enhances photo-generated carrier separation. Consequently, WO$_3$/TiO$_2$ heterostructures exhibit improved photocatalytic activity compared to pure WO$_3$ nanosheets and TiO$_2$ nanoribbons upon visible light irradiation. WO$_3$/TiO$_2$-25 possesses the highest photocatalytic activity and can remove 92.8% of RhB pollutants in 120 min. Both further increase and decrease in the growth density of WO$_3$ nanosheets result in an obvious reduction in photocatalytic activity. The kinetic studies confirmed that the photocatalytic degradation of RhB follows the kinetics of the pseudo-first-order model. The present study demonstrates that the prepared WO$_3$/TiO$_2$ 2D/1D heterostructures are promising materials for photocatalytic removal of organic pollutants to produce clean water.

Keywords: WO$_3$/TiO$_2$; 2D/1D heterojunction; photocatalytic activity; water purification

1. Introduction

The demand for fresh water increases exponentially with global population growth and industrial development [1–3]. Clean water scarcity is one of the most pressing issues for human beings to confront nowadays, especially in developing countries [4]. According to World Wildlife, two-thirds of the world's population will face water shortages by 2025 [5]. To address this challenge, many advanced technologies have been developed to remove aqueous pollutants for clean water production [6,7], including adsorption [8,9], membrane filtration [10], and semiconductor photocatalytic technology [11]. Among various techniques, semiconductor photocatalysis offers a green and sustainable water purification technology because harmful pollutants can be completely decomposed to CO$_2$, H$_2$O, and other small molecules without the involvement of any chemicals [11–14]. Mainly, photocatalysts absorb photons as an energy source to (1) generate electron−hole pairs, (2) separate the photoexcited charges, (3) transfer electron and hole to the photocatalyst's surface, and (4) utilize the active charges on the photocatalyst's surface for catalytic decomposition of pollutants [12–17].

Heterojunction photocatalysts, particularly systems consisting of a low-dimensional 1D and 2D semiconductor, are recognized as prospective building blocks for next-generation advanced photon harvesting and conversion technologies [18,19]. The considerable potential of 1D/2D heterojunctions arises from high charge carrier mobility along the 1D nanostructures and effective charge recombination prevention ability of 2D nanostructures due to the strong electron confinement effect in atomic-thick layers. Among four different configurations based on interfacial contact, heterojunctions with vertically aligned 2D nanosheets on 1D nanostructures are highly desirable for photocatalytic systems [19]. That is because the vertically aligned 2D nanosheets on 1D nanostructures not only allow maximum exposure of the entire surface, thereby maximally providing the edge active sites, but also create better electronic contacts and optimized electron transport pathways, leading to enhanced photo-generated electron–hole separation efficiency [18,19]. This unique 2D array on 1D structures widely exists in nature, such as leaf array on a tree branch and the bony plates along stegosaur backs, and their advancement in light absorption and conversion have been proven in nature for millions of years.

Since the discovery of the Honda–Fujishima effect in 1972 [20], TiO_2 has been considered to be the most promising photocatalyst because of its high photosensitivity, excellent stability, nontoxic nature, and low cost [13,21,22]. In particular, 1D TiO_2 nanoribbons have received much more attention for advanced photocatalyst design because of the following features [22–24]. First, the well-defined 1D geometry facilitates fast and long-distance electron transport, leading to long-time photocatalytic stability. Second, 1D nanoribbons possess larger specific surface areas than corresponding bulk materials, providing more active catalytic sites. Third, the large length-to-diameter ratio of 1D nanoribbons increases light absorption and scattering, enhancing light use efficiency [19,20]. Beside TiO_2, WO_3 is another extensively investigated photocatalyst driven by its narrow band gap and visible light absorption ability [25–27]. Among various morphologies, 2D WO_3 nanosheets possess an extremely high percentage of exposed surfaces and a strong quantum confinement effect in the thinnest dimension, leading to abundant active sites and enhanced light conversion efficiency, which are highly desirable for high-performance photocatalysts [25–30]. Inspired by the booming progress in 1D TiO_2 nanoribbon and 2D WO_3 nanosheet photocatalysts, the rational design of 2D/1D WO_3/TiO_2 multidimensional heterojunction photocatalysts is expected to integrate the merits of both 1D and 2D nanogeometry and lead to enhanced photocatalytic performance. Particularly, a 2D WO_3 array on a 1D TiO_2 heterojunction is of special interest for highly efficient photocatalyst design. Coupling narrow band gap WO_3 (2.4–2.8 eV) can not only broaden the absorption wavelength of TiO_2 from UV to visible light regions, resulting in the formation of visible light-responded heterojunction photocatalysts, but also effectively suppresses electron−hole pair recombination, thus improving the photocatalytic activity [31]. To date, substantial efforts have been devoted to construct WO_3/TiO_2 heterojunction systems, including WO_3 nanorod/TiO_2 nanofiber composite [32], WO_3 nanoparticle/TiO_2 nanoparticle composite [33,34], heterostructured TiO_2/WO_3 porous microspheres [35,36], a WO_3 nanoparticle/TiO_2 nanotubes system [37–39], heterostructured WO_3/TiO_2 nanosheets [40,41], and WO_3 nanosheet/TiO_2 nanoparticle composite [42]. However, to the best of our knowledge, 2D/1D heterojunction photocatalysts composed of TiO_2 nanoribbons and WO_3 nanosheets have not been reported. Demonstrating the vertical growth of WO_3 nanosheets on TiO_2 nanoribbons for 2D/1D heterojunction photocatalysts is especially challenging.

Herein, we report the vertical growth of WO_3 nanosheets on TiO_2 nanoribbons as 1D/2D heterojunction photocatalysts with improved photocatalytic performance under visible light. The WO_3/TiO_2 heterojunctions were constructed through the introduction of TiO_2 nanoribbons into the hydrothermal growth system of WO_3 nanosheets. The as-obtained WO_3/TiO_2 sample was characterized by using XRD, SEM, TEM, XPS, UV–vis, and PL analysis. WO_3 nanosheets vertically arrayed on the surface of TiO_2 nanoribbons lead to the formation of 1D/2D heterojunctions with maximum exposure of the entire surface, which not only broadens the light adsorption spectrum into the visible region but also inhibits the recombination of photoinduced carriers, resulting in enhanced photocatalytic degradation of aqueous pollutants.

2. Results and Discussion

2.1. XRD Analysis

The hydrothermal growth process is accompanied by an obvious color change in TiO_2 from white to yellow-green (Figure 1a), indicating the successful and uniform growth of WO_3 on TiO_2. In order to confirm WO_3 growth, XRD patterns of the samples before and after hydrothermal reaction were recorded (Figure 1b). Before growth of WO_3 nanosheets, all the diffraction peaks located at 25.3°, 37.8°, 48.0°, 53.9°, 55.0°, 62.1°, 62.7°, and 68.8° can be indexed to anatase TiO_2 (JCPDS file no. 21-1272), which is in good agreement with previous reports [43,44]. After the growth of the WO_3 nanosheets, except for the peaks originating from TiO_2 nanoribbons (marked with red boxes), new peaks (marked with blue circles) centered at 16.5°, 25.6°, 30.5°, 33.4°, 34.1°, 34.8°, 38.9°, 44.1°, 45.9°, 46.3°, 49.6°, 52.6°, 56.2°, 57.2°, 58.4°, 61.2°, 64.3°, and 66.0° were observed, which match well with $WO_3 \bullet H_2O$ (JCPDS file no. 43-0679) [45], indicating the formation of WO_3/TiO_2-25 composites [32–34].

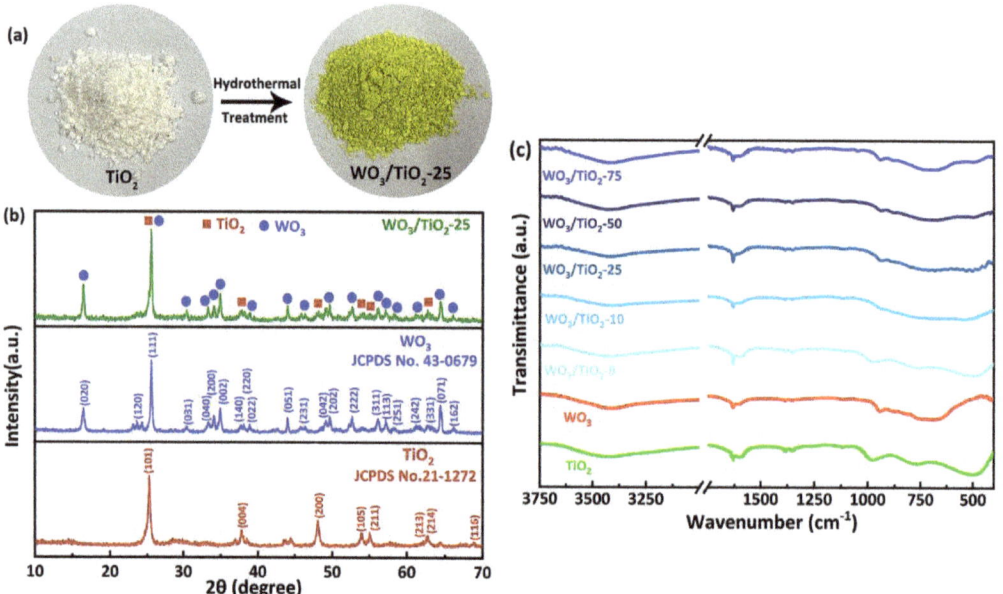

Figure 1. (a) Photograph of TiO_2 before and after hydrothermal growth of WO_3. (b) XRD patterns of TiO_2, WO_3, and WO_3/TiO_2−25. (c) FTIR spectra of TiO_2, WO_3, and WO_3/TiO_2 composites with different ratios.

2.2. FTIR Analysis

In order to further confirm the formation of WO_3/TiO_2 composites, FTIR spectra of all samples have been collected and shown in Figure 1c. For the spectrum of pure TiO_2 nanoribbons (green line), the broad adsorption band centered at 3430 cm^{-1} could be ascribed to the stretching vibrations of the surface hydroxyl groups and molecularly chemisorbed water [46]. The peak at 1630 cm^{-1} results from -OH bending of molecularly physisorbed water [46]. The large absorption in the range 492 cm^{-1} can be attributed to the strong stretching vibrations of the Ti-O bond in TiO_2 nanoribbons [46]. For the spectrum of pure WO_3 nanosheets (red line), the broad adsorption band center at 3410 cm^{-1} is related to the stretch region of the surface hydroxyl groups with hydrogen bonds and crystal water of $WO_3 \bullet H_2O$, while the peak centered at 1633 cm^{-1} is from O–H bending of physisorbed water. The weak band observed around 600–750 cm^{-1} is attributed to the O–W–O stretching modes of WO_3, and the peak located at 941 cm^{-1} is associated with W=O stretching [46]. For WO_3/TiO_2 composites with different ratios, the broad peak around 3412–3420 cm^{-1} of each sample can be ascribed to the stretching vibrations of -OH groups [46]. These characteristic absorption peaks of WO_3 and TiO_2 located in the range of 400–1000 cm^{-1} are all observed in the spectra of WO_3/TiO_2 composites, indicating the successful combination of WO_3 and TiO_2. From WO_3/TiO_2-5 to WO_3/TiO_2-75, the intensity of peaks from TiO_2 nanoribbons decreases, and the intensity of peaks ascribed to WO_3 nanosheets increases, indicating the increasing growth density of WO_3 nanosheets on TiO_2 nanoribbons, which is in good agreement with the raw material proportioning.

2.3. SEM and TEM Analysis

In order to reveal the configuration of WO_3/TiO_2 heterojunctions, SEM and TEM were applied to observe the pure TiO_2 nanoribbons and the WO_3/TiO_2-25 heterostructure. Before WO_3 growth, pure TiO_2 nanoribbons exhibit typical 1D morphology with a length from several tens to several hundreds of micrometers (Figure 2a). TEM images (Figure 2b–d) of an individual nanoribbon show that the width of the nanoribbons is around 100 nm and the surface of the nanoribbons are flat and clean, which provides comfortable platforms for the nucleation and growth of WO_3 nanosheets. After WO_3 growth, the 1D typical morphology is still maintained, while the surface of the nanoribbons is vertically arrayed with tetragonal nanosheets at a length of 100 nm to 400 nm and thickness of 40 nm (Figure 2e). A similar structure has also been reported in $BiOBr/TiO_2$ systems [43]. A TEM image (Figure 2f) of a single heterostructured nanoribbon demonstrates that 2D nanosheets were grown on both sides of the nanoribbons, leading to the formation of vertically aligned 2D/1D heterostructures with maximum exposure of junctions and the entire surface. High-magnification TEM images (Figure 2g,h) show very clear crystalline planes with a d-spacing of 0.347 nm, which closely matched the inter-planar spacing of the (111) facets of WO_3 [45], further confirming the successful formation of WO_3/TiO_2 heterojunctions. Figure 2i–l show HAADF-STEM-EDS mapping images of a typical nanoribbon arrayed with sheet structures. Ti and O elements can be found in the nanoribbons, while W and O elements existed in the nanosheets, which directly proves that the nanoribbons were TiO_2 and the nanosheets arrayed on nanoribbons were WO_3. These results visually verified that the flat TiO_2 nanoribbons served as the backbones and the WO_3 nanosheets with a tetragonal profile were arrayed on them, leading to the formation of typical vertically aligned 2D on 1D heterostructures. This unique 2D array on 1D structures widely exist in nature and their advancement in light absorption and conversion have been proved in nature for millions of years. For example, trees adopt a leaf array on branch structures to obtain enough access to maximum light. Stegosaurs possessed bony plates along their backs to absorb more heat from the sun to warm their blood on cool days. Therefore, 2D WO_3/1D TiO_2 heterostructures are anticipated to be highly desired for the design of advanced photocatalytic systems.

The growth density of WO_3 nanosheets can be regulated simply via control over the precursor concentration. When the concentration of the precursors was decreased to 5 mM, few nanosheets could be observed on the surfaces of the nanoribbons (Figure S1a). At 10 mM precursor concentration, only a small amount of nanoribbons was decorated with WO_3 nanosheets (Figure S1b). On increasing the precursor concentration to 50 mM, a much higher density of nanosheets could be grown on the nanoribbon backbones, and some isolate nanosheets (Figure S1c) were found between nanoribbons. Further increasing the precursor concentration to 75 mM would lead to a large amount of free nanosheets (Figure S1d), indicating the formation of the mixture of WO_3/TiO_2 heterojunctions and WO_3 nanosheets (Figure S2).

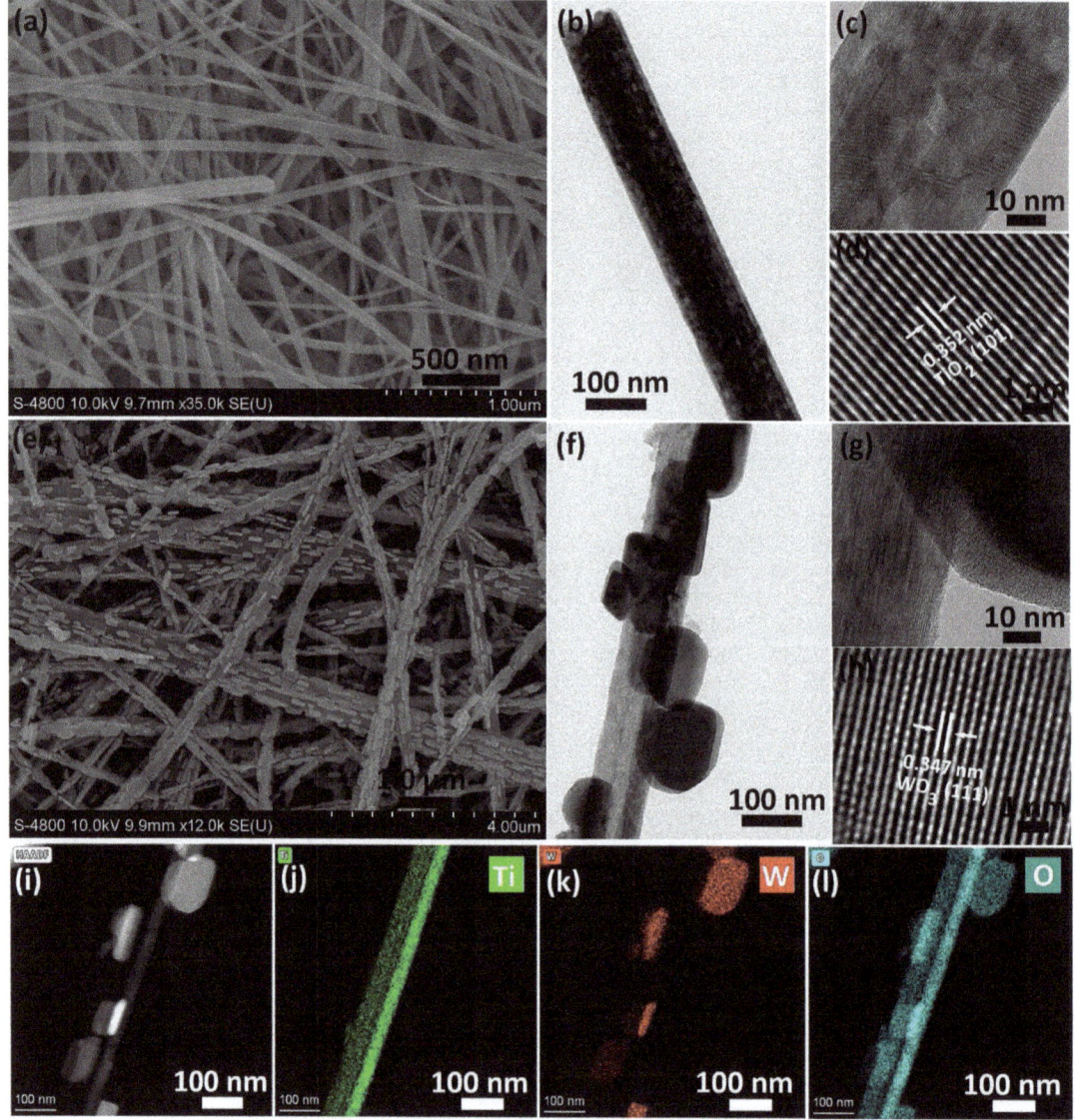

Figure 2. (a) SEM image of TiO$_2$ nanoribbons. (b–d) TEM images of TiO$_2$ nanoribbons. (e) SEM image of WO$_3$/TiO$_2$-25 heterostructure. (f–h) TEM images of WO$_3$/TiO$_2$-25 heterostructure. (i–l) HAADF-STEM micrograph of WO$_3$/TiO$_2$-25 heterostructure and the corresponding EDS mapping of Ti, W, and O.

2.4. XPS Analysis

The element composition and valence of WO$_3$/TiO$_2$-25 composite were analyzed by X-ray photoelectron spectroscopy (XPS). Figure 3a shows the Ti 2p spectra of pure TiO$_2$ nanoribbons and WO$_3$/TiO$_2$-25 heterojunctions. The pure TiO$_2$ nanoribbons exhibit two XPS peaks at 458.6 eV and 464.3 eV, which correspond to Ti 2p$_{1/2}$ and Ti 2p$_{3/2}$, respectively. The energy gap between Ti 2p$_{1/2}$ and Ti 2p$_{3/2}$ peaks is 5.7 eV, revealing the oxidation state of Ti in TiO$_2$ nanoribbons is +4 [36,47], which is also consistent with the XRD result. After the growth of WO_3 nanosheets on TiO_2, both the Ti $2p_{1/2}$ and Ti $2p_{3/2}$ peaks exhibit obvious red shifts from 458.6 to 459.1 eV and from 464.3 to 464.8 eV, respectively. This result suggests the conversion of Ti-O-Ti bonds to Ti-O-W bonds, which confirms the strong interaction between WO_3 and TiO_2 in the WO_3/TiO_2-25 heterojunctions [34,36]. Furthermore, the +4 oxidation state of Ti is unchanged after the formation of WO_3/TiO_2-25 heterojunctions as the energy gap between the peaks at 459.1 and 464.8 eV remains 5.7 eV. The W 4f XPS spectra of pure WO_3 nanosheets and WO_3/TiO_2-25 heterojunctions are shown in Figure 3b. The two XPS peaks of pure WO_3 nanosheets at 35.7 eV and 37.8 eV are attributed to W $4f_{7/2}$ and W $4f_{5/2}$, respectively. The energy gap between W $4f_{7/2}$ and W $4f_{5/2}$ peaks is 2.1 eV, which suggests that the W element existed in the form of W^{6+} in WO_3 nanosheets [34,36]. For WO_3/TiO_2-25 heterojunctions, the binding energies of W $4f_{7/2}$ and W $4f_{5/2}$ peaks were both shifted to low energy by 0.1 eV, while the energy gap between these two peaks is unchanged, indicating the W^{6+} oxidation state in WO_3/TiO_2-25 heterojunctions [34,36,48], which is in good agreement with the XRD data. The high-resolution O 1s XPS spectrum of the samples is shown in Figure 3c,d. It can be seen that the Ti-O bond peak in TiO_2 nanoribbons is located at 530.0 eV (Figure 3c), while the W-O bond peak in WO_3 nanosheets is centered at 530.5 eV (Figure 3d). For WO_3/TiO_2-25 heterojunctions, the binding energy at 530.4 eV corresponded to the lattice oxygen of Ti^{4+}-O or W^{6+}-O, indicating that W-O and Ti-O shared the orbital O 1s in the W-O-Ti bond [32,43,44]. The peak exhibited a 0.4 eV up-shift from Ti-O bonds (from 530.0 to 530.4 eV) and 0.1 eV down-shift from W-O bonds (from 530.5 to 530.4 eV). The above XPS results show that the binding energy of Ti 2p shifted to high energy, while the binding energy of W 4f and O 1s shifted to low energy after the formation of WO_3/TiO_2-25 heterojunctions, confirming the electron density decrease on TiO_2 and electron density increase on WO_3, respectively [36,49,50]. The electron density change indicates that the photo-generated carriers have been successfully transferred between WO_3 and TiO_2 in the composite, which further proved the formation of WO_3/TiO_2-25 heterojunction structures [34,36].

2.5. Optical Analysis

The UV–vis diffuse reflection spectra of TiO_2, WO_3, and WO_3/TiO_2 heterojunctions were recorded to understand the light absorption property and are shown in Figure 4a. The TiO_2 nanoribbons show strong absorbance in the ultraviolet region, and the optical absorption edge was found to be around 400 nm owing to the large band gap of anatase TiO_2 [21,22]. After the combination of WO_3, the optical absorption edge exhibits a 130 nm red shift compared to pure TiO_2 nanoribbons. Thus, the hybrid WO_3/TiO_2 heterostructures can utilize a larger fraction of the solar spectrum for photocatalytic reactions. In order to quantitatively determine their band gaps, $(\alpha h \nu)^{1/2}$ vs. photon energy ($h\nu$) are generated from the diffuse reflectance spectra (Figure 4b). Therein, α is the Kubelka–Munk function of the diffuse reflectance spectra ($\alpha = (1 - R)^2/2R$, and R is the reflectance). The apparent band gaps of pristine TiO_2 and WO_3 were calculated to be 3.12 eV and 2.26 eV, respectively, while the estimated band gap of WO_3/TiO_2 heterostructures was calculated to be 2.30 eV. The band gap reduction in TiO_2 nanoribbons after WO_3 sheet growth suggests the strong interaction between TiO_2 and WO_3 in WO_3/TiO_2 heterostructures, which can effectively hinder the recombination of e-h+ through the WO_3/TiO_2 heterojunction [27,29,35].

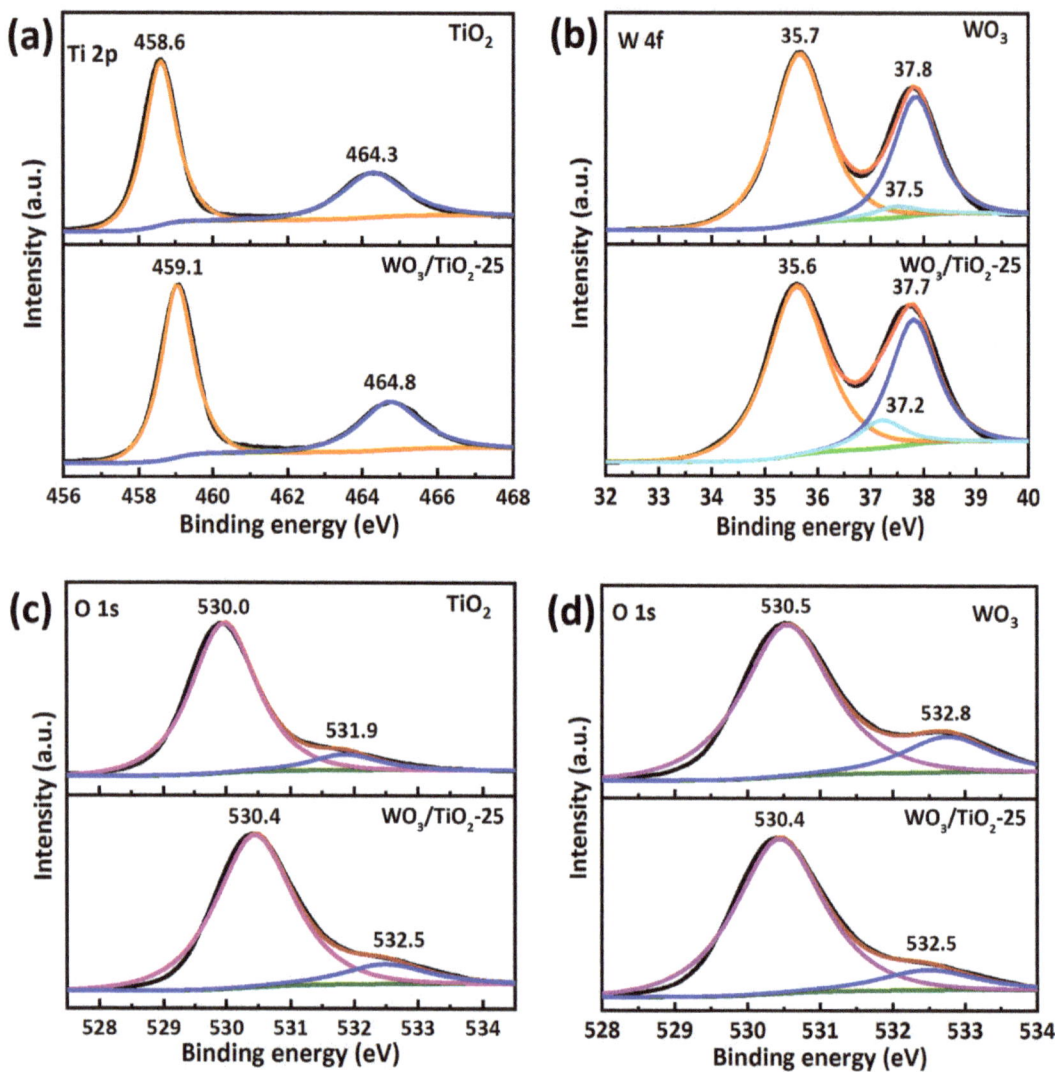

Figure 3. XPS spectra: (a) Ti 2p spectra of TiO_2 and WO_3/TiO_2-25 heterojunctions. The orange and blue curves are fitting curves corresponding to the Ti $2p_{1/2}$ and Ti $2p_{3/2}$, respectively. The black line is the raw data. (b) W 4f spectra of WO_3 and WO_3/TiO_2-25 heterojunctions. The orange, cyan, blue and green curves are fitting curves corresponding to the W $4f_{7/2}$ and W $4f_{5/2}$, respectively. The black represent the raw data and the red line is the fitting line. (c) O 1s spectra of TiO_2 and WO_3/TiO_2-25 heterojunctions. The magenta, blue and olive curves are fitting curves corresponding to the Ti-O bond. The black represent the raw data and the red line is the fitting line. (d) O 1s spectra of WO_3 and WO_3/TiO_2-25 heterojunctions. The magenta, blue and olive curves are fitting curves corresponding to the W-O bond. The black represent the raw data and the red line is the fitting line.

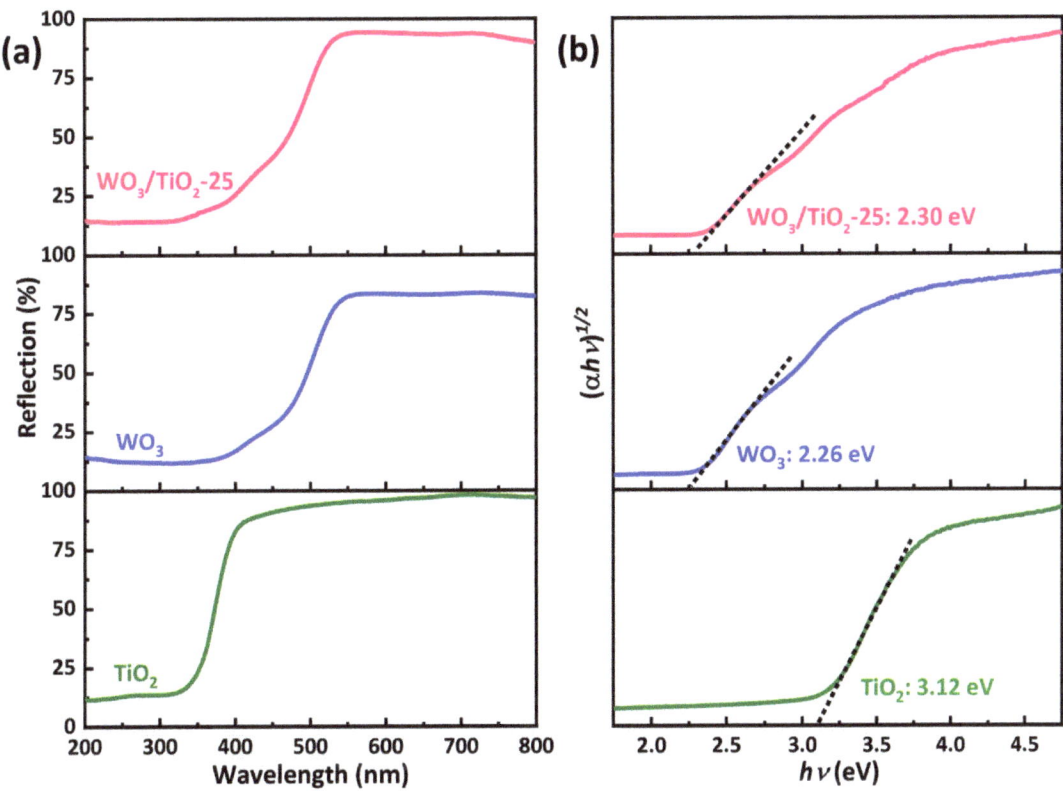

Figure 4. (a) Diffuse reflectance spectra for the TiO$_2$ nanoribbons, WO$_3$ nanosheets, and WO$_3$/TiO$_2$-25 heterojunction. (b) Plot of $(\alpha h\nu)^{1/2}$ as a function of photon energy (hν).

2.6. Photoluminescence Analysis

Photoluminescence measurements were further undertaken to unveil the charge recombination behavior of TiO$_2$ and WO$_3$/TiO$_2$-25 heterostructures. It is widely known that fluorescence emission signals are derived mainly from the recombination of photogenerated electron–hole pairs, and a lower fluorescence intensity signifies a higher photocatalytic efficiency [12,31,43]. Figure 5a shows the fluorescence spectra of TiO$_2$ and WO$_3$/TiO$_2$-25 heterostructures in the wavelength range of 475–725 nm. It can be seen clearly that the emission peaks of TiO$_2$ and WO$_3$/TiO$_2$-25 heterostructures are in similar shapes, and the fluorescence emission intensity of TiO$_2$ shows an obvious reduction after modification with WO$_3$ nanosheets, indicating that growth of WO$_3$ nanosheets on TiO$_2$ nanoribbons can effectively suppress recombination of electron–hole pairs and increase the lifetime of the charge carriers, which provides solid basis for photocatalytic activity improvement. In order to further confirm the enhanced photo-generated electron–hole separation efficiency, photocurrent responses of all samples have been recorded and shown in Figure S4. The photocurrent of all the WO$_3$/TiO$_2$ composite samples is significantly higher than that of the pure TiO$_2$ and WO$_3$, further confirming the higher charge separation induced by WO$_3$ nanosheet growth, which is consistent with photoluminescence spectra. It is worth noting that WO$_3$/TiO$_2$-25 exhibits the highest photocurrent density of 7.9 µA·cm^{-2}, which is roughly 7.1, 6.6 times higher than that of pure TiO$_2$ nanowires and pure WO$_3$ nanosheets. The best photocurrent response manifests that the WO$_3$/TiO$_2$-25 heterostructure composite possesses the highest photoelectrons–holes transfer efficiency and thus should be a promising candidate as a high-performance photocatalyst.

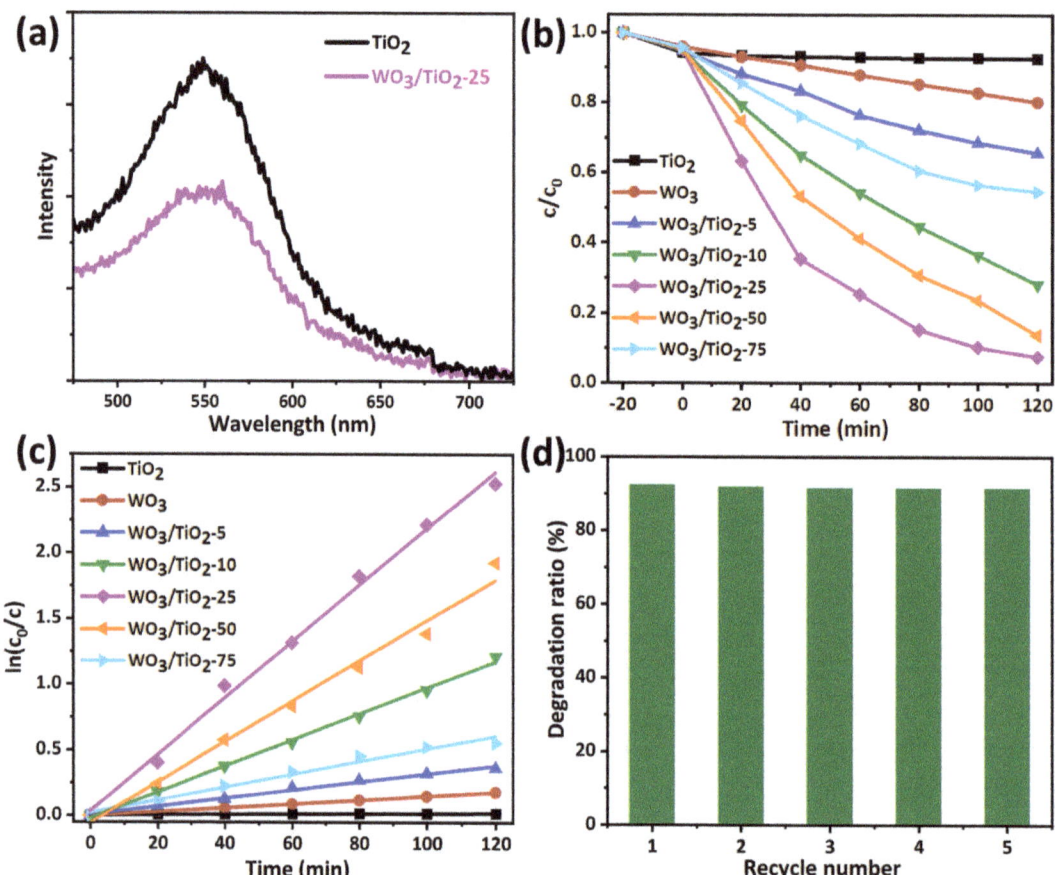

Figure 5. (**a**) Fluorescence spectrum of TiO$_2$ nanoribbons and WO$_3$/TiO$_2$−25 heterojunction. (**b**) Photocatalytic degradation of RhB over various photocatalysts under sunlight irradiation, where C$_0$ is the initial concentration of the pollutant, and C is the concentration of the pollutant at irradiation time t. (**c**) First-order reaction kinetic curves for the RhB degradation reaction over various photocatalysts. (**d**) Curve of the degradation ratio of RhB versus reuse times of WO$_3$/TiO$_2$-25.

2.7. Photocatalytic Performance

UV–vis diffuse reflection, photoluminescence, and photocurrent measurements demonstrate that construction of 2D/1D heterojunctions through vertical growth WO$_3$ nanosheets on TiO$_2$ nanoribbons not only broaden the light utilization region but also effectively suppresses electron–hole pair recombination; thus, theoretically, WO$_3$/TiO$_2$ heterostructures should possess higher photocatalytic activity than pure TiO$_2$ and WO$_3$ [11,12,31]. Figure 5b shows the photocatalytic degradation behavior of RhB over various photocatalysts under visible light irradiation. In the dark condition, all the samples exhibit similar RhB removal efficiency, owing to the similar BET surface area (Figure S3). All five kinds of WO$_3$/TiO$_2$ heterostructures prepared under different precursor concentrations show higher photocatalytic activity than pure WO$_3$ nanosheets and TiO$_2$ nanoribbons. Of the five heterojunctions, the WO$_3$/TiO$_2$-25 system shows the highest photocatalytic activity, which can remove 92.8% of

RhB pollutants in 120 min. The photocatalytic degradation ration of RhB over WO_3/TiO_2-50 and WO_3/TiO_2-10 heterojunctions is 86.3% and 71.7%, respectively, indicating that too much higher and lower growth density of WO_3 nanosheets causes photocatalytic activity decrease. The reason is that over-dense WO_3 nanosheets result in lower exposure of junctions, while sparse growth density leads to few junctions. Figure 5c shows the first-order reaction kinetic curves for the RhB degradation reaction over various photocatalysts. It can be seen that the slope of WO_3/TiO_2 heterostructures is bigger than that of pure WO_3 nanosheets and TiO_2 nanoribbons, suggesting the enhanced photocatalytic rate constant after growth of WO_3 nanosheets on TiO_2 nanoribbons. The enhanced photocatalytic activity of the WO_3/TiO_2 heterostructures is attributed to the synergetic effects of the improved visible light utilization and enhanced electron–hole separation, which have been proven by UV–vis diffuse reflection and photoluminescence spectra. Over five consecutive cycles, there was no notable change for the apparent photocatalytic degradation ratio, indicating the excellent durability of WO_3/TiO_2 heterostructures (Figure 5d).

The effect of WO_3/TiO_2-25 catalyst loading on the photocatalytic degradation ratio of RhB is shown in Figure 6a. The photodegradation ratio increased with the increase in catalyst loading in the range of 5–20 mg and reached a maximum when 20 mg of the catalyst was used. Further increase in catalyst loading from 20 mg to 40 mg led to a decline in the photodegradation ratio from 92.8% to 82.6%. The observation can be explained as follows. Increasing catalyst loading from 5 mg to 20 mg would provide more active photodegradation sites, thus leading to an enhancement of the degradation ratio. When the catalyst was increased to 40 mg, the photodegradation reaction system became thick colloid, reducing the light penetration depth, which means only the catalysts in the solution surface layer can be photo activated as degradation sites, while the catalyst in the solution bottom layer make little contribution to RhB photodegradation, thus resulting in a decline in the degradation ratio [51]. Another reason might be due to catalyst aggregation resulting from high catalyst concentration, which would lead to a decrease in total surface area available for degradation, reduction in site density for surface holes and electrons, and an increase in the diffusion path length [51].

In order to reveal the photocatalytic mechanism, several trapping agents, including ascorbic acid (AC), ammonium oxalate (AO), isopropanol (IPA), and hydrogen peroxide (H_2O_2), were applied as scavengers for probing the active radicals in photocatalytic degradation. The scavengers AC, AO, IPA, and H_2O_2 functioned as trapping agents for superoxide anions (O_2^-), holes (h^+), hydroxyl radicals (OH), as well as electrons (e^-), respectively [52]. As shown in Figure 6b, various sacrificial agents have a great impact on the RhB degradation efficiency. After adding AO, AC, and IPA, the RhB photocatalytic degradation efficiency decreased obviously compared with the blank. The photodegradation ratios of RhB corresponding to AC, AO, and IPA were 40.2%, 20.7%, and 32.2%, respectively, indicating that h^+ and OH constituted the major active species for RhB photodegradation [52]. The photodegradation ratio of RhB reached nearly 100% in the presence of H_2O_2. The addition of H_2O_2 resulted in the consumption of e^- in the conduction band (CB) and the enhancement of photoinduced e^-/h^+ pair separation. Furthermore, H_2O_2 was reduced by e^- to generate OH, as indicated by $H_2O_2 + e^- \rightarrow OH + OH^-$ [53]. The generation of extra ·OH further promoted photocatalytic degradation. Therefore, active species detection reveals that photoinduced h^+ and ·OH were the main active species in the RhB photodegradation process.

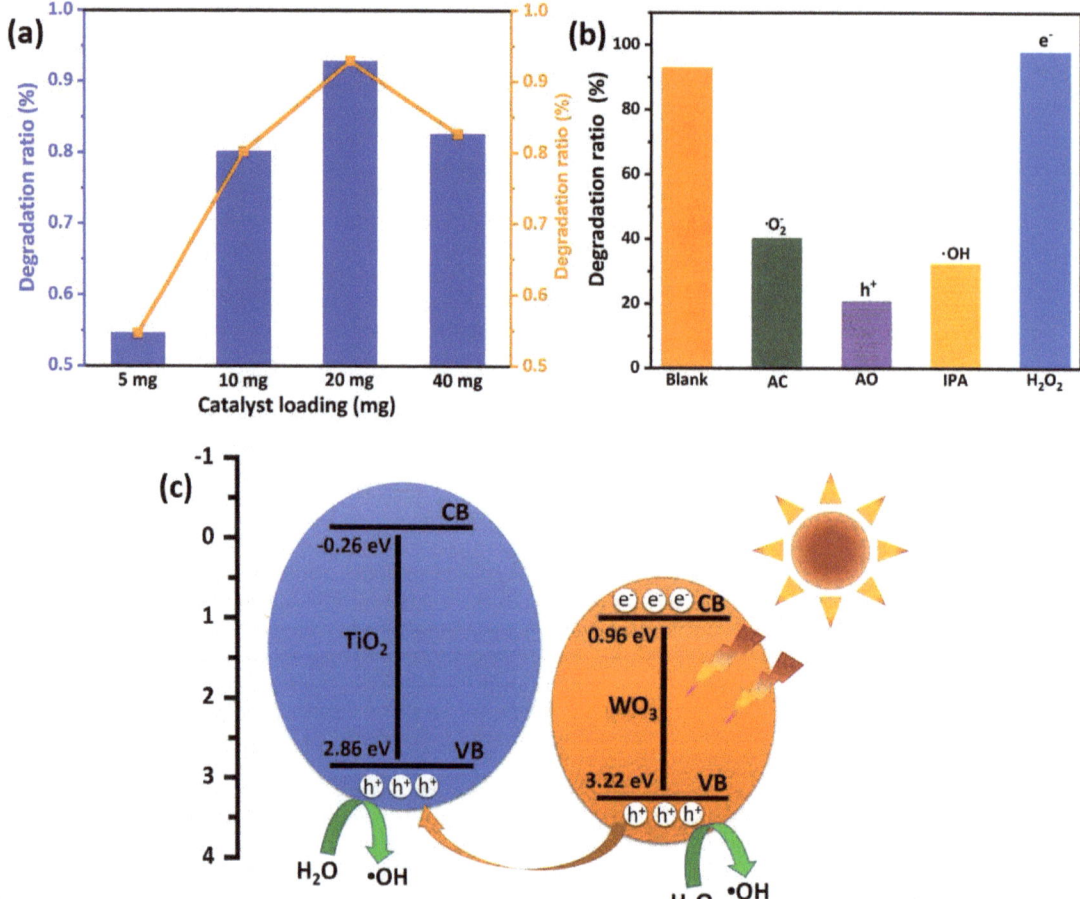

Figure 6. (a) The effect of WO_3/TiO_2-25 catalyst loading on RhB photodegradation ratio. (b) Trapping test for active species during RhB photodegradation with the WO_3/TiO_2-25 photocatalyst. (c) Proposed photocatalytic mechanism of WO_3/TiO_2 photocatalyst.

The conduction band (CB) and the valence band (VB) edge level potentials of TiO_2 and WO_3 were measured by applying the following formula [54]: $E_{VB} = X - E_c + 0.5E_g$, $E_{CB} = E_{VB} - E_g$. Where E_{VB} is the VB edge level potential and E_{CB} is the CB edge level potential, and X is the absolute electronegativity of the semiconductor material (electronegativity of TiO_2 and WO_3 is 5.8 eV and 6.59 eV, respectively) [54]. E_c is the energy of free electrons (4.5 eV vs. NHE), and E_g is the band gap of TiO_2 and WO_3. The CB and VB edge level potentials of TiO_2 were determined as −0.26 eV and 2.86 eV, respectively, while the CB and VB edge level potentials of WO_3 were calculated to be 0.96 eV and 3.22 eV, respectively. The photocatalytic degradation mechanism of the WO_3/TiO_2 heterojunctions is schematically demonstrated in Figure 6c. When WO_3/TiO_2 heterojunction photocatalysts are exposed to visible light irradiation, the photo-generated electrons in the WO_3 valence band will be excited to the conduction band, generating holes in the valence band. However, it cannot occur for TiO_2 due to the band gap energy of TiO_2 (3.12 eV) being higher than the visible photon energy. The valence band of WO_3 is higher than that of TiO_2 (3.22 eV vs. 2.86 eV), so the photo-generated holes can be easily transferred to the valence band of TiO_2 through the 2D/1D junction interface [31–36]. The photo-generated electrons on the WO_3 conduction band cannot react with O_2 to produce $•O_2^-$ radicals because the conduction edge potential of WO_3 was higher than the redox potential of O_2 to $•O_2^-$ (0.96 eV vs. −0.33 eV). Thus, the radicals of $•O_2^-$ can hardly participate in the photodegradation reaction, which is in good agreement with the scavenger experiment results. Furthermore, photo-generated holes (h^+) can react with adsorbed H_2O or OH^- to generate $•OH$ radicals, which are the main active species for organic pollutant photo-degradation confirmed by the scavenger experiment. The above results showed TiO_2 acted as a photo-generated hole acceptor in the heterojunction, effectively reducing the recombination of photo-generated carriers and enhancing photocatalytic performance, and h^+ and $•OH$ are responsible for the photodegradation of RhB.

3. Materials and Methods

3.1. Preparation of WO_3/TiO_2

TiO_2 nanoribbons were synthesized according to our previous report [55,56]. Then, WO_3 nanosheets were vertically grown on TiO_2 nanoribbons using a hydrothermal process. Typically, 0.1 g of TiO_2 nanoribbons were dispersed in 30 mL of distilled water dissolved by 0.25 g of $Na_2WO_4•2H_2O$ (25 mM), 2 mL of 3 mol/L HCl aqueous solution, and 0.3 g of citric acid. Then, the resulting suspension was transferred into a 50 mL Teflon-lined stainless steel autoclave and kept at 120 °C for 24 h. After the hydrothermal reaction was completed, the resulting products were collected by centrifugation, washed several times with deionized water, and dried at 60 °C overnight. For comparison, WO_3/TiO_2 composites with different ratios were prepared in a similar manner. The samples were denoted as WO_3/TiO_2-5, WO_3/TiO_2-10, WO_3/TiO_2-25, and WO_3/TiO_2-50, WO_3/TiO_2-75 in which the number represents the concentration of $Na_2WO_4·2H_2O$.

3.2. Material Characterization

The morphology and structure of the photocatalysts were examined by a HITACHIS-4800 field emission scanning electron microscope (SEM, Hitachi, Tokyo, Japan). Transmission electron microscope (TEM, Thermo Fisher Scientific, Waltham, MA, USA) and high-resolution TEM (HRTEM, Thermo Fisher Scientific, Waltham, MA, USA) images were taken with an FEI Tecnai G20 electron microscope. For TEM samples, the photocatalyst was dispersed in ethanol using ultrasonic treatment and the TEM samples were prepared by depositing a drop of diluted suspension on a carbon-coated copper grid. UV-vis-near-infrared (NIR, Agilent, Polo Alto, CA, USA) reflection spectra were recorded on an Agilent Australia Carry-5000 spectrophotometer. The UV–Visible absorption spectra were recorded on a Shimadzu UV-2550 spectrophotometer. The X-ray photoelectron spectroscopy (XPS, Thermo Fisher Scientific, Waltham, MA, USA) analysis was carried out on an Escalab 250Xi spectrometer with monochromatic Al Kα radiation (hν = 1486.7 eV). Phase identification of the photocatalysts was measured by a Shimadzu powder difractometer with Cu-Kα radiation. The solid-state fluorescence spectra were characterized using a FLS980 fluorescence spectrophotometer.

3.3. Photodegradation and Photocurrent Test

Rhodamine B (RhB) solution with a concentration of 9 mg/L was used as the model waste water. An amount of 20 mg of samples were dispersed in 100 mL of RhB solution. The solution was kept in the dark for 20 min and then exposed to visible light irradiation. A 500 W halogen lamp was used as the visible light source, and the average light intensity was 60 mW•cm^{-2}. The distance between the lamp and the solution was 10 cm. During the measurement process, a certain amount of the solution was sampled at certain time intervals and centrifuged (8000 rpm, 3 min) to remove the photocatalyst particles. The supernatant was analyzed to measure the concentration of RhB by using a Shimadzu UV-2550 UV–vis spectrometer (peak center: 554 nm). To analyze the photocatalytic degradation mechanism, ascorbic acid (AC), ammonium oxalate (AO), isopropanol (IPA), and hydrogen peroxide (H_2O_2) reagents were selected as scavengers for anions (O_2^-), holes (h^+), hydroxyl radicals ($\cdot OH$), as well as electrons (e^-), respectively. Next, 5 mg of AC, 5 mg of AO, 0.2 mL of IPA, and 0.2 mL of H_2O_2 were added into the above photodegradation system while keeping other conditions unchanged.

Photocurrent studies were performed on a CHI 660D electrochemical workstation using a three-electrode configuration where ITO electrodes were deposited with the samples as a working electrode, Pt as a counter electrode, and a saturated calomel electrode as reference. The electrolyte was 0.35 M/0.25 M Na_2S-Na_2SO_3 aqueous solution. For the fabrication of the working electrode, 0.25 g of the sample was grinded with 0.06 g polyethylene glycol (PEG, molecular weight: 20,000) and 0.5 mL ethanol to make a slurry. The slurry was spread onto a 1 cm × 4 cm ITO glass using the doctor blade technique and then allowed to air-dry. A 500 W halogen lamp was used as the visible light source and the average light intensity was 60 mW•cm^{-2}.

4. Conclusions

In conclusion, a novel WO_3/TiO_2 2D/1D heterojunction photocatalyst was constructed via the introduction of TiO_2 nanoribbons into a WO_3 nanosheet growth system. Two-dimensional WO_3 nanosheets were vertically arrayed on the surface of TiO_2 nanoribbons, and the growth density could be easily controlled by adjusting the concentration of the precursors. The UV–vis diffuse reflection result reveals that vertical growth of WO_3 nanosheets on TiO_2 nanoribbons successfully decreases the band gap energy of TiO_2 from 3.12 to 2.30 eV and broadens the photoresponse range from the UV region to the visible light region. Furthermore, the photoluminescence measurement demonstrates that construction of WO_3/TiO_2 heterojunctions significantly reduces electron–hole pair recombination. Consequently, WO_3/TiO_2 heterostructures with different WO_3 nanosheet growth density show higher photocatalytic activity than pure WO_3 nanosheets and TiO_2 nanoribbons. WO_3/TiO_2-25 possesses the highest photocatalytic activity, which can remove 92.8% of RhB pollutants in 120 min and maintain high removal efficiency after five consecutive cycles. The present study demonstrates that the prepared WO_3/TiO_2 2D/1D heterostructures are promising materials for photocatalytic removal of organic pollutants to purify water.

Supplementary Materials: The following supporting information can be downloaded at: https://www.mdpi.com/article/10.3390/catal13030556/s1, Figure S1: SEM images of WO_3/TiO_2 heterostructures prepared under different precursor concentrations: (a) 5 mM, (b) 10 mM, (c) 50 mM, (d) 75 mM. Figure S2: SEM image of pure WO_3 nanosheets. Figure S3: Nitrogen adsorptione-desorption isotherm: (a) TiO_2 nanoribbons; (b) WO_3 nanosheets; (c) WO_3/TiO_2 heterostructures and Figure S4. Photocurrent responses of all samples.

Author Contributions: Conceptualization, L.Z.; methodology, L.Z., L.W., K.X. and H.T.; validation, L.W. and H.T.; formal analysis, L.W., K.X. and H.T.; investigation, L.Z.; data curation, L.W., K.X. and L.W.; writing—original draft preparation, L.Z.; writing—review and editing, L.Z. and L.W.; visualization, L.W. and K.X.; supervision, L.Z.; project administration, L.Z.; funding acquisition, L.Z. All authors have read and agreed to the published version of the manuscript.

Funding: This research was funded by Jiaxing City Public Welfare Research Project of China, grant number 2021AY10059 and Innovation Jiaxing Excellent Talents Support Program, grant number 84321005.

Data Availability Statement: The data presented in this study are available on request from the corresponding authors.

Conflicts of Interest: The authors declare no conflict of interest.

References

1. McDonald, R.I.; Green, P.; Balk, D.; Montgomery, M. Urban growth, climate change, and freshwater availability. *Proc. Natl. Acad. Sci. USA* **2011**, *108*, 6312–6317. [CrossRef] [PubMed]
2. Oki, T.; Kanae, S. Global hydrological cycles and world water resources. *Science* **2006**, *313*, 1068–1072. [CrossRef] [PubMed]
3. Vörösmarty, C.J.; McIntyre, P.B.; Gessner, M.O.; Dudgeon, D.; Prusevich, A.; Green, P.; Glidden, S.; Bunn, S.E.; Sullivan, C.A.; Liermann, C.R.; et al. Global threats to human water security and river biodiversity. *Nature* **2010**, *467*, 555–561. [CrossRef] [PubMed]
4. Gadgil, A. Drinking Water in Developing Countries. *Annu. Rev. Energy Environ.* **1998**, *23*, 253–286. [CrossRef]
5. Water Scarcity Threats. Available online: https://www.worldwildlife.org/threats/water-scarcity (accessed on 12 January 2023).
6. Shannon, M.A.; Bohn, P.W.; Elimelech, M.; Georgiadis, J.G.; Mariñas, B.J.; Mayes, A.M. Science and technology for water purification in the coming decades. *Nature* **2008**, *452*, 301–310. [CrossRef]
7. Macedonio, F.; Drioli, E.; Gusev, A.A.; Bardow, A.; Semiatf, R.; Kuriharag, M. Efficient technologies for worldwide clean water supply. *Chem. Eng. Process.* **2012**, *51*, 2–17. [CrossRef]
8. Tee, G.T.; Gok, X.Y.; Yong, W.F. Adsorption of pollutants in wastewater via biosorbents, nanoparticles and magnetic biosorbents: A review. *Environ. Res.* **2022**, *212*, 113248. [CrossRef]
9. Ouachtak, H.; Akhouairi, S.; Haounati, R.; Addi, A.A.; Jada, A.; Taha, M.L.; Douch, J. 3,4-Dihydroxybenzoic acid removal from water by goethite modified natural sand column fixed-bed: Experimental study and mathematical modeling. *Desalin. Water Treat.* **2020**, *194*, 439–449. [CrossRef]
10. Pendergast, M.T.M.; Eric, M.V.; Hoek, E.M.V. A review of water treatment membrane nanotechnologies. *Energy Environ. Sci.* **2011**, *4*, 1946–1971. [CrossRef]
11. Andrew, M.; Davies, R.H.; Worsley, D. Water purification by semiconductor photocatalysis. *Chem. Soc. Rev.* **1993**, *22*, 417–425.
12. Parvulescu, V.I.; Epron, F.; Garcia, H.; Granger, P. Recent progress and prospects in catalytic water treatment. *Chem. Rev.* **2021**, *122*, 2981–3121. [CrossRef] [PubMed]
13. Schneider, J.; Matsuoka, M.; Takeuchi, M.; Zhang, J.; Horiuchi, Y.; Anpo, M.; Bahnemann, D.W. Understanding TiO$_2$ Photocatalysis: Mechanisms and Materials. *Chem. Rev.* **2014**, *114*, 9919–9986. [CrossRef] [PubMed]
14. Kumar, A.; Choudhary, P.; Kumar, A.; Camargo, P.H.; Krishnan, V. Recent advances in plasmonic photocatalysis based on TiO$_2$ and noble metal nanoparticles for energy conversion, environmental remediation, and organic synthesis. *Small* **2022**, *18*, 2101638. [CrossRef] [PubMed]
15. Weng, B.; Lu, K.Q.; Tang, Z.C.; Chen, H.M.; Xu, Y.J. Stabilizing ultrasmall Au clusters for enhanced photoredox catalysis. *Nat. Commun.* **2018**, *9*, 1543. [CrossRef] [PubMed]
16. Pan, X.; Chen, W.J.; Cai, H.; Li, H.; Sun, X.J.; Weng, B.; Yi, Z. A redox-active support for the synthesis of Au@SnO$_2$ core-shell nanostructure and SnO$_2$ quantum dots with efficient photoactivities. *RSC Adv.* **2020**, *10*, 33955–33961. [CrossRef] [PubMed]
17. Haounati, R.; Alakhras, F.; Ouachtak, H.; Saleh, T.A.; Al-Mazaideh, G.; Alhajri, E.; Jada, A.; Hafid, N.; Addi, A.A. Synthesized of Zeolite@Ag$_2$O Nanocomposite as Superb Stability Photocatalysis toward Hazardous Rhodamine B Dye from Water. *Arab. J. Sci. Eng.* **2023**, *48*, 169–179. [CrossRef]
18. Nawaz, A.; Goudarzi, S.; Asghari, M.A.; Pichiah, S.; Selopal, G.S.; Rosei, F.; Wang, Z.M.; Zarrin, H. Review of Hybrid 1D/2D Photocatalysts for Light-Harvesting Applications. *ACS Appl. Nano Mater.* **2021**, *4*, 11323–11352. [CrossRef]
19. Hou, H.; Zhang, X. Rational design of 1D/2D heterostructured photocatalyst for energy and environmental applications. *Chem. Eng. J.* **2020**, *395*, 125030. [CrossRef]
20. Fujishima, A.; Honda, K. Electrochemical photolysis of water at a semiconductor electrode. *Nature* **1972**, *238*, 37–38. [CrossRef]
21. Ren, Y.; Dong, Y.; Feng, Y.; Xu, J. Compositing two-dimensional materials with TiO$_2$ for photocatalysis. *Catalysts* **2018**, *8*, 590. [CrossRef]
22. Guo, Q.; Zhou, C.; Ma, Z.; Yang, X. Fundamentals of TiO$_2$ photocatalysis: Concepts, mechanisms, and challenges. *Adv. Mater.* **2019**, *31*, 1901997. [CrossRef] [PubMed]
23. Feng, T.; Feng, G.S.; Yan, L. One-dimensional nanostructured TiO$_2$ for photocatalytic degradation of organic pollutants in wastewater. *Int. J. Photoenergy* **2014**, *2014*, 563879. [CrossRef]
24. Ge, M.; Cao, C.; Huang, J. A review of one-dimensional TiO$_2$ nanostructured materials for environmental and energy applications. *J. Mater. Chem. A* **2016**, *4*, 6772–6801. [CrossRef]
25. Szilágyi, I.M.; Fórizs, B.; Rosseler, O.; Szegedi, Á.; Németh, P.; Király, P.; Tárkányi, G.; Vajna, B.; Katalin, V.-J.; László, K. WO$_3$ photocatalysts: Influence of structure and composition. *J. Catal.* **2012**, *294*, 119–127. [CrossRef]
26. Dutta, V.; Sharma, S.; Raizada, P.; Thakurb, V.K.; Parvaz, K.A.A.; Sainie, V.; Asiri, A.M.; Singha, P. An overview on WO$_3$ based photocatalyst for environmental remediation. *J. Environ. Chem. Eng.* **2021**, *9*, 105018. [CrossRef]
27. Wang, F.; Valentin, C.D.; Pacchioni, G. Rational band gap engineering of WO$_3$ photocatalyst for visible light water splitting. *Mater. Sci. ChemCatChem* **2012**, *4*, 476–478. [CrossRef]
28. Liu, D.; Zhang, S.; Wang, J.; Peng, T.; Li, R. Direct Z-scheme 2D/2D photocatalyst based on ultrathin g-C$_3$N$_4$ and WO$_3$ nanosheets for efficient visible-light-driven H$_2$ generation. *ACS Appl. Mater. Interfaces* **2019**, *11*, 27913–27923. [CrossRef]

29. Lei, B.; Cui, W.; Chen, P.; Chen, L.; Li, J.; Dong, F. C-doping induced oxygen-vacancy in WO_3 nanosheets for CO_2 activation and photoreduction. *ACS Catal.* **2022**, *12*, 9670–9678. [CrossRef]
30. Liang, Y.; Yang, Y.; Zou, C.; Xu, K.; Luo, X.; Luo, T.; Li, J.; Yang, Q.; Shi, P.; Yuan, C. 2D ultra-thin WO_3 nanosheets with dominant {002} crystal facets for high-performance xylene sensing and methyl orange photocatalytic degradation. *J. Alloy. Compd.* **2019**, *783*, 848–854. [CrossRef]
31. Lai, C.W. WO_3-TiO_2 Nanocomposite and its applications: A review. *Nano Hybrids Compos.* **2018**, *20*, 1–26. [CrossRef]
32. Zhang, L.; Qin, M.; Yu, W.; Zhang, Q.; Xie, H.; Sun, Z.; Shao, Q.; Guo, X.; Hao, L.; Zheng, Y.; et al. Heterostructured TiO_2/WO_3 nanocomposites for photocatalytic degradation of toluene under visible light. *J. Electrochem. Soc.* **2017**, *164*, H1086–H1090. [CrossRef]
33. Liu, Y.; Xie, C.; Li, J.; Zou, T.; Zeng, D. New insights into the relationship between photocatalytic activity and photocurrent of TiO_2/WO_3 nanocomposite. *Appl. Catal. A* **2012**, *433–434*, 81–87. [CrossRef]
34. Prabhu, S.; Cindrella, L.; Kwon, O.J.; Mohanrajubet, K. Photoelectrochemical and photocatalytic activity of TiO_2-WO_3 heterostructures boosted by mutual interaction. *Mater. Sci. Semicond. Process.* **2018**, *88*, 10–19. [CrossRef]
35. Yang, J.; Zhang, X.; Liu, H.; Wang, C.; Liu, S.; Sun, P.; Wang, L.; Liu, Y. Heterostructured TiO_2/WO_3 porous microspheres: Preparation, characterization and photocatalytic properties. *Catal. Today* **2013**, *201*, 195–202. [CrossRef]
36. Wang, Q.; Zhang, W.; Hu, X.; Xua, L.; Chen, G.; Li, X. Hollow spherical WO_3/TiO_2 heterojunction for enhancing photocatalytic performance in visible-light. *J. Water Process. Eng.* **2021**, *40*, 101943. [CrossRef]
37. Zhang, Y.; Xu, M.; Li, H.; Ge, H.; Bian, Z. The enhanced photoreduction of Cr (VI) to Cr (III) using carbon dots coupled TiO_2 mesocrystals. *Appl. Catal. B* **2018**, *226*, 213–219. [CrossRef]
38. Song, Y.Y.; Gao, Z.D.; Wang, J.H.; Xia, X.H.; Lynch, R. Multistage coloring electrochromic device based on TiO_2 nanotube arrays modified with WO_3 nanoparticles. *Adv. Funct. Mater.* **2011**, *21*, 1941–1946. [CrossRef]
39. Paramasivam, I.; Nah, Y.C.; Das, C.; Shrestha, N.K.; Schmuki, P. WO_3/TiO_2 nanotubes with strongly enhanced photocatalytic activity. *Chem. Eur. J.* **2010**, *16*, 8993–8997. [CrossRef]
40. Xu, T.; Wang, Y.; Zhou, X.; Zheng, X.; Xu, Q.; Chen, Z.; Ren, Y.; Yan, B. Fabrication and assembly of two-dimensional TiO_2/WO_3·H_2O heterostructures with type II band alignment for enhanced photocatalytic performance. *Appl. Surf. Sci.* **2017**, *403*, 564–571. [CrossRef]
41. Lee, J.Y.; Jo, W.K. Heterojunction-based two-dimensional N-doped TiO_2/WO_3 composite architectures for photocatalytic treatment of hazardous organic vapor. *J. Hazard. Mater.* **2016**, *314*, 22–31. [CrossRef]
42. Liu, J.; Yang, L.; Li, C.; Chen, Y.; Zhang, Z. Optimal monolayer WO_3 nanosheets/TiO_2 heterostructure and its photocatalytic performance under solar light. *Chem. Phys. Lett.* **2022**, *804*, 139861. [CrossRef]
43. Mei, Y.; Su, Y.; Li, Z.; Bai, S.; Yuan, M.; Li, L.; Yan, Z.; Wu, J.; Zhu, L.W. BiOBr nanoplates@TiO_2 nanowires/carbon fiber cloth as a functional water transport network for continuous flow water purification. *Dalton Trans.* **2017**, *46*, 347–354. [CrossRef] [PubMed]
44. Chen, J.Z.; Ko, W.Y.; Yen, Y.C.; Chen, P.H.; Lin, K.J. Hydrothermally processed TiO_2 nanowire electrodes with antireflective and electrochromic properties. *ACS Nano* **2012**, *6*, 6633–6639. [CrossRef] [PubMed]
45. Li, F.; Li, C.; Zhu, L.; Guo, W.; Shen, L.; Wen, S.; Ruan, S. Enhanced toluene sensing performance of gold-functionalized WO_3·H_2O nanosheets. *Sens. Actuators B Chem.* **2016**, *223*, 761–767. [CrossRef]
46. Liu, W.; Du, T.; Ru, Q.; Zuo, S.; Cai, Y.; Yao, C. Preparation of graphene/WO_3/TiO_2 composite and its photocathodic protection performance for 304 stainless steel. *Mater. Res. Bull.* **2018**, *102*, 399–405. [CrossRef]
47. Idriss, H.; Barteau, M.A. Characterization of TiO_2 surfaces active for novel organic syntheses. *Catal. Lett.* **1994**, *26*, 123–139. [CrossRef]
48. Chen, S.; Chen, L.; Gao, S.; Ca, G. The preparation of coupled WO_3/TiO_2 photocatalyst by ball milling. *Powder Technol.* **2005**, *160*, 198–202.
49. Li, S.; Wang, P.; Wang, R.; Liu, Y.; Jing, R.; Zi, Z.; Meng, L.; Liu, Y.; Zhang, Q. One-step co-precipitation method to construct black phosphorus nanosheets/ZnO nanohybrid for enhanced visible light photocatalytic activity. *Appl. Surf. Sci.* **2019**, *497*, 143682. [CrossRef]
50. Zhang, Z.; Liu, K.; Feng, Z.; Bao, Y.; Dong, B. Hierarchical sheet-on-sheet $ZnIn_2S_4$/g-C_3N_4 heterostructure with highly efficient photocatalytic H_2 production based on photoinduced interfacial charge transfer. *Sci. Rep.* **2016**, *6*, 19221. [CrossRef]
51. Paul, D.R.; Sharma, R.; Nehra, S.P.; Sharma, A. Effect of calcination temperature, pH and catalyst loading on photodegradation efficiency of urea derived graphitic carbon nitride towards methylene blue dye solution. *RSC Adv.* **2019**, *9*, 15381–15391. [CrossRef]
52. Gao, J.; Hu, J.; Wang, Y.; Zheng, L.; He, G.; Deng, J.; Liu, M.; Li, Y.; Liu, Y.; Zhou, H. Fabrication of Z-Scheme TiO_2/SnS_2/MoS_2 ternary heterojunction arrays for enhanced photocatalytic and photoelectrochemical performance under visible light. *J. Solid State Chem.* **2022**, *307*, 122737. [CrossRef]
53. Zhu, B.; Xia, P.; Li, Y.; Ho, W.; Yu, J. Fabrication and photocatalytic activity enhanced mechanism of direct Z-scheme g-C_3N_4/Ag_2WO_4 photocatalyst. *Appl. Surf. Sci.* **2017**, *391*, 175–183. [CrossRef]

54. Vembuli, T.; Thiripuranthagan, S.; Sureshkumar, T.; Erusappan, E.; Kumaravel, S.; Kasinathan, M.; Sivakumar, A. Degradation of Harmful Organics Using Visible Light Driven N-TiO_2/rGO Nanocomposite. *J. Nanosci. Nanotechnol.* **2021**, *21*, 3081–3091. [CrossRef] [PubMed]
55. Zhu, L.; Gu, L.; Zhou, Y.; Cao, S.; Cao, X. Direct production of a free-standing titanate and titania nanofiber membrane with selective permeability and cleaning performance. *J. Mater. Chem.* **2011**, *21*, 12503–12510. [CrossRef]
56. Zhou, Y.; Zhu, L.; Gu, L.; Cao, S.; Wang, L.; Cao, X. Guided growth and alignment of millimetre-long titanate nanofibers in solution. *J. Mater. Chem.* **2012**, *22*, 16890–16896. [CrossRef]

Disclaimer/Publisher's Note: The statements, opinions and data contained in all publications are solely those of the individual author(s) and contributor(s) and not of MDPI and/or the editor(s). MDPI and/or the editor(s) disclaim responsibility for any injury to people or property resulting from any ideas, methods, instructions or products referred to in the content.

Article

Fabrication of Porous Hydrophilic CN/PANI Heterojunction Film for High-Efficiency Photocatalytic H₂ Evolution

Xiaohang Yang [1,*], Yulin Zhang [2], Jiayuan Deng [2], Xuyang Huo [1], Yanling Wang [1] and Ruokun Jia [2,*]

1. Department of Biomedical Engineering, Jilin Medical University, Jilin 132013, China
2. College of Chemical Engineering, Northeast Electric Power University, Jilin 132012, China
* Correspondence: xhyang16@jlu.edu.com (X.Y.); jiaruokun@163.com (R.J.)

Abstract: The modulation of surface wettability and morphology are essential to optimize the photocatalytic H_2 evolution activity of graphitic carbon nitride (CN)-based photocatalysts. In this work, the porous hydrophilic CN/PANI heterojunction film was prepared via interfacial polymerization and loaded on a porous PCL substrate. The construction of the type-II CN/PANI heterojunction enabled an overall spectrum response and the efficient separation and transportation of photoexcited charge carriers. The fabricated CN/PANI solid-state film in comparison with its powder counterpart elevated the utilization efficiency and maintained the long-term stability of photocatalyst. The porous morphology and hydrophilic surface increased the surface area and enhanced the surface wettability, favoring water-molecule adsorption and activation. The as-prepared CN/PANI heterojunction film exhibited photocatalytic H_2 production activity up to 3164.3 $\mu mol \cdot h^{-1} \cdot g^{-1}$, which was nearly 16-fold higher than that of pristine CN (569.1 $\mu mol \cdot h^{-1} \cdot g^{-1}$).

Keywords: CN/PANI heterojunction; porous solid-state film; hydrophilic surface; photocatalytic H_2 evolution

Citation: Yang, X.; Zhang, Y.; Deng, J.; Huo, X.; Wang, Y.; Jia, R. Fabrication of Porous Hydrophilic CN/PANI Heterojunction Film for High-Efficiency Photocatalytic H₂ Evolution. *Catalysts* **2023**, *13*, 139. https://doi.org/10.3390/catal13010139

Academic Editors: Yongming Fu and Qian Zhang

Received: 2 December 2022
Revised: 25 December 2022
Accepted: 29 December 2022
Published: 6 January 2023

Copyright: © 2023 by the authors. Licensee MDPI, Basel, Switzerland. This article is an open access article distributed under the terms and conditions of the Creative Commons Attribution (CC BY) license (https://creativecommons.org/licenses/by/4.0/).

1. Introduction

Photocatalytic H_2 evolution via water splitting—that is, the application of solar energy to drive water reduction—has developed over the past decades years into an appealing route for the green production of hydrogen energy. This technique, through which incident photons are adsorbed by semiconductor photocatalysts to generate charge carriers participating in redox reaction, conforms to the concept of green chemistry and could help achieve the sustainable development goals [1–3]. Its productivity is mainly determined by the utilization range of the visible light spectrum and the photoelectric conversion efficiency of the photocatalysts [4]. A wide range of semiconductor materials are applied for photocatalytic H_2 generation, primarily including metal-based oxides, sulfides, and nitrides [5,6]. Non-metallic based polymers are introduced for photocatalysis owing to their easy modification, tunable band structures, and satisfactory visible light response [7]. Among them, graphitic carbon nitride (CN) has rapidly developed into one of the most promising photocatalysts for H_2 evolution from water due to its visible light-harvesting capacity ($\lambda < 460$ nm), strong reduction ability and environmentally friendly characteristics [8]. However, there are drawbacks related to pristine CN, including an insufficient adsorption range of the visible light spectrum and the poor separation and migration of the photoexcited charge carriers [9]. Heterojunction fabrication offers a feasible strategy to overcome these problems by integrating another semiconductor photocatalyst with a suitable band structure and superior electronic properties [10,11].

Polyaniline (PANI), a conductive polymer with an extended π-conjugated system, possesses a high absorption coefficient in the visible light range and high mobility of charge carriers [12]. It is selected as a preferred candidate to fabricate CN-based heterojunctions attributed to its wide spectral response, appropriate energy band edge, and high conductivity [13]. Ge et al. synthesized the CN/PANI heterojunction via in situ deposition oxidative polymerization. The obtained CN/PANI heterojunction showed enhanced photocatalytic H_2 evolution activity due to the effectively improved carrier separation [14]. Zhang et al. presented the PANI nanorod array grown on the CN nanosheet by cryogenic dilution polymerization. The as-prepared CN/PANI heterojunction boosted photocatalytic H_2 production attributed to enhanced charge carrier transportation [15]. However, these reported CN/PANI heterojunctions were generally developed as a suspended powder in the photocatalytic system. There are some shortcomings in the use of the powder heterojunctions, such as difficult separation and segregation after the reaction [16]. The large losses and low utilization efficiency induced by the particulate properties of the heterojunctions could significantly constrain the photocatalytic H_2 generation performance. Meanwhile, the aggregation of powder heterojunctions would reduce the surface active sites, leading to decreased photocatalytic activity.

The membrane immobilization of powder photocatalyst is beneficial for long-term maintenance for photocatalytic activity and stability [17]. Generally, photocatalyst precursors that are dispersed into a solvent could be cast into solid state films through various methods, including spin coating, inkjet printing, and spray coating [18]. The porous membranes have been recognized as a suitable substrate for powder photocatalyst immobilization [19]. Herein, porous PCL film was selected as the substrate to obtain porous heterojunction film. In the presence of aniline monomer and pristine CN, the CN/PANI heterojunction was synthesized by interfacial polymerization and loaded on the porous PCL substrate. The obtained porous CN/PANI film exhibited a high surface area and hydrophilicity, thereby exposing abundant reactive sites and enhancing the water molecule adsorption. As a result, the as-prepared porous hydrophilic CN/PANI heterojunction membrane showed a nearly 16-fold increase in photocatalytic H_2 generation rate compared to that of the pristine CN.

2. Results and Discussion

2.1. Morphology and Structure Characterization

The morphology of pristine CN, PANI, and the CN/PANI heterojunction film were observed by scanning electron microscopy (SEM). Pristine CN showed a two-dimensional layer-like architecture with a size range of several micrometers (Figure 1a). Bare PANI exhibited agglomerated particles with an irregular shape (Figure 1b). The CN/PANI heterojunction film deposited on the PCL substrate showed large numbers of macropores and a rough surface (Figure 1c,d). As a comparison, the SEM image of the film made of CN/PANI heterojunction powder was shown in Figure S1. No regular porous structures were observed on its surface.

The FTIR spectra of the pristine CN, PANI, and the CN/PANI film was shown in Figure 2a. For pristine CN, the peak located at 813 cm^{-1} was assigned to the bending vibration of heptazine rings. A series of peaks ranging between 1250 and 1630 cm^{-1} were ascribed to the typical stretching modes of CN heterocycles. The broad absorption peak at around 3200 cm^{-1} originated from the stretching vibration of the N H bond [20]. For bare PANI, the peaks at 806 cm^{-1} and 1132 cm^{-1} were the out-of-plane and in-plane stretching vibration of the C-H bond. The peak at 1302 cm^{-1} corresponded to the stretching vibration of the C-N bond. The peaks at 1485 cm^{-1} and 1568 cm^{-1} were ascribed to the vibration peaks of benzene ring and quinone ring, respectively [21–25]. For the CN/PANI heterojunction film, the characteristic peaks of CN and PANI were simultaneously observed.

The X-ray diffraction (XRD) patterns of pristine CN, PANI, and CN/PANI heterojunction film are shown in Figure 2b. For the pristine CN, two strong diffraction peaks at 13.1° and 27.4° were observed, corresponding to the (100) and (002) crystal plane of graphitic carbon nitride (JCPDS 87-1526) [26]. For bare PANI, the diffraction peaks at 14.94°, 20.22°, and 25.52° were assigned to the (011), (020), and (200) crystal planes of the emerald salt (ES) form of polyaniline [27]. The XRD pattern of the CN/PANI heterojunction film showed characteristic peaks of CN and PANI. The FTIR and XRD result indicated the simultaneous presence of CN and PANI within the film, implying the successful fabrication of the

CN/PANI heterojunction film. X-ray photoelectron spectroscopy (XPS) measurements were conducted to identify the elemental state of as-synthesized CN/PANI heterojunction film.

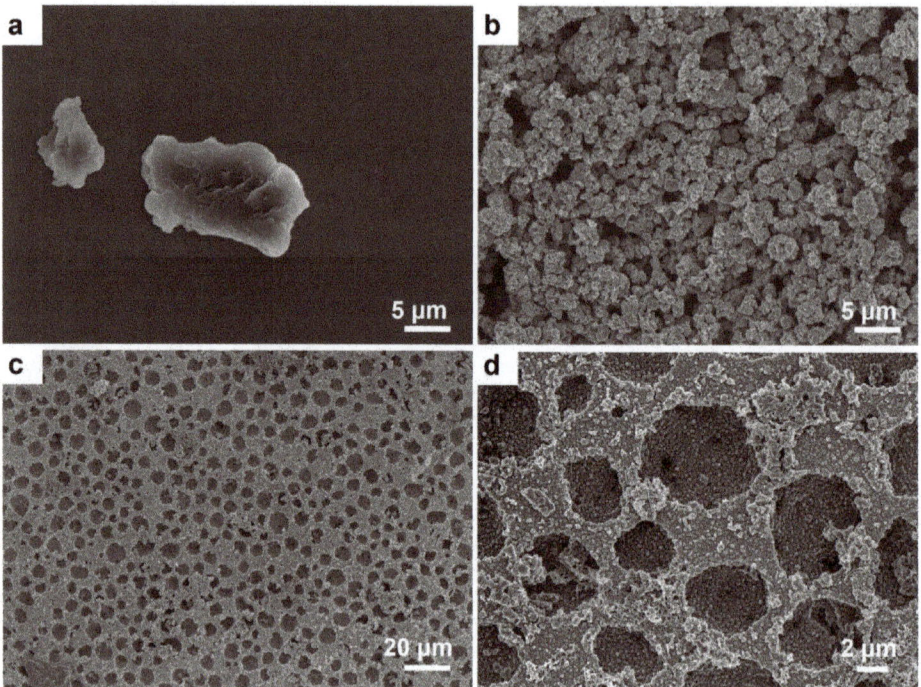

Figure 1. SEM images of (**a**) pristine CN, (**b**) pure PANI, and (**c**,**d**) CN/PANI heterojunction film.

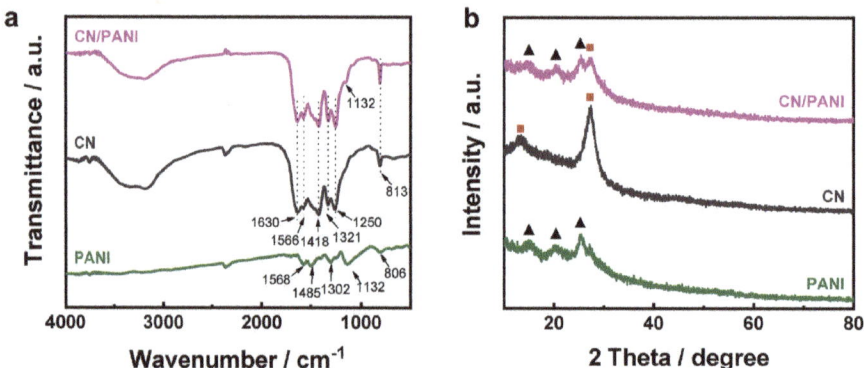

Figure 2. (**a**) FTIR spectra and (**b**) XRD patterns of pristine CN, PANI, and CN/PANI heterojunction film.

The survey XPS spectra of the CN/PANI heterojunction film in Figure 3a demonstrated the coexistence of C, N, and O elements. The high-resolution C 1s spectra deconvoluted into three peaks was exhibited in Figure 3b. The peaks centered at 288.4 and 284.8 eV were ascribed to N-C=N and C-C groups of CN [26]. The peak at binding energy of 286.4 eV was assigned to the C-N group of PANI [21]. The high-resolution N 1s spectra also confirmed the simultaneous presence of CN and PANI. The N 1s spectra showed in Figure 3c could be deconvoluted into four peaks centered at 398.6, 399.2, 400.1, and 401.1 eV, respectively.

The peaks at 398.6, 400.1, and 401.1 eV were associated with C-N=C, N-(C)$_3$, and -NHx groups of CN, while the peak at 399.2 eV was ascribed to C-N group of PANI [21,26]. The high-resolution O 1s spectra is given in Figure 3d. The peak at 532.2 eV was assigned to surface-adsorbed oxygen species of CN [20]. The XPS result further confirmed the successful construction of the CN/PANI heterojunction film.

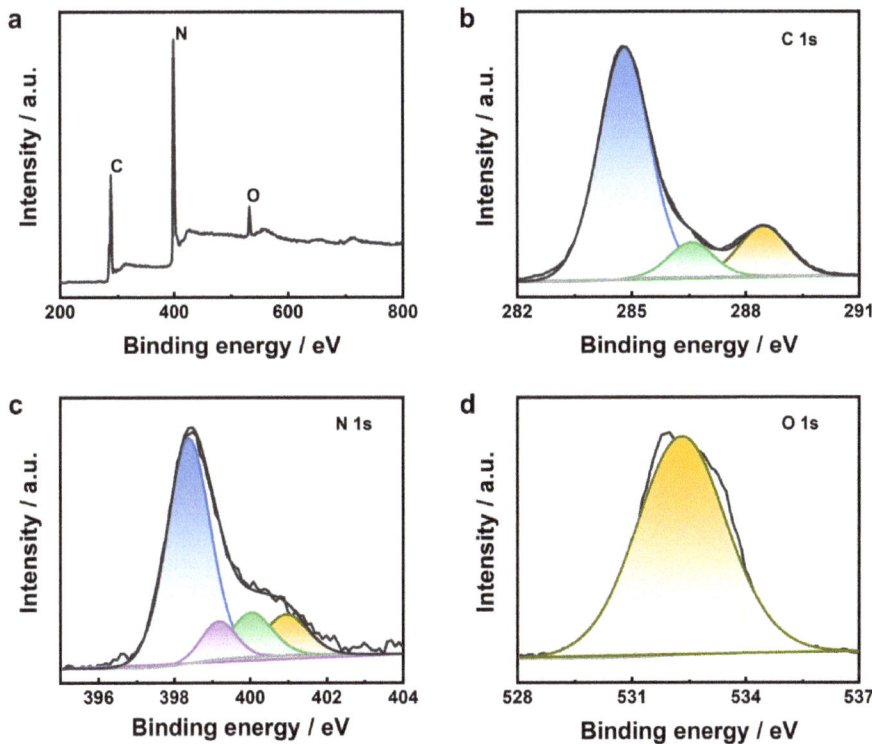

Figure 3. (a) Survey XPS spectra and (b–d) high-resolution XPS spectra of the CN/PANI heterojunction film: (b) C 1s, (c) N 1s, and (d) O 1s.

2.2. Photocatalytic Performance Evaluation

The photocatalytic H$_2$ evolution rate of the pristine CN, bare PANI, and CN/PANI heterojunction film were evaluated. As shown in Figure 4a, the bare PANI showed a minimum H$_2$ generation rate of 308.2 μmol·h^{-1}·g^{-1}, and the pristine CN exhibited an H$_2$ production rate of 569.1 μmol·h^{-1}·g^{-1}. The as-prepared CN/PANI heterojunction film showed a significantly enhanced H$_2$ evolution rate (3164.3 μmol·h^{-1}·g^{-1}) compared to the single component, indicating that the construction of the heterojunction film was favorable to boost the photocatalytic performance. The photocatalytic H$_2$ generaction activity of the CN/PANI heterojunction powder was measured for comparison (Figure S2). Its low H$_2$ evolution rate (1674.5 μmol·h^{-1}·g^{-1}) revealed that the the porous membrane fabrication was crucial to facilitate photocatalytic H$_2$ production. The photocatalytic H$_2$ evolution acticities of other similar CN-based and PANI-based heterojunctions were exhibited in Table S1 [28–33], suggesting the superior photocatalytic H$_2$ generation performance of the as-synthesized porous hydrophilic CN/PANI heterojunction film. The photocatalytic recycling tests of the CN/PANI heterojunction film was depicted in Figure 4b. Its photocatalytic H$_2$ generation activity was preserved after four cycles, confirming the excellent photostability of the as-prepared CN/PANI heterojunction film. To understand the enhanced photocatalytic performance of the CN/PANI heterojunction film, the surface wettability, porosity characteristics, and optical and photoelectric conversion properties were characterized.

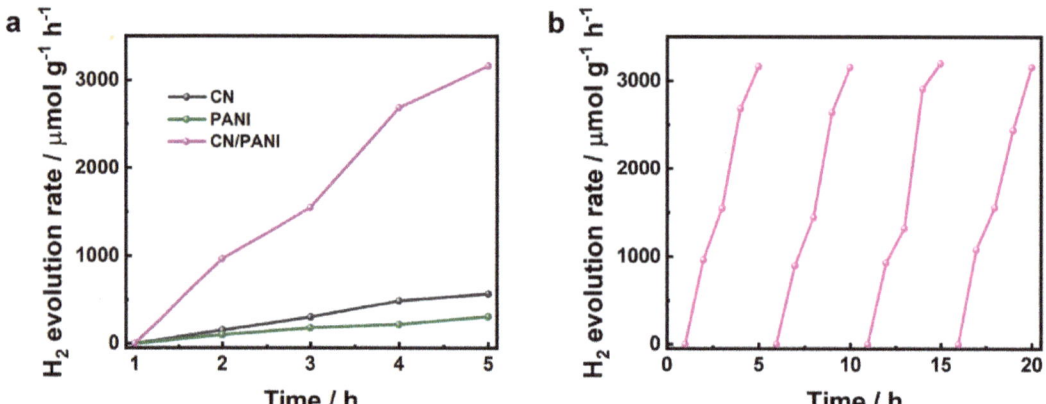

Figure 4. (a) Photocatalytic H_2 evolution rate of pristine CN, bare PANI, and the CN/PANI heterojunction film. (b) Cyclic H_2 production tests of the CN/PANI heterojunction film.

2.3. Mechanism Analysis

Figure 5 presents the N_2 adsorption/desorption isotherms of the pristine CN, PANI, and CN/PANI heterojunction film. The BET surface area, pore size, and pore volume of the as-prepared photocatalysts were displayed in Table 1. It could be seen that all of the photocatalysts show the typical type IV isotherms with a H3 hysteresis loop, proving their predominant mesoporous structures [34]. The measured BET surface areas of the pristine CN and PANI were only 9.47 and 5.75 $m^2 \cdot g^{-1}$, much lower than the fabricated CN/PANI heterojunction film (137.87 $m^2 \cdot g^{-1}$). The higher surface area of the CN/PANI film provided more catalytic active sites and a shorted diffusion distance for charge carrier, which was one reason for its improved photocatalytic activity.

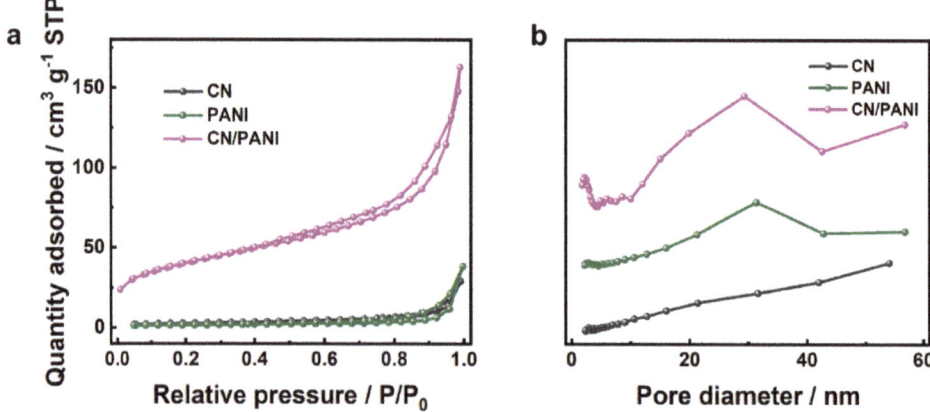

Figure 5. (a) N_2 adsorption–desorption isotherms of the pristine CN, PANI, and CN/PANI heterojunction film. (b) The corresponding pore size distributions calculated from BJH method.

The surface wettability of pristine CN, bare PANI, and CN/PANI heterojunction film were tested via the sessile drop method. As shown in Figure 6, the contact angle of the CN/PANI heterojunction film (35.0°) was evidently smaller than that of pristine CN photocatalyst (50.2°), indicating the improved hydrophilicity of the obtained heterojunction film. The enhanced surface wettability of the CN/PANI heterojunction film was beneficial for the adsorption of water molecules and the escape promotion of gaseous product (H_2) [35], which was desirable for photocatalytic H_2 evolution enhancement.

Table 1. BET surface area, pore size, and pore volume of the pristine CN, PANI, and CN/PANI heterojunction film.

Photocatalyst	BET Surface Area ($m^2 \cdot g^{-1}$)	Pore Size (nm)	Pore Volume ($cm^3 \cdot g^{-1}$)
CN	9.47	9.6	0.057
PANI	5.75	15.9	0.045
CN/PANI heterojunction film	137.87	9.4	0.177

Figure 6. Water-droplet contact angles of (**a**) the pristine CN, (**b**) bare PANI, and (**c**) CN/PANI heterojunction film.

The visible light response capacity and photoelectric conversion efficiency of the as-prepared photocatalysts were characterized by the UV-vis diffuse reflectance spectra, PL spectra, and electrochemical measurements. The pristine CN showed an adsorption band edge of 460 nm due to the typical semiconductor band–band transition (Figure 7a) [28]. Bare PANI exhibited strong light adsorption in both the ultraviolet and visible spectrum regions attributed to the π→π* transition [36]. The CN/PANI heterojunction film exhibited an overall spectrum response, which was conducive to utilize more photons to generate charge carriers for photocatalytic reaction.

The separation efficiency of photogenerated charge carriers of the as-synthesized photocatalysts was characterized by the intensity of PL emission spectra. As illustrated in Figure 7b, the pristine CN showed the highest PL emission intensity due to the high recombination rate of photoexcited charge carriers. Weak photoluminescence effect was detected for the bare PANI. In comparison with pristine CN, the PL emission intensity of the CN/PANI heterojunction film dramatically decreased, demonstrating the supriority of porous solid-state membrane fabrication on the separation promotion of photoexcited charge carriers. The migration resistance of photogenerated charge carriers for the as-prepared photocatalysts was measured by EIS spectra. The typical π-conjugated skeleton structure of conductive polymers (CPs) enabled good conductivity of the PANI molecular chain. Therefore, the CN/PANI heterojunction film showed significantly decreased slop of Nyquist plots in comparison with pristine CN (Figure 7c), implying the reduced transportation resistance at the interface [37]. The photocurrent density indicated the migration efficiency of photoinduced charge carriers of the as-prepared photocatalysts, which was displayed in Figure 7d. The pristine CN showed the lowest photocurrent density attributed to the high transportation resistance of charge carriers induced by its inherent π-deficient conjugated system [26]. The CN/PANI film showed evidently increased photocurrent density, suggesting the heterojunction construction effectively promoted the migration of photogenerated charge carriers. The optical and electrochemical measurement revealed the extended visible light harvesting ability and optimized photoelectric conversion efficiency of the as-prepared CN/PANI heterojunction film, which was responsible for its improved photocatalytic H_2 generation performance.

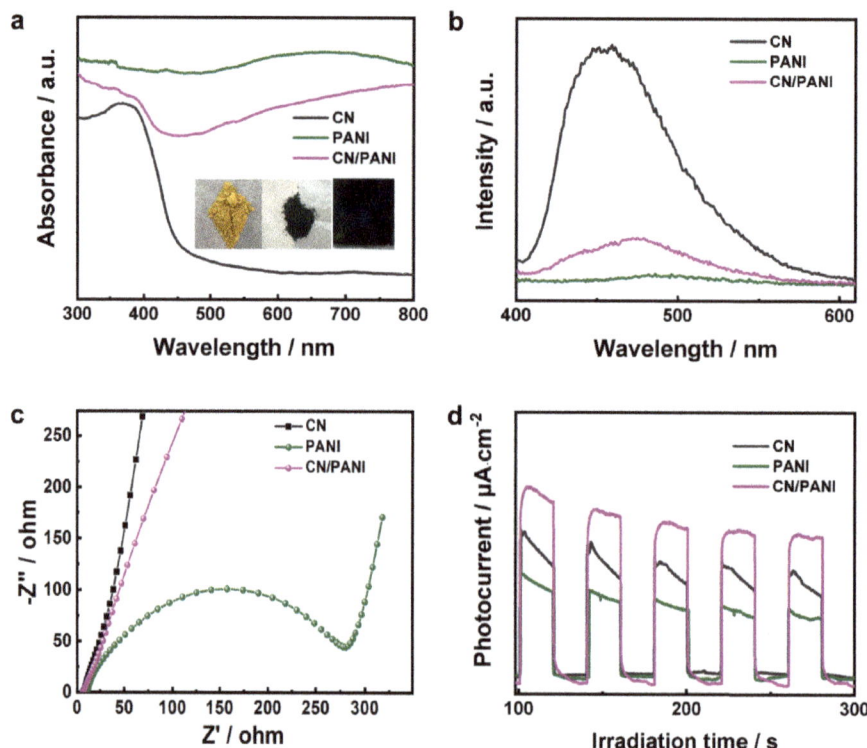

Figure 7. (a) UV-Vis diffuse reflectance spectra and optical pictures (inserted ones); (b) PL emission spectra; (c) EIS spectra; and (d) periodic on/off photocurrent response for the pristine CN, PANI, and CN/PANI heterojunction film.

The energy band edge potentials of CN and PANI were computed to elucidate the enhancement of the photocatalytic activity enhancement mechanism. The highest occupied molecular orbital (HOMO) and lowest unoccupied molecular orbital (LUMO) of PANI were composed of bonded π orbitals and anti-bonded π* orbitals. It was known that the HOMO and LUMO of PANI were 0.62 and −2.14 V, respectively. The band gap between orbitals was approximately 2.76 eV [14]. The band gap and conduction band minimum (CBM) potential of CN were determined to be 2.59 eV and −1.8 V according to its Tauc plot and Mott–Schottky plot (Figures S3 and S4). Then, the valence band minimum (VBM) potential was derived to be 0.79 V. The schematic illustration for the energy-band structure of the CN/PANI film was displayed in Figure 8.

The mechanism for improved photocatalytic H_2 production of the CN/PANI heterojunction film was proposed. After interfacial polymerication of aniline, a tight interface was formed between CN and PANI. Due to the relatively low CB and VB edge potential of CN, a type-II heterostructure was constructed and a charge carriers channel was generated at the interface of two phases. Under visible light irradiation, PANI absorbed photons to induce a π→π* transition, and the excited electrons were transferred to the π* orbital. At the same time, CN could be excited and thus create photogenerated electronhole pairs. Following the standard type II heterojunction transfer path of photogenerated carriers, the electrons on the LUMO of PANI were able to inject into the CB of CN and the holes on the VB of CN could transfer to the HOMO of PANI, which effectively promoted the separation and directional transportation of photogenerated charge carriers. The electrons accumulated on the CB of CN participate in the water-reduction reaction, and the holes accumulated on the

HOMO of PANI take part in the TEOA oxidation reaction. Consequently, the CN/PANI heterojunction film showed enhanced photocatalytic H_2 generation performance attributed to increased surface area, improved surface wettability, an extended visible light response, and the separation and migration promotion of photoexcited charge carriers.

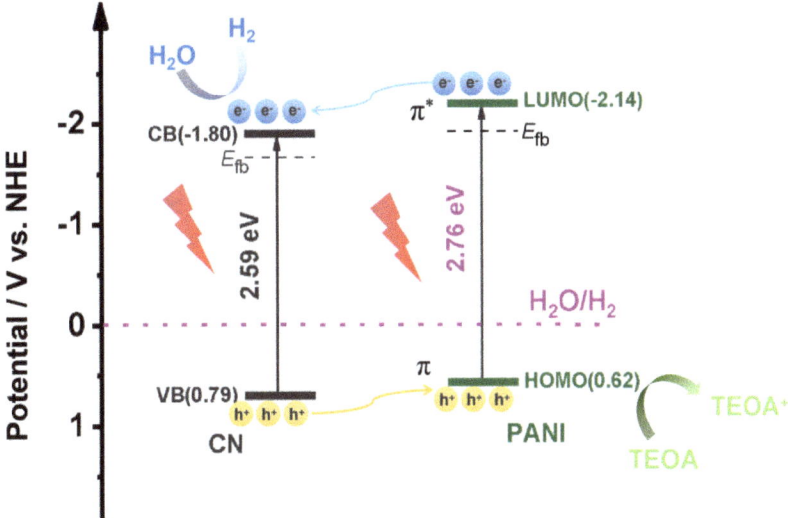

Figure 8. The schematic illustration for electronic energy-band structure and photocatalytic H_2 evolution improvement mechanism of CN/PANI heterojunction film under visible light irradiation.

3. Materials and Methods

3.1. Chemicals

Urea (CH_4N_2O), sodium sulfate (Na_2SO_4), and Nafion were purchased from Beijing Chemical Works, Beijing, China. Aniline monomer (An), hydrochloric acid (HCl), ammonium persulfate (($NH_4)_2S_2O_8$, APS), and ethanol were supplied by Yongda Chemical Reagent, Tianjin, China. All of the chemicals were analytic-grade and used without purification. Fluoride tin oxide (FTO) glass was purchased from Xiangcheng Technology, Hunan, China. Deionized water (18.2 MΩ·cm) was used throughout all of the experiments.

3.2. Photocatalysts Preparation Procedure

3.2.1. Synthesis of CN

The pristine graphitic carbon nitride (CN) was prepared by direct thermal polymerization using urea as starting material. Typically, 10 g of urea was put out in a 50 mL ceramic crucible with a cover and calcined at 520 °C for 2 h in a murfle furnace with a heating rate of 5 °C/min. The obtained yellow product was washed several times by water and ethanol and dried in an oven for further use.

3.2.2. Synthesis of CN/PANI Composites and CN/PANI@PCL Film

The ordered porous polycaprolactam lactone (PCL) film substrate was prepared by the solvent evaporation self-organization method. The detailed process had been described in our previous study [38]. The CN/PANI@PCL composite films were prepared by impregnation adsorption of aniline (An) monomer and CN on PCL substrate, followed by in situ polymerization. Typically, 40 mL of the HCl solution containing 0.01 mol of An, the PCL film substrate (2.5 cm × 4 cm), and 0.5 g of CN was added to a reactor. The mixture was stirred in an ice-water bath for 1 h to obtain a homogeneous solution. A cooled HCl solution of APS was added dropwise into the homogeneous solution to initiate the in situ polymerization of aniline. After the complete polymerization, the film was washed three times by water and ethanol and dried at 80 °C in vacuum oven for 12 h to obtain the CN/PANI composite loaded on the PCL substrate. For comparison, the CN/PANI composite was synthesized by the same procedure without adding a PCL film substrate.

3.3. Characterizations

The morphology of the as-prepared samples was observed on a Hitachi S4800 scanning electron microscopy (SEM). Fourier transformed infrared (FTIR) spectra were recorded using a Thermo Nicolet iS50 FTIR spectrometer. Powder X-ray diffraction (XRD) patterns were collected with Bruker D8 Advance diffractometer with Cu Kα radiation at a scanning rate of 5°/min. Surface areas and pore size distributions of the photocatalysts were acquired by nitrogen physisorption at 77 K on a Nova 1200e Surface Area and Porosity Analyzer. X-ray photoelectron spectroscopy (XPS) measurements were performed on PHI Quantera II spectrometer using nonmonochromatized Al-Kα X-ray as an excitation source. The water contact angle was measured with a JY-82B Kruss DSA video-based contact angle goniometer. The average CA value was obtained by measuring the sample at four different positions. UV-vis diffuse reflection spectra was recorded in the spectral region of 200–800 nm on a Shimadzu UV-2550 spectrophotometer with $BaSO_4$ as the reference substance. Room temperature photoluminescence (PL) spectra were monitored at a PerkinElmer LS-55 Luminesence Spectrometer under an excitation wavelength of 365 nm. Conductivity measurements were conducted on a Chenjing ET3000 Hall system.

3.4. Photoelectrochemical Tests

Photoelectrochemical tests were performed on a Chenhua CHI760E electrochemical workstation with a standard three-electrode photoelectrochemical cell. A gauze platinum, Ag/AgCl (saturated KCl), and FTO glass coated with the CN/PANI membrane were used as the counter electrode, reference electrode, and working electrode, respectively. The three electrodes were immersed in a sodium sulfate (Na_2SO_4) electrolyte solution (0.5 M, pH 6.8). A 300 W Xe lamp (Perfect Light PLS-SXE300, Beijing, China) equipped with a 420 nm UV-cutoff filter was used as irradiation light resource. The solution was continuously in an N_2-purged flow to remove O_2 before photoelectrochemical measurements. The working electrode was illuminated to record the electrochemical impedance spectra (EIS) under the perturbation signal of 5 mV and the frequency ranged from 0.1 Hz to 100 kHz. Periodic on/off light was applied to monitor the transient photocurrent response in 400 s.

3.5. Evaluation of Photocatalytic Performance

A 300 W Xe lamp (Perfect Light PLS-SXE300, Beijing, China) equipped with a 420 nm UV-cutoff filter was used as irradiation resource for photocatalytic H_2 evolution measurements. For the photocatalytic film system, four pieces of as-prepared film (2.5 cm × 4 cm) were immersed into 300 mL aqueous solution containing 10% TEOA and 3 wt.% Pt. Before light irradiation, the reactor was bubbled with N_2 to remove dissolved O_2. Then, the Xe lamp was turned on to initiate the photocatalytic H_2 generation test and the gaseous products were analyzed on a gas chromatography (Fuli GC9790II(PLF-01), Taizhou, China). The cyclic stability measurements of the CN/PANI film were evaluated under the same procedure.

4. Conclusions

In summary, we successfully synthesized the porous and hydrophilic CN/PANI solid-state film loaded on the porous PCL substrate. The formed type II CN/PANI heterojunction achieved an overall spectrum response and high-efficient separation and migration of photogenerated charge carriers. The porous morphology and hydrophilic surface contributed to an increased surface area and improved surface wettability, enhancing the adsorption and activation of reactants (H_2O). Meanwhile, the design of solid film was conducive to maintain the long-term stability of the photocatalyst. Thereby, the obtained porous hydrophilic CN/PANI heterojunction membrane showed the significantly enhanced photocatalytic H_2 evolution performance. Such work will provide new ideas for further improving the feasibility of CN-based photocatalysts in the practical application of hydrogen energy production.

Supplementary Materials: The following supporting information can be downloaded at https://www.mdpi.com/article/10.3390/catal13010139/s1, Figure S1: The SEM image of pristine CN/PANI heterojunction film; Figure S2: Photocatalytic H_2 evolution rate of the CN/PANI heterojunction powder; Figure S3: Tauc plot of pristine CN. Figure S4: Mott–Schottky plot of pristine CN; Table S1: Comparison in photocatalytic H2 evolution activities of similar CN based heterojunctions.

Author Contributions: Conceptualization, X.Y. and Y.Z.; methodology, J.D. and R.J.; formal analysis, X.H.; investigation, X.Y.; data curation, X.Y. and Y.Z.; writing—original draft preparation, X.Y. and Y.Z.; writing—review and editing, Y.W. and R.J.; visualization, X.H.; supervision, Y.Z.; project administration, X.H.; and funding acquisition, X.Y. and R.J. All authors have read and agreed to the published version of the manuscript.

Funding: This research was funded by the Jilin Province Science and Technology Development Project (20220101248JC), the Jilin Provincial Education Department Project (JJKH20230538KJ), and the Doctoral Research Start-up Fund of Jilin Medical University (JYBS2021032LK).

Data Availability Statement: The data presented in this study are available on request from the corresponding author.

Acknowledgments: The authors would like to thank Shiyanjia Lab (www.shiyanjia.com, accessed on 1 January 2021) for their help on material characterizations.

Conflicts of Interest: The authors declare no conflict of interest.

References

1. Lu, Y.; Liu, X.L.; He, L.; Zhang, Y.X.; Hu, Z.-Y.; Tian, G.; Cheng, X.; Wu, S.-M.; Li, Y.-Z.; Yang, X.-H.; et al. Spatial heterojunction in nanostructured TiO_2 and its cascade effect for efficient photocatalysis. *Nano Lett.* **2020**, *20*, 3122–3129. [CrossRef] [PubMed]
2. Zhang, J.; Chen, X.; Takanabe, K.; Maeda, K.; Domen, K.; Epping, J.D.; Fu, X.; Antonietti, M.; Wang, X. Synthesis of a carbon nitride structure for visible-light catalysis by copolymerization. *Angew. Chem. Int. Ed. Engl.* **2010**, *49*, 441–444. [CrossRef] [PubMed]
3. Lu, Y.; Liu, Y.X.; He, L.; Wang, L.Y.; Liu, X.L.; Liu, J.-W.; Li, Y.-Z.; Tian, G.; Zhao, H.; Yan, X.-H.; et al. Interfacial co-existence of oxygen and titanium vacancies in nanostructured Tio_2 for enhancement of carrier transport. *Nanoscale* **2020**, *12*, 8364–8370. [CrossRef] [PubMed]
4. Zhang, G.; Lan, Z.A.; Lin, L.; Lin, S.; Wang, X. Overall water splitting by $Pt/g-C_3N_4$ photocatalysts without using sacrificial agents. *Chem. Sci.* **2016**, *7*, 3062–3066. [CrossRef] [PubMed]
5. Brown, K.A.; Harris, D.F.; Wilker, M.B.; Rasmussen, A.; Khadka, N.; Hamby, H.; Keable, S.; Dukovic, G.; Peters, J.W.; Seefeldt, L.C.; et al. Light-driven dinitrogen reduction catalyzed by a CdS:nitrogenase MoFe protein biohybrid. *Science* **2016**, *352*, 448–450. [CrossRef]
6. Wang, L.; Dong, Y.; Yan, T.; Hu, Z.; Jelle, A.A.; Meira, D.M.; Duchesne, P.N.; Loh, J.Y.Y.; Qiu, C.; Storey, E.E.; et al. Black indium oxide a photothermal CO_2 hydrogenation catalyst. *Nat. Commun.* **2020**, *11*, 2432–2440. [CrossRef]
7. Wang, L.; Zheng, X.; Chen, L.; Xiong, Y.; Xu, H. Van Der Waals heterostructures comprised of ultrathin polymer nanosheets for efficient Z-scheme overall water splitting. *Angew. Chem. Int. Ed. Engl.* **2018**, *57*, 3454–3458. [CrossRef]
8. Wang, X.; Maeda, K.; Thomas, A.; Takanabe, K.; Xin, G.; Carlsson, J.M.; Domen, K.; Antonietti, M. A metal-free polymeric photocatalyst for hydrogen production from water under visible light. *Nat. Mater.* **2009**, *8*, 76–80. [CrossRef]
9. Barrio, J.; Volokh, M.; Shalom, M. Polymeric carbon nitrides and related metal-free materials for energy and environmental applications. *J. Mater. Chem. A* **2020**, *8*, 11075–11116. [CrossRef]
10. She, X.; Wu, J.; Xu, H.; Zhong, J.; Wang, Y.; Song, Y.; Nie, K.; Liu, Y.; Yang, Y.; Rodrigues, M.T.F.; et al. High efficiency photocatalytic water splitting using 2D $\alpha-Fe_2O_3/g-C_3N_4$ Z-scheme catalysts. *Adv. Energy Mater.* **2017**, *7*, 1700025–1700032. [CrossRef]
11. Fu, Y.; Ren, Z.; Wu, J.; Li, Y.; Liu, W.; Li, P.; Xing, L.; Ma, J.; Wang, H.; Xue, X. Direct Z-scheme heterojunction of ZnO/MoS_2 nanoarrays realized by flowing-induced piezoelectric field for enhanced sunlight photocatalytic performances. *Appl. Catal. B Environ.* **2021**, *285*, 119785–119795. [CrossRef]
12. Zhou, Q.; Zhao, D.; Sun, Y.; Sheng, X.; Zhao, J.; Guo, J.; Zhou, B. $g-C_3N_4$ and polyaniline-co-modified TiO_2 nanotube arrays for significantly enhanced photocatalytic degradation of tetrabromobisphenol A under visible light. *Chemosphere* **2020**, *252*, 126468–126475. [CrossRef]

13. Wu, H.; Chang, C.; Lu, D.; Maeda, K.; Hu, C. Synergistic effect of hydrochloric acid and phytic acid doping on polyaniline-coupled g-C_3N_4 nanosheets for photocatalytic Cr(VI) reduction and dye degradation. *ACS Appl. Mater. Interfaces* **2019**, *11*, 35702–35712. [CrossRef]
14. Ge, L.; Han, C.; Liu, J. In situ synthesis and enhanced visible light photocatalytic activities of novel PANI-g-C_3N_4 composite photocatalysts. *J. Mater. Chem.* **2012**, *22*, 11843–11850. [CrossRef]
15. Zhang, S.; Zhao, L.; Zeng, M.; Li, J.; Xu, J.; Wang, X. Hierarchical nanocomposites of polyaniline nanorods arrays on graphitic carbon nitride sheets with synergistic effect for photocatalysis. *Catal. Today* **2014**, *224*, 114–121. [CrossRef]
16. Zhang, M.; Bao, Y.; Hou, L.; Gao, K.; Yang, Y. Will the photocatalytic ceramic membrane be the solution for the next generation of photocatalysis?—A comprehensive comparison between g-C_3N_4 powder and g-C_3N_4 modified ceramic membrane. *Sep. Purif. Technol.* **2023**, *305*, 122440–122451. [CrossRef]
17. Jia, C.; Yang, L.; Zhang, Y.; Zhang, X.; Xiao, K.; Xu, J.; Liu, J. Graphitic carbon nitride films: Emerging paradigm for versatile applications. *ACS Appl. Mater. Interfaces* **2020**, *12*, 53571–53591. [CrossRef] [PubMed]
18. Lin, E.; Qin, N.; Wu, J.; Yuan, B.; Kang, Z.; Bao, D. $BaTiO_3$ nanosheets and caps grown on TiO_2 nanorod arrays as thin-film catalysts for piezocatalytic applications. *ACS Appl. Mater. Interfaces* **2020**, *12*, 14005–14015. [CrossRef] [PubMed]
19. Liu, M.; Li, J.; Guo, Z. Polyaniline coated membranes for effective separation of oil-in-water emulsions. *J. Colloid Interface Sci.* **2016**, *467*, 261–270. [CrossRef]
20. Yang, X.; Bian, X.; Yu, W.; Wang, Q.; Huo, X.; Teng, S. Organosilica-assisted superhydrophilic oxygen doped graphitic carbon nitride for improved photocatalytic H_2 evolution. *Int. J. Hydrog. Energy* **2022**, *47*, 34444–34454. [CrossRef]
21. Naciri, Y.; Hsini, A.; Bouziani, A.; Tanji, K.; El Ibrahimi, B.; Ghazzal, M.N.; Bakiz, B.; Albourine, A.; Navío, J.A.; et al. Z-scheme WO_3/PANI Heterojunctions with enhanced photocatalytic activity under visible light: A depth experimental and DFT studies. *Chemosphere* **2022**, *292*, 133468–133481. [CrossRef]
22. Qin, N.; Pan, A.; Yuan, J.; Ke, F.; Wu, X.; Zhu, J.; Liu, J.; Zhu, J. One-step construction of a hollow Au@bimetal-organic framework core-shell catalytic nanoreactor for selective Alcohol oxidation reaction. *ACS Appl. Mater. Interfaces* **2021**, *13*, 12463–12471. [CrossRef]
23. Qin, L.; Li, Y.; Liang, F.; Li, L.; Lan, Y.; Li, Z.; Lu, X.; Yang, M.; Ma, D. A microporous 2D cobalt-based MOF with pyridyl sites and open metal sites for selective adsorption of CO_2. *Microporous Mesoporous Mater.* **2022**, *341*, 112098–112106. [CrossRef]
24. Qin, L.; Liang, F.; Li, Y.; Wu, J.; Guan, S.; Wu, M.; Xie, S.; Luo, M.; Ma, D. A 2D porous zinc-organic framework platform for loading of 5-fluorouracil. *Inorganics* **2022**, *10*, 202. [CrossRef]
25. Jin, J.C.; Wu, X.R.; Luo, Z.D.; Deng, F.Y.; Wu, X.; Luo, Z. Luminescent sensing and photocatalytic degradation properties of an uncommon (4, 5, 5)-connected 3D MOF based on 3, 5-di (3′, 5′-dicarboxylphenyl) benzoic acid. *CrystEngComm* **2017**, *19*, 4368–4377. [CrossRef]
26. Jia, R.; Zhang, Y.; Yang, X. High efficiency photocatalytic CO_2 reduction realized by Ca^{2+} and HDMP group co-modified graphitic carbon nitride. *Int. J. Hydrog. Energy* **2021**, *46*, 32893–32903. [CrossRef]
27. Poulain, M. Significantly improving the performance and dispersion morphology of porous g-C_3N_4/PANI composites by an interfacial polymerization method. *e-Polymers* **2015**, *15*, 95–101. [CrossRef]
28. Yu, C.F.; Tan, L.; Shen, S.J.; Fang, M.H.; Yang, L.; Fu, X.; Dong, S.; Sun, J. In situ preparation of g-C_3N_4/polyaniline hybrid composites with enhanced visible-light photocatalytic performance. *J. Environ. Sci.* **2021**, *104*, 317–325. [CrossRef]
29. Li, Y.M.; Zhong, J.B.; Li, J.Z. Reinforced photocatalytic H_2 generation behavior of S-scheme NiO/g-C_3N_4 heterojunction photocatalysts with enriched nitrogen vacancies. *Opt. Mater.* **2023**, *135*, 113296–113304. [CrossRef]
30. Sun, N.; Zhang, Y.; Li, X.; Jing, Y.; Zhang, Z.; Gao, Y.; Liu, J.; Tan, H.; Cai, X.; Cai, J. Ultrathin g-PAN/PANI-encapsulated Cu nanoparticles decorated on $SrTiO_3$ with high stability as an efficient photocatalyst for the H_2 evolution and degradation of 4-nitrophenol under visible-light irradiation. *Catal. Sci. Technol.* **2022**, *12*, 2482–2489. [CrossRef]
31. Li, T.; Cui, J.D.; Gao, L.M.; Lin, Y.-Z.; Li, R.; Xie, H.; Zhang, Y.; Li, K. Competitive self-assembly of PANI confined MoS_2 boosting the photocatalytic activity of the graphitic carbon nitride. *ACS Sustain. Chem. Eng.* **2020**, *8*, 13352–13361. [CrossRef]
32. Dong, P.; Zhang, A.; Cheng, T.; Pan, J.; Song, J.; Zhang, L.; Guan, R.; Xi, X.; Zhang, J. 2D/2D S-scheme heterojunction with a covalent organic framework and g-C_3N_4 nanosheets for highly efficient photocatalytic H_2 evolution. *Chin. J. Catal.* **2022**, *43*, 2592–2605. [CrossRef]
33. Sk, S.; Vennapoosa, C.S.; Tiwari, A.; Abraham, B.M.; Pal, U. Polyaniline encapsulated Ti-MOF/CoS for efficient photocatalytic hydrogen evolution. *Int. J. Hydrog. Energy* **2022**, *47*, 33955–33965. [CrossRef]
34. Yang, X.; Guo, Z.; Zhang, X.; Han, Y.; Xue, Z.; Xie, T.; Yang, W. The effect of indium doping on the hydrogen evolution performance of g-C_3N_4 based photocatalysts. *New J. Chem.* **2021**, *45*, 544–550. [CrossRef]
35. Zhu, H.; Cai, S.; Liao, G.; Gao, Z.F.; Min, X.; Huang, Y.; Jin, S.; Xia, F. Recent advances in photocatalysis based on bioinspired superwettabilities. *ACS Catal.* **2021**, *11*, 14751–14771. [CrossRef]
36. Shang, M.; Wang, W.; Sun, S.; Ren, J.; Zhou, L.; Zhang, L. Efficient visible light-induced photocatalytic degradation of contaminant by spindle-like PANI/$BiVO_4$. *J. Phys. Chem. C* **2009**, *113*, 20228–20233. [CrossRef]

37. Hung, S.F.; Xiao, F.X.; Hsu, Y.Y.; Suen, Y.-Y.; Yang, N.-T.; Chen, H.-B.; Liu, H.M.; Chen, B.H.M. Iridium oxide-assisted plasmon-induced hot carriers: Improvement on kinetics and thermodynamics of hot carriers. *Adv. Energy Mater.* **2016**, *6*, 1501339–1501351. [CrossRef]
38. Moura, N.K.d.; Siqueira, I.A.W.B.; Machado, J.P.d.B.; Kido, H.W.; Avanzi, I.R.; Rennó, A.C.M.; Trichês, E.d.S.; Passador, F.R. Production and characterization of porous polymeric membranes of PLA/PCL blends with the addition of hydroxyapatite. *J. Compos. Sci.* **2019**, *3*, 45. [CrossRef]

Disclaimer/Publisher's Note: The statements, opinions and data contained in all publications are solely those of the individual author(s) and contributor(s) and not of MDPI and/or the editor(s). MDPI and/or the editor(s) disclaim responsibility for any injury to people or property resulting from any ideas, methods, instructions or products referred to in the content.

Article

Solvothermal Synthesis of g-C$_3$N$_4$/TiO$_2$ Hybrid Photocatalyst with a Broaden Activation Spectrum

Amit Imbar, Vinod Kumar Vadivel and Hadas Mamane *

Environmental Engineering Program, School of Mechanical Engineering, Faculty of Engineering, Tel Aviv University, Tel Aviv 69978, Israel
* Correspondence: hadasmg@tauex.tau.ac.il

Abstract: A solvothermal self-made composite of graphitic carbon nitride (g-C$_3$N$_4$) and commercially available titanium dioxide (TiO$_2$) demonstrated the removal of commercial acid green-25 (AG-25) textile dye in a saline water matrix when activated by ultraviolet (UV) and visible light. The g-C$_3$N$_4$-TiO$_2$ composite was characterized by X-ray diffraction (XRD), Nitrogen sorption–desorption recording and modeling by the Brunauer–Emmett–Teller (BET) theory, scanning electron microscopy (SEM), energy-dispersive X-ray spectroscopy (EDX), diffuse reflectance spectroscopy (DRS), X-ray photoelectron spectroscopy (XPS), photoluminescence (PL), and electron spin resonance (ESR). The solvothermal process did not modify the crystalline structure of the g-C$_3$N$_4$ and TiO$_2$ but enhanced the surface area by interlayer delamination of g-C$_3$N$_4$. Under a simulated solar spectrum (including UVA/B and vis wavelengths), the degradation rate of AG-25 by the composite was two and four times higher than that of TiO$_2$ and pure g-C$_3$N$_4$, respectively (0.04, 0.02, and 0.01 min^{-1}). Unlike TiO$_2$, the g-C$_3$N$_4$-TiO$_2$ composite was activated with visible light (the UV portion of the solar spectrum was filtered out). This work provides insight into the contribution of various reactive oxidative species (ROS) to the degradation of AG-25 by the composite.

Keywords: acid dyes; textile dyes; UV–vis activated photocatalyst; graphitic carbon nitride; TiO$_2$-g-C$_3$N$_4$ photocatalyst; solvothermal coupling

Citation: Imbar, A.; Vadivel, V.K.; Mamane, H. Solvothermal Synthesis of g-C$_3$N$_4$/TiO$_2$ Hybrid Photocatalyst with a Broaden Activation Spectrum. *Catalysts* 2023, 13, 46. https://doi.org/10.3390/catal13010046

Academic Editors: Yongming Fu and Qian Zhang

Received: 13 November 2022
Revised: 19 December 2022
Accepted: 20 December 2022
Published: 26 December 2022

Copyright: © 2022 by the authors. Licensee MDPI, Basel, Switzerland. This article is an open access article distributed under the terms and conditions of the Creative Commons Attribution (CC BY) license (https://creativecommons.org/licenses/by/4.0/).

1. Introduction

The textile industry is one of the world's largest industries, providing jobs in low-income countries and playing a vital role in their economy. Over 100,000 synthetic dyes produced annually generate 7 × 10^5 tons of dyes [1,2]. The textile industry consumes large amounts of water and mainly contributes to wastewater production [3]. Due to process inefficiency, 10–15% of the dyes are washed out into sewer or water reservoir [4]. Synthetic dyes in water reservoirs pose a toxic threat to living organisms and vegetation [5] and hinder photosynthesis by blocking light penetration and increasing biological oxygen demand (BOD) [6,7]. Physical separation methods used in wastewater treatment (such as filtration and sorption) transfer the dyes from the water to a solid phase [8], biodegradation is generally inefficient in dye removal, and the use of oxidants (such as chlorine and ozone) is expensive [4,8]. Advanced oxidation processes (AOPs) are based on reactive oxygen treatment technologies aimed at degrading recalcitrant organic compounds in water through reaction with highly reactive species formed from O$_2$ as hydroxyl radical (OH·) and superoxide (O$_2^-$) that breakdown the organic compounds [9]. Specifically, photocatalysis is of high potential due to the complete removal of organic compounds [10] and the capability to avoid by-products and can be executed without consuming chemicals. The main drawback of photocatalysis is the requirement for intensive energy and light for activation which plays a crucial role in the photocatalytic efficiency and applicability.

The use of TiO$_2$ in water treatment has been extensively reported in the literature; it is inexpensive, non-toxic, and chemically stable [11–15]. Its main drawbacks are the high recombination rate of electron-hole (e^-/h^+) pairs and wide band gap (BG), ≥ 3.0 eV. This BG allows electron excitation only by UV wavelengths, making solar light inefficient as only 4% of the solar spectrum lies in the UV spectrum [16,17], making the process energy intensive [18,19].

Coupling photocatalysts to form heterojunctions is a known method for improving photocatalytic activity (PCA) and energy efficiency [20]. Graphitic carbon nitride (g-C_3N_4) gained interest due to its photocatalytic (PC) properties (BG of 2.7 eV), biocompatibility, low cost and easy production. g-C_3N_4 combined with TiO_2 showed improved PCA [21–27] while allowing a broader range of solar spectrum use.

In the past decades, many reports of photocatalysts have been published. However, they mainly focused on pure contaminants in de-ionized (DI) water matrices and pure dyes. These conditions are far from the purity of the dyes used in textile dyeing and the water quality of the effluent discharged. At the same time, the existence of other ions is proved to be harmful in photocatalysis [28]. This study demonstrates the solvothermal coupling method of TiO_2 and g-C_3N_4 for PC degradation of a commercial AG-25, a textile dye used for fabric coloring in a saline water matrix, imitating the salinity and electrical conductivity of dye house effluents.

2. Results and Discussion

2.1. X-ray Diffraction (XRD) Analysis

X-ray diffraction (XRD) analysis is a technique used in materials science to determine the crystallographic structure of a material. The prepared PC composite (g-C_3N_4/TiO_2) was analyzed by an XR diffractometer. The clear, sharp peaks of the diffraction pattern and low background noise indicate that the obtained product is of high purity. The resulting sharp peaks in Figure 1 reveal major intensive diffraction peaks for TiO_2 at 25.6°, corresponding to the (111) planes, confirming that the crystalline structure matches the anatase phase of TiO_2, as analyzed using JCPDS [29]. The graphitic carbon nitride has major intensive peaks at 12.3° and 26.7°, corresponding to the (100) and (200) planes as reported by Miranda et al. [24]. These are attributed to the planar, which repeats the cell structure and sandwich stack reflection of g-C_3N_4. After compositing with TiO_2, the significant peaks of g-C_3N_4 still had decreased peak intensity. Thus, the generation of g-C_3N_4/TiO_2 nanostructures can confirm by these structural characterizations.

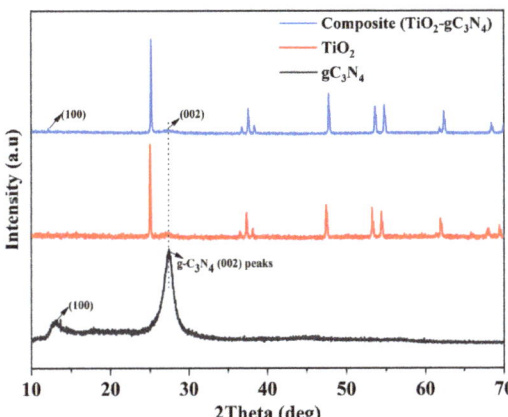

Figure 1. XRD patterns of the composite, TiO_2 and g-C_3N_4.

2.2. Brunauer-Emmett-Teller (BET) Analysis

Adsorption and desorption experiments were carried out at 77 K, and N_2 isotherms were applied to calculate the specific surface area. The surface area, pore size, and volume distribution of the g-C_3N_4/TiO_2 composite were measured by the multi-point BET and BJH (Barrett, Joyner, and Halenda) method. As shown in Figure S3, mesoporous g-C_3N_4/TiO_2 nanostructures are evident with an estimated pore volume of 0.039 cc/g, pore radius of ~2.038 nm, and pore size of 4.08 nm. As shown in Figure S4, the g-C_3N_4/TiO_2 composite shows a hysteresis loop with a type H3 shape and shifts at high relative pressures (P/P_0) between 0.8 and 1.0, which suggests a mesoporous structure. Mesoporosity is typical of wrinkled, sheetlike particles, and this result is consistent with SEM images [30]. The surface area was 15.43 $m^2\ g^{-1}$, higher than obtained in our previous study (Kumar et al. [31]) for sole g-C_3N_4 after solvothermal treatment. The different results are possibly rooted in the large amount of TiO_2 in the composite, which is smaller than g-C_3N_4, as shown in Figure 2.

Figure 2. SEM images of the composite. (**a**) magnification of 10^4 and (**b**) magnification of 2500.

2.3. Scanning Electron Microscopy (SEM)

EDX analysis was performed to obtain the composite composition. Figures S5–S7 show that the sheet-shaped particles are primarily composed of carbon and nitrogen, while the sphere-shaped particles are composed mainly of titanium and oxygen. The composition and the shape of the sheet-shaped particles both confirm that these particles are of g-C_3N_4. In contrast, the composition of the spherical particles is that of TiO_2.

The SEM images of as-synthesized g-C_3N_4/TiO_2 structures are presented in Figure 2. After compositing TiO_2 with g-C_3N_4, TiO_2 nanoparticles uniformly adhered to g-C_3N_4. However, g-C_3N_4 has a wrinkled sheetlike morphology, as reported in former publications [31–33], whereas TiO_2 has smaller and sphere-shaped particles. g-C_3N_4 has a crystalline structure similar to the solvothermal-treated g-C_3N_4 and contrary to the untreated g-C_3N_4 [31]. This combination could promote the charge transfer between g-C_3N_4 and TiO_2 by the Z-scheme route [30], as demonstrated in Figure 3.

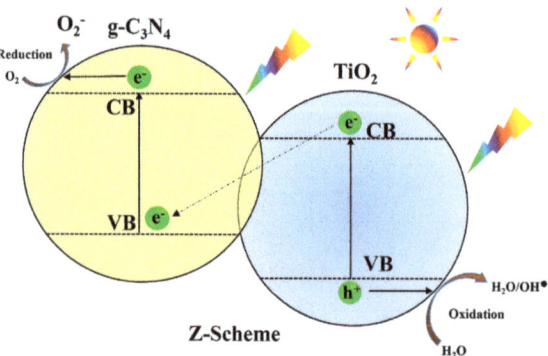

Figure 3. Z–scheme of TiO_2 and g-C_3N_4.

2.4. Absorption Spectrum and DRS Analysis

UV–vis diffuse reflectance spectra were recorded to probe the prepared photocatalysts' optical properties. As shown in Figure S8b, the absorption edges of g-C_3N_4, TiO_2 and the composite are located at approximately 389, 334 and 340 nm, respectively. Compared to TiO_2, the absorption edge positions of the composite exhibited a red shift toward the visible range alongside an extension of the absorbance spectrum toward ~430 nm wavelengths due to the g-C_3N_4 introduction with TiO_2 [34]. Furthermore, the band gaps of photocatalysts could be determined by the following equation:

$$\alpha h\nu = A(h\nu - E_g)^{n/2} \qquad (1)$$

where α, A, h, ν, and E_g represent the absorption coefficient, proportionality constant, Planck constant, light frequency, and band gap energy, respectively [35]. A plot based on the Kubelka–Munk function (Equation (1)) versus the energy of light is shown in Figure S8a. The estimated band gap values of the photocatalysts are about 2.80, 2.86, and 3.19 eV for g-C_3N_4, composite, and TiO_2, respectively. The composite shows a narrower BG than TiO_2 and is slightly wider than g-C_3N_4. As the band gap increases, the recombination of e^-/h^+ pairs decreases, enhancing the ROSs production, and the narrowing of the BG (compared with TiO_2) enables activation by the visible-light region. Therefore, the UV–vis diffuse reflectance spectroscopy (DRS) results indicated more photogenerated charges when activating the composite are excited under UV–vis spectrum irradiation, which enhances their photocatalytic performance [36].

2.5. X-ray Photoelectron Spectroscopy (XPS)

Figure 4 shows the XPS spectra of g-C_3N_4, TiO_2 and the composite. The full XPS spectrum presented all the peaks (C 1s, N 1s, Ti 2p, O 1s) of g-C_3N_4, TiO_2 plus composite (TiO_2/g-C_3N_4) nanomaterials. Thus, the generation of g-C_3N_4/TiO_2 nanostructures can be confirmed by these structural characterizations and well supports the XRD analysis in Figure 1. However, the in the valence band (VB) of TiO_2 and g-C_3N_4 is directly excited to its conduction band (CB) by absorbing photons with energy higher than or equal to its energy gap (E_g) and meanwhile resulting in the generation of a positive hole in the VB. Compared to typical photocatalysts such as TiO_2, g-C_3N_4 has the most negative CB value of -1.69 eV versus the normal hydrogen electrode (NHE) and a neutral band gap (~3.19 eV), summarized in Table S2. Furthermore, a detailed analysis of the XPS valence band spectra confronted with band structure calculations was achieved for the three photocatalysts TiO_2, g-C_3N_4 and composite (TiO_2/g-C_3N_4). Furthermore, through valence band spectra analysis, conduction band offset (-0.63 eV) and valence band offset (2.23 eV) of g-C_3N_4–TiO_2 heterojunction were estimated (Figure S9). The band gap (E_{CB}) and valence band (E_{VB}) positions of the photocatalyst materials can be calculated by the following equations:

$$E_{CB} = X - E_C - \frac{1}{2}E_g \qquad (2)$$

$$E_{VB} = E_{CB} + E_g \qquad (3)$$

where E_{VB} and E_{CB} represent the valence band and conduction band position of the semiconductor material, E_C is the hydrogen electron free energy (4.5 eV), and X represents the electronegativity of the semiconductor material.

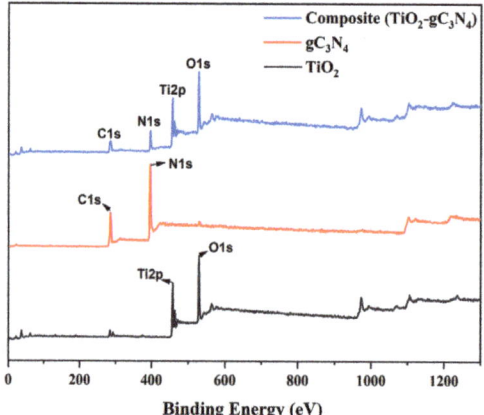

Figure 4. Full XPS spectrum presented all the peaks (C 1s, N 1s, Ti 2p, O 1s) of g-C_3N_4, TiO_2 and the composite (TiO_2/g-C_3N_4) nanomaterials.

2.6. Photo Luminescence (PL)

Figure 5 shows PL of g-C_3N_4, TiO_2 and the composite. The spectrum shows emission at 350 nm corresponding to the recombination of e^-/h^+ pairs, with the intensity directly proportional to the photogenerated e^-/h^+ recombination in the catalyst. The composite's intensity is between the two substances of origin, a decrease in intensity was obtained with the composite compared with g-C_3N_4 and increased intensity compared with TiO_2.

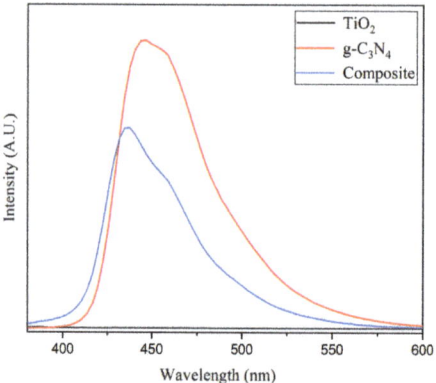

Figure 5. PL spectra of g-C_3N_4, TiO_2 and the composite.

2.7. Electron Spin Resonance Spectroscopy (ESR)

The production of ROSs by the g-C_3N_4, TiO_2 and the composite under solar light was studied by ESR spin trap experiments. Aqueous suspensions of each catalyst were irradiated separately under solar light in the presence of 5-methyl-1-pyrroline N-oxide (BMPO), which traps both hydroxyl and superoxide radicals. To distinguish the signal corresponding to hydroxyl radical, dimethyl sulfoxide (DMSO), a hydroxyl radical scavenger, was added to the catalysts-BMPO suspensions.

The resulting signals are shown in Figure 6. When only BMPO is present, the signals show the formation of ROSs (either OH· or O_2^-) by the irradiation of all three substances. The composite's signals almost utterly correspond to those of BMPO-OOH, implying that it produces superoxide radicals. The consistent intensity of the signal suggests that the composite does not produce hydroxyl radicals.

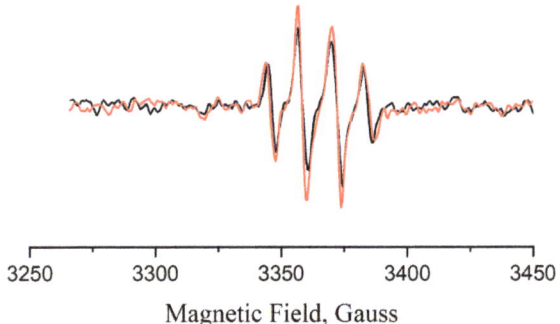

Figure 6. BMPO-OOH with (red line) and without (black line) DMSO signal from the suspension of the composite.

2.8. Photocatalytic Degradation of AG-25 under Solar Irradiation in Saline Water

10-ppm AG-25 in saline water matrix was used as a target pollutant to study the degradation by the catalysts. The catalyst was stirred in the solution for 30 min (time −30 to 0) under dark conditions for sorption control, followed by simulated solar irradiation for 90 min (time 0 to 90). Under dark conditions, 16, 12, and 11% removal of AG-25 were observed by the composite, g-C_3N_4, and TiO_2, respectively. The composite has an increased surface area, as mentioned in Section 3.2, promoting higher sorption over the catalyst.

Figure 7 shows the degradation rate of AG-25 by the three catalysts. AG-25 total removal by g-C_3N_4, TiO_2, and the composite were 59, 86, and 98%, respectively, implying that all three catalysts can degrade AG-25. The degradation rate of AG-25 using the composite was twice as high as commercially available TiO_2 and four times higher than that of pure g-C_3N_4. The higher results obtained by TiO_2 and the composite can be explained by the type of radicals formed by the catalyst. TiO_2 primarily produces hydroxyl radicals [37,38], which have higher oxidation potential than superoxide radicals, the main ROS produced by g-C_3N_4. Moreover, the PL results show significant recombination in g-C_3N_4 compared with TiO_2. However, coupling the two catalysts in the composite can form a heterojunction that suppresses the recombination rate and improves photon energy usage, as shown in PL spectra in Figure 5 and the broadened absorption spectrum demonstrated in Figure S8, resulting in enhanced PCA [39].

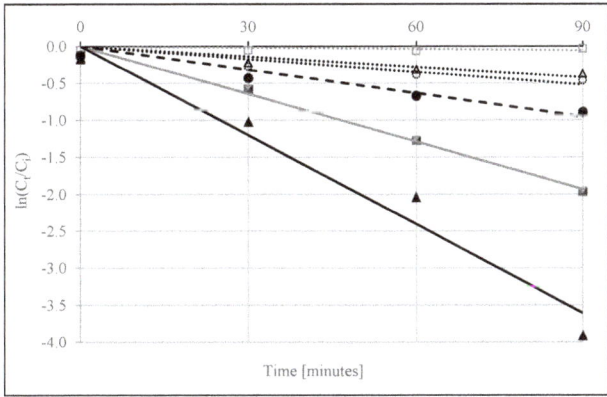

Figure 7. Photocatalytic degradation rate of 100 mL AG−25 10 ppm in brine for 90 min under solar irradiation with (filled shapes) and without (hollow shapes) UV spectrum cutoff by the composite (▲, △), g-C_3N_4 (●, ○), and TiO_2 (■, □).

Table 1 summarizes the degradation rate and total degradation of each catalyst. Adly et al. [40] degraded AG-25 with graphene oxide/titanium dioxide composites (GO-TiO_2) under UV–vis irradiation, reporting 40% removal (vs. 98% in our work) of AG-25 after 90 min by a similar concentration of catalysts but a higher concentration of AG-25.

Table 1. Total degradation (after 90 min) and degradation rate of AG-25 by coupled and uncoupled g-C_3N_4 -TiO_2 under solar irradiation, with and without LP400 filter.

Catalyst	Percent Degradation of AG-25 (%)	Rate of AG-25 Degradation (min^{-1}) $\times 10^{-3}$
g-C_3N_4	59%	10.6
TiO_2	86%	21.5
Composite	98%	40.1
g-C_3N_4—LP400	37%	5.8
TiO_2—LP400	0%	0
Composite—LP400	31%	4.7

2.9. Photocatalytic Degradation of AG-25 under Solar Irradiation in Saline Water with a UV Cutoff

The role of UV and visible light in PCA was examined by degrading 10-ppm AG-25 in saline water under simulated solar irradiation, similar to Section 2.8, with a UV filter for UV cutoff. Figure 8 presents the degradation rate of AG-25 by the composite, g-C_3N_4, and TiO_2 under the visible spectrum. AG-25 degradation by g-C_3N_4, TiO_2, and the composite after 90 min was 37%, 0%, and 31%, respectively, demonstrating that TiO_2 is a UV-driven catalyst. Moreover, a higher g-C_3N_4 ratio in the catalysts resulted in higher visible-based PCA in the absence of UV wavelengths. These results show the composite's versatility as both a UV and a visible light-activated catalyst. In both cases (with and without the LP400 filter), the composite is superior to the commercial TiO_2, while the cutoff results show increased degradation (37% compared with 31% removal) by g-C_3N_4. However, comparing the full spectrum results (without the LP400 filter) shows a clear advantage of the composite over pure g-C_3N_4. Generally, the composite's versatile activation spectrum and improved results under total spectrum irradiation are advantageous.

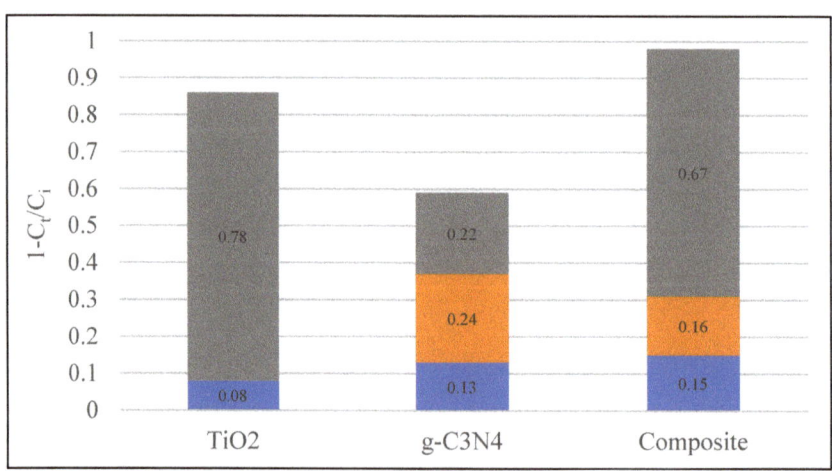

Figure 8. Segmentation of sorption (blue background), visible light (orange background), and UV (grey background) contribution to the total degradation of 100 mL AG-25 10 ppm by TiO_2, g-C_3N_4 and the composite after 90 min irradiation.

3. Materials and Methods

3.1. Materials

Melamine and Titanium dioxide (99% purity) were purchased from Sigma-Aldrich, Darmstadt, Germany.

Acid Green-25 Textile Dye

AG-25, a light green textile acid dye, was provided by Colourtex Industries Ltd., Hamburg, Germany, and has a molecular formula of $C_{28}H_{20}N_2Na_2O_8S_2$. AG-25 stock solution of 300-ppm AG-25 was prepared and diluted with deionized (DI) water to reach a concentration of 10 ppm used for the experiments. The conductivity was adjusted to 36 mS cm^{-1} using NaCl to simulate a real textile dye effluent. Due to AG-25 physical properties, the concentration of AG-25 could be determined by spectroscopic measurement of the visible light spectrum. The maximum absorbance wavelength value of the dye solutions was 614 nm (Evolution 220 UV–Visible spectrophotometer, Thermo Scientific, Waltham, MA, USA) to determine the AG-25 concentration.

The removal of AG-25 was calculated using Equation (4).

$$\text{AG-25 Removal} = (C_i - C_t)/C_i \qquad (4)$$

Here, C_i stands for the initial concentration, and C_t stands for concentrating the samples.

3.2. Characterization

The $TiO_2/g\text{-}C_3N_4$ powder crystalline structure was characterized by a D8 Advance diffractometer (Bruker AXS, Karlsruhe, Germany) with a secondary graphite monochromator, 2° Soller slits, and a 0.2 mm receiving slit. Low-background quartz sample holders were carefully filled with the powdered samples. The nanoparticle morphology and average particle size were further investigated by a Quanta 200 FEG environmental scanning electron microscope equipped with a field-emission electron gun (FEG). The samples were coated with a conductive carbon grid and imaged in a Hitachi S3200N SEM-EDS system at 20 kV accelerating voltage. The surface area, pore size and volume distribution of the modified and unmodified $g\text{-}C_3N_4$ were measured by N_2 adsorption–desorption and the BJH (Barrett, Joyner, and Halenda) method, respectively. To probe the optical properties of the prepared photocatalysts, UV–visible diffuse reflectance spectra were recorded using a UV–Vis–NIR spectrophotometer (Cary 5000 + UMA, Agilent, Santa Clara, CA, USA).

3.3. Preparation of $g\text{-}C_3N_4$

$g\text{-}C_3N_4$ was synthesized via the thermal-condensation method. In short, 4 g melamine (Sigma-Aldrich, 99%) was heated to 540 °C in a packed aluminum crucible under ambient pressure in the air for 2 h with a heating rate of 15 °C min^{-1} in a muffle furnace. The resultant powder reached room temperature and was washed with acetone.

3.4. Solvothermal Method

TiO_2 and $g\text{-}C_3N_4$ (named composite) were coupled via the solvothermal method. The $TiO_2\text{-}g\text{-}C_3N_4$ mixture (2:1 ratio, 1 g) was stirred for 3 h in ethanol (70%) in a 50 mL PTFE pressure-tight liner at ~65 °C. Subsequently, the PTFE liner was placed in a stainless-steel jacket and heated to 180 °C for 10 h. The resulting catalyst was filtered and dried before usage.

3.5. Photocatalytic Degradation of AG-25 in Saline Water

The PC experiments were conducted under a 300-W solar simulator, ozone-free xenon arc lamp (Newport full-spectrum, 50.8 mm × 50.8 mm, Irvine, CA USA) equipped with a 1.5 Global Air Mass filter to remove infrared light, as shown in Figure S1. The xenon lamp's incident spectral irradiance and photon fluence are recorded by a spectroradiometer (International Light, ILT 900R, Peabody, MA, USA), as in Figure S2. The integrated incident irradiance was 2.865 and 62.737 mW cm^{-2} for the UV (280–400 nm) and visible (400–1251 nm) ranges, respectively.

The PC experiments were conducted using AG-25 (100 mL, 10 ppm) in a crystalline cylinder (70 × 50 mm). The powdered catalyst (500 ppm) was sonicated and added to the AG-25 solution under stirring. The solution was constantly mixed under dark conditions for 30 min, followed by 90 min under solar-like irradiation. Samples of the slurry solution were taken in 30 min intervals and centrifuged (14600 RPM, 10 min) to separate the catalyst from the solution. The samples were analyzed in a spectroradiometer (Evolution 220 UV–Visible spectrophotometer, Thermo Scientific) to determine the AG-25 concentration.

3.6. Photocatalytic Degradation of AG-25 with UV Cutoff—LP400

Distinguishing between PCA driven by visible light and UV irradiation was performed similarly to as in Section 2.5, using a UV filter (400 HLP/130 mm, OMEGA OPTICAL, Brattleboro, VT, USA) to block UV wavelengths (i.e., λ < 400 nm) from reaching the slurry.

4. Conclusions

The composite showed improved photocatalytic activity compared with TiO_2 and g-C_3N_4 under solar irradiation. As demonstrated by the BET results, the solvothermal method increased the surface area resulting in increased photocatalytic activity relative to the origin catalysts, reflected in the enhanced degradation rate of AG-25. The UV cutoff experiments showed a versatile spectrum ranging from UV to the visible spectrum, allowing composite activation at various wavelengths. This study provides a simple synthesis method of a composite that can efficiently remove acid-dye effluents in a versatile spectrum of wavelengths.

Supplementary Materials: The following supporting information can be downloaded at: https://www.mdpi.com/article/10.3390/catal13010046/s1, Figure S1: Newport solar simulator, Figure S2: Spectral incident irradiance, Figure S3: Pore size distribution and volume curves, Figure S4: BET sorption\desorption curves, Figures S5–S7: EDX elemental analysis, Figure S8: UV–visible DRS spectra, Figure S9: XPS, S10:ESR, Table S1: XRD, Table S2: XPS.

Author Contributions: Supervision, reviewing and project administration, H.M.; methodology, V.K.V.; investigation, V.K.V. and A.I.; writing—original draft, A.I.; writing—review and editing, H.M., V.K.V. and A.I. All authors have read and agreed to the published version of the manuscript.

Funding: This research received no external funding.

Data Availability Statement: The data presented in this study are available on request from the corresponding author.

Conflicts of Interest: The authors declare no conflict of interest.

References

1. Ajmal, A.; Majeed, I.; Malik, R.N.; Idriss, H.; Nadeem, M.A. Principles and mechanisms of photocatalytic dye degradation on TiO2 based photocatalysts: A comparative overview. *RSC Adv.* **2014**, *4*, 37003–37026. [CrossRef]
2. Ananthashankar, A.G.R. Production, Characterization and Treatment of Textile Effluents: A Critical Review. *J. Chem. Eng. Process Technol.* **2013**, *5*, 1–18. [CrossRef]
3. Arami, M.; Limaee, N.Y.; Mahmoodi, N.M. Evaluation of the adsorption kinetics and equilibrium for the potential removal of acid dyes using a biosorbent. *Chem. Eng. J.* **2008**, *139*, 2–10. [CrossRef]
4. Natarajan, S.; Bajaj, H.C.; Tayade, R.J. Recent advances based on the synergetic effect of adsorption for removal of dyes from waste water using photocatalytic process. *J. Environ. Sci.* **2018**, *65*, 201–222. [CrossRef] [PubMed]
5. Chung, K.T. Azo dyes and human health: A review. *J. Environ. Sci. Heal.—Part C Environ. Carcinog. Ecotoxicol. Rev.* **2016**, *34*, 233–261. [CrossRef]
6. Gita, S.; Hussan, A.; Choudhury, T.G. Impact of Textile Dyes Waste on Aquatic Environments and its Treatment. *Environ. Ecol.* **2017**, *35*, 2349–2353.
7. Holkar, C.R.; Jadhav, A.J.; Pinjari, D.V.; Mahamuni, N.M.; Pandit, A.B. A critical review on textile wastewater treatments: Possible approaches. *J. Environ. Manag.* **2016**, *182*, 351–366. [CrossRef]

8. Rajeshwar, K.; Osugi, M.E.; Chanmanee, W.; Chenthamarakshan, C.R.; Zanoni, M.V.B.; Kajitvichyanukul, P.; Krishnan-Ayer, R. Heterogeneous photocatalytic treatment of organic dyes in air and aqueous media. *J. Photochem. Photobiol. C Photochem. Rev.* **2008**, *9*, 171–192. [CrossRef]
9. Thiruvenkatachari, R.; Vigneswaran, S.; Moon, I.S. A review on UV/TiO$_2$ photocatalytic oxidation process. *Korean J. Chem. Eng.* **2008**, *25*, 64–72. [CrossRef]
10. Liu, C.; Mao, S.; Shi, M.; Wang, F.; Xia, M.; Chen, Q.; Ju, X. Peroxymonosulfate activation through 2D/2D Z-scheme CoAl-LDH/BiOBr photocatalyst under visible light for ciprofloxacin degradation. *J. Hazard. Mater.* **2021**, *420*, 126613. [CrossRef]
11. Ajmal, A.; Majeed, I.; Malik, R.N.; Iqbal, M.; Nadeem, M.A.; Hussain, I.; Yousaf, S.; Zeshan; Mustafa, G.; Zafar, M.I.; et al. Photocatalytic degradation of textile dyes on Cu$_2$O-CuO/TiO$_2$ anatase powders. *J. Environ. Chem. Eng.* **2016**, *4*, 2138–2146. [CrossRef]
12. Chauhan, A.; Sharma, M.; Kumar, S.; Thirumalai, S.; Kumar, R.V.; Vaish, R. TiO$_2$@C core@shell nanocomposites: A single precursor synthesis of photocatalyst for efficient solar water treatment. *J. Hazard. Mater.* **2020**, *381*, 2019. [CrossRef] [PubMed]
13. Mortazavian, S.; Saber, I.; James, D.E. Optimization of Photocatalytic Degradation of Acid Suspension System: Application of Response Surface. *Catalysts* **2019**, *9*, 360. [CrossRef]
14. Rizzo, L.; della Sala, A.; Fiorentino, A.; Puma, G.L. ScienceDirect Disinfection of urban wastewater by solar driven and UV lamp e TiO$_2$ photocatalysis: Effect on a multi drug resistant Escherichia coli strain. *Water Res.* **2014**, *53*, 145–152. [CrossRef]
15. Teh, C.M.; Mohamed, A.R. Roles of titanium dioxide and ion-doped titanium dioxide on photocatalytic degradation of organic pollutants (phenolic compounds and dyes) in aqueous solutions: A review. *J. Alloys Compd.* **2011**, *509*, 1648–1660. [CrossRef]
16. Wang, H.; Zhang, L.; Chen, Z.; Hu, J.; Li, S. Semiconductor heterojunction photocatalysts: Design, construction, and photocatalytic. *Chem. Soc. Rev.* **2014**, *43*, 5234–5244. [CrossRef]
17. Junwang, S. Visible-light driven heterojunction photocatalysts for water splitting—A critical review. *Energy Environ. Sci.* **2015**, *8*, 731–759. [CrossRef]
18. Gong, S.; Jiang, Z.; Zhu, S. enhanced visible-light photocatalytic activities heterojunctions with enhanced visible-light photocatalytic activities. *J. Nanopart. Res.* **2018**, *20*, 310. [CrossRef]
19. Hou, W.; Cronin, S.B. A review of surface plasmon resonance-enhanced photocatalysis. *Adv. Funct. Mater.* **2013**, *23*, 1612–1619. [CrossRef]
20. Liu, C.; Mao, S.; Wang, H.; Wu, Y.; Wang, F.; Xia, M.; Chen, Q. Peroxymonosulfate-assisted for facilitating photocatalytic degradation performance of 2D/2D WO$_3$/BiOBr S-scheme heterojunction. *Chem. Eng. J.* **2022**, *430*, 2021. [CrossRef]
21. Caudillo-Flores, U.; Muñoz-Batista, M.J.; Luque, R.; Fernández-García, M.; Kubacka, A. g-C$_3$N$_4$/TiO$_2$ composite catalysts for the photo-oxidation of toluene: Chemical and charge handling effects. *Chem. Eng. J.* **2019**, *378*, 122228. [CrossRef]
22. Jiang, X.H.; Xing, Q.J.; Luo, X.B.; Li, F.; Zou, J.P.; Liu, S.S.; Li, X.; Wang, X.K. Simultaneous photoreduction of Uranium(VI) and photooxidation of Arsenic(III) in aqueous solution over g-C$_3$N$_4$/TiO$_2$ heterostructured catalysts under simulated sunlight irradiation. *Appl. Catal. B Environ.* **2018**, *228*, 29–38. [CrossRef]
23. Luan, S.; Qu, D.; An, L.; Jiang, W.; Gao, X.; Hua, S.; Miao, X.; Wen, Y.; Sun, Z. Enhancing photocatalytic performance by constructing ultrafine TiO$_2$ nanorods/g-C$_3$N$_4$ nanosheets heterojunction for water treatment. *Sci. Bull.* **2018**, *63*, 683–690. [CrossRef]
24. Miranda, C.; Mansilla, H.; Yáñez, J.; Obregón, S.; Colón, G. Improved photocatalytic activity of g-C$_3$N$_4$/TiO$_2$ composites prepared by a simple impregnation method. *J. Photochem. Photobiol. A Chem.* **2013**, *253*, 16–21. [CrossRef]
25. Kočí, K.; Reli, M.; Troppová, I.; Šihor, M.; Kupková, J.; Kustrowski, P.; Praus, P. Photocatalytic decomposition of N$_2$O over TiO$_2$/g-C$_3$N$_4$ photocatalysts heterojunction. *Appl. Surf. Sci.* **2017**, *396*, 1685–1695. [CrossRef]
26. Li, J.; Liu, Y.; Li, H.; Chen, C. Fabrication of g-C$_3$N$_4$/TiO$_2$ composite photocatalyst with extended absorption wavelength range and enhanced photocatalytic performance. *J. Photochem. Photobiol. A Chem.* **2016**, *317*, 151–160. [CrossRef]
27. Zhou, B.; Hong, H.; Zhang, H.; Yu, S.; Tian, H. Heterostructured Ag/g-C$_3$N$_4$/TiO$_2$ with enhanced visible light photocatalytic performances. *J. Chem. Technol. Biotechnol.* **2019**, *94*, 3806–3814. [CrossRef]
28. Liu, C.; Mao, S.; Shi, M.; Hong, X.; Wang, D.; Wang, F.; Xia, M.; Chen, Q. Enhanced photocatalytic degradation performance of BiVO$_4$/BiOBr through combining Fermi level alteration and oxygen defect engineering. *Chem. Eng. J.* **2022**, *449*, 137757. [CrossRef]
29. Ranjithkumar, R.; Lakshmanan, P.; Devendran, P.; Nallamuthu, N.; Sudhahar, S.; Kumar, M.K. Investigations on effect of graphitic carbon nitride loading on the properties and electrochemical performance of g-C$_3$N$_4$/TiO$_2$ nanocomposites for energy storage device applications. *Mater. Sci. Semicond. Process.* **2021**, *121*, 105328. [CrossRef]
30. Liu, Y.; Wu, S.; Liu, J.; Xie, S.; Liu, Y. Synthesis of g-C$_3$N$_4$/TiO$_2$nanostructures for enhanced photocatalytic reduction of U(vi) in water. *RSC Adv.* **2021**, *11*, 4810–4817. [CrossRef]
31. Kumar, V.; Avisar, D.; Betzalel, Y.; Mamane, H. Rapid visible-light degradation of EE2 and its estrogenicity in hospital wastewater by crystalline promoted g-C$_3$N$_4$. *J. Hazard. Mater.* **2020**, *398*, 122880. [CrossRef]
32. Zhao, S.; Chen, S.; Yu, H.; Quan, X. G-C$_3$N$_4$/TiO$_2$ hybrid photocatalyst with wide absorption wavelength range and effective photogenerated charge separation. *Sep. Purif. Technol.* **2012**, *99*, 50–54. [CrossRef]
33. Zhang, Y.; Thomas, A.; Antonietti, M.; Wang, X. Activation of carbon nitride solids by protonation: Morphology changes, enhanced ionic conductivity, and photoconduction experiments. *J. Am. Chem. Soc.* **2009**, *131*, 50–51. [CrossRef] [PubMed]
34. Li, C.; Sun, Z.; Xue, Y.; Yao, G.; Zheng, S. A facile synthesis of g-C$_3$N$_4$/TiO$_2$ hybrid photocatalysts by sol-gel method and its enhanced photodegradation towards methylene blue under visible light. *Adv. Powder Technol.* **2016**, *27*, 330–337. [CrossRef]

35. Li, J.; Wu, X.; Pan, W.; Zhang, G.; Chen, H. Vacancy-Rich Monolayer BiO$_{2-x}$ as a Highly Efficient UV, Visible, and Near-Infrared Responsive Photocatalyst. *Angew. Chemie—Int. Ed.* **2018**, *57*, 491–495. [CrossRef] [PubMed]
36. Tang, Q.; Meng, X.; Wang, Z.; Zhou, J.; Tang, H. In Situ Microwave-Assisted Synthesis of Porous N-TiO$_2$/g-C$_3$N$_4$ Heterojunctions with Enhanced Visible-Light Photocatalytic Properties. *Appl. Surf. Sci.* **2018**, *430*, 253–262. [CrossRef]
37. Ren, H.T.; Liang, Y.; Han, X.; Liu, Y.; Wu, S.H.; Bai, H.; Jia, S.Y. Photocatalytic oxidation of aqueous ammonia by Ag$_2$O/TiO$_2$ (P25): New insights into selectivity and contributions of different oxidative species. *Appl. Surf. Sci.* **2020**, *504*, 2019. [CrossRef]
38. Ishibashi, K.I.; Fujishima, A.; Watanabe, T.; Hashimoto, K. Detection of active oxidative species in TiO$_2$ photocatalysis using the fluorescence technique. *Electrochem. Commun.* **2000**, *2*, 207–210. [CrossRef]
39. Wen, J.; Li, X.; Liu, W.; Fang, Y.; Xie, J.; Xu, Y. Photocatalysis fundamentals and surface modification of TiO$_2$ nanomaterials. *Cuihua Xuebao Chin. J. Catal.* **2015**, *36*, 2049–2070. [CrossRef]
40. Adly, M.S.; El-Dafrawy, S.M.; El-Hakam, S.A. Application of nanostructured graphene oxide/titanium dioxide composites for photocatalytic degradation of rhodamine B and acid green 25 dyes. *J. Mater. Res. Technol.* **2019**, *8*, 5610–5622. [CrossRef]

Disclaimer/Publisher's Note: The statements, opinions and data contained in all publications are solely those of the individual author(s) and contributor(s) and not of MDPI and/or the editor(s). MDPI and/or the editor(s) disclaim responsibility for any injury to people or property resulting from any ideas, methods, instructions or products referred to in the content.

Article

Synergistic Effect of Amorphous Ti(IV)-Hole and Ni(II)-Electron Cocatalysts for Enhanced Photocatalytic Performance of Bi₂WO₆

Chenjing Sun [1,†], Kaiqing Zhang [1,†], Bingquan Wang [2] and Rui Wang [1,*]

[1] School of Environmental Science and Engineering, Shandong University, Qingdao 266237, China
[2] School of Chemistry and Molecular Engineering, Qingdao University of Science and Technology, Qingdao 266042, China
* Correspondence: wangrui@sdu.edu.cn
† These two authors contributed equally and share co-first authorship.

Abstract: Bi₂WO₆ has become a common photocatalyst due to its advantages of simple synthesis and high activity. However, the defects of pure Bi₂WO₆ such as low light reception hinder its application in photocatalysis. In this study, based on the modification of Bi₂WO₆ with Ti(IV) as a cavity co-catalyst, new Ni- and Ti-doped nanosheets of Bi₂WO₆ (Ni/Ti-Bi₂WO₆) were prepared by a one-step wet thermal impregnation method and used for the photocatalytic degradation of tetracycline. The experimental results showed that the photocatalytic activity of Ni/Ti-Bi₂WO₆ modified by the two-component catalyst was significantly better than those of pure Bi₂WO₆ and Ti-Bi₂WO₆ modified with Ti(IV) only. The photocatalytic effect of Ni/Ti-Bi₂WO₆ with different Ni/Ti molar ratios was investigated by the degradation of TC. The results showed that 0.4Ni/Ti-Bi₂WO₆ possessed the best photocatalytic performance, with a degradation rate of 92.9% at 140 min TC. The results of cycling experiments showed that the catalyst exhibited high stability after five cycles. The scavenger experiment demonstrated that the h^+ and O_2^- were the main reactive species. The enhanced photocatalytic activity of Bi₂WO₆ could be attributed to the synergistic effect between the Ti(IV) as a hole cocatalyst and Ni(II) as an electron cocatalyst, which effectively promoted the separation of photogenerated carriers.

Keywords: Bi₂WO₆ nanoflake; Ni/Ti dual cocatalyst; visible light; degradation of antibiotics

Citation: Sun, C.; Zhang, K.; Wang, B.; Wang, R. Synergistic Effect of Amorphous Ti(IV)-Hole and Ni(II)-Electron Cocatalysts for Enhanced Photocatalytic Performance of Bi₂WO₆. *Catalysts* **2022**, *12*, 1633. https://doi.org/10.3390/catal12121633

Academic Editors: Yongming Fu and Qian Zhang

Received: 13 September 2022
Accepted: 8 December 2022
Published: 13 December 2022

Publisher's Note: MDPI stays neutral with regard to jurisdictional claims in published maps and institutional affiliations.

Copyright: © 2022 by the authors. Licensee MDPI, Basel, Switzerland. This article is an open access article distributed under the terms and conditions of the Creative Commons Attribution (CC BY) license (https://creativecommons.org/licenses/by/4.0/).

1. Introduction

Since the emergence of penicillin in the 1920s, many antibiotics have been used in pharmaceuticals, agriculture, and aquaculture [1]. Among them, tetracycline, as one of the most widely used antibiotics, would cause adverse impacts on human health and environmental safety [2]. Therefore, it is imperative to develop an efficient method for the treatment of TC. In recent years, various traditional and emerging techniques, including biological treatment [3], the Fenton method [4], and photochemistry [5], have been developed to remove TC from water. Among them, photocatalysis has attracted considerable attention owing to the following advantages: mild reaction conditions, high efficiency and stability, environmental friendliness, and so on [6–8]. Semiconductor photocatalysts represented by TiO₂ have been widely used in the field of photocatalytic decontamination of environmental pollution due to the advantages of low cost, high stability, and no environmental hazards [9,10]. However, due to the limitations of the material itself, TiO₂ still has problems of the fast compounding of photogenerated electron–hole pairs, wide band gap, and narrow absorption range [11,12]. In addition to modifying TiO₂ materials to improve their photocatalytic activity, the search for other semiconductor photocatalytic materials is also an important way to synthesize high-performance photocatalysts [13–15].

Among a series of developed visible-light reactive photocatalysts, Bi_2WO_6 has attracted much attention in degrading TC in wastewater due to its unique band structure, non-toxicity, and high stability [16,17]. However, pure Bi_2WO_6 suffers from the rapid binding of photogenerated carriers, inefficient absorption of visible light at wavelengths less than 450 nm, and the low number of active surface sites, and these drawbacks greatly limit the potential applications of Bi_2WO_6 in environmental remediation [18,19]. Therefore, many modifications of Bi_2WO_6 have been explored to enhance its photocatalytic performance, such as noble metal deposition [20–22], construction of semiconductor heterojunctions [23–26], ion doping [27–30], and co-catalyst modification [31–33]. Among the above improvement methods, co-catalyst modification is a promising method that can effectively promote the separation of photogenerated electrons and holes [34–36].

Co-catalysts can be generally classified as cavity co-catalysts and electron co-catalysts [37,38]. For the photocatalytic degradation of organic pollutants, the rapid transfer of photogenerated holes to the catalyst surface and participation in the oxidation reaction are generally required. Cavity co-catalysts can improve photocatalytic performance by rapidly trapping interfacial holes and facilitating the oxidation reaction [39]. For example, Yu et al. modified several Ag-based materials (AgCl, AgBr, AgI, and Ag_2O) with Ti(IV) as a hole co-catalyst. The synthesized Ti(IV)/Ag-based photocatalysts were all found to exhibit enhanced photocatalytic performance for the degradation of phenol, indicating that Ti(IV) can be used as a general cavity co-catalyst to effectively improve the photocatalytic performance of various Ag-based materials [40]. In addition to Ti(IV), other cavity co-catalysts, such as RuO_2, PdS, CoO_x, and $B_2O_{3-x}N_x$, have been widely developed and applied in photocatalysis [41–44]. On the other hand, electron co-catalysts, such as noble metal nanoparticles (Pt, Pd, etc.), are generally used to capture photogenerated electrons [45,46]. For example, Yan et al. could greatly enhance the photocatalytic performance of CdS by loading PdS as a hole catalyst and Pt as an electron co-catalyst on the CdS photocatalyst [47]. However, precious metals are expensive and rare, so it is essential to develop efficient and economical electronic co-catalyst materials.

In this work, we successfully synthesized Ti-Bi_2WO_6 composites loaded with Ti(IV) hole co-catalysts on Bi_2WO_6. However, the rapid transfer and trapping of photogenerated holes by the Ti(IV) hole co-catalyst led to the accumulation of a large number of photogenerated electrons on the conduction band (CB) of Bi_2WO_6. This resulted in Ti-Bi_2WO_6 exhibiting a limited enhancement of photocatalytic activity. To improve the photocatalytic performance of Ti-Bi_2WO_6, the surface was further loaded with the Ni(II) electron catalyst. A series of Ni/Ti-Bi_2WO_6 composites loaded with different Ni/Ti molar ratios were prepared. At this time, Ti(IV) and Ni(II) were effective co-catalysts for the fast transfer of photogenerated holes and photogenerated electrons, respectively. The photocatalytic activity of Ni/Ti-Bi_2WO_6 is expected to be further improved because the dual co-catalysts can simultaneously promote the transfer rate of photogenerated electrons and holes to reach the specific reaction sites of the photocatalyst. Then, the photocatalytic activity and stability of the synthesized catalysts were investigated by TC degradation under visible-light irradiation. The properties of the prepared samples were characterized by scanning electron microscopy (SEM), transmission electron microscopy (TEM), X-ray diffraction (XRD), X-ray photoelectron spectroscopy (XPS), and UV diffuse reflectance spectroscopy (DRS). Among them, the 0.4Ni/Ti-Bi_2WO_6 photocatalyst exhibited a high degradation efficiency of 92.9% under visible light. In addition, the photocatalytic mechanism of tetracycline degradation was investigated and discussed on the basis of experiments and different characterization methods. This work may provide new insights for the development of low-cost and efficient photocatalytic materials.

2. Results and Discussion

2.1. Characterization of Ni/Ti-Bi$_2$WO$_6$

2.1.1. SEM, TEM, and EDS Analysis

SEM and EDS characterized various Bi$_2$WO$_6$ photocatalysts to investigate the detailed morphology and microstructure. As shown in Figure 1a, it can be found that the surface of pure Bi$_2$WO$_6$ was relatively smooth with a layered structure. In Figure 1b, Bi$_2$WO$_6$ modified with Ti(IV) (Ti-Bi$_2$WO$_6$) showed a similar structure to the Bi$_2$WO$_6$ sample and fine particles of about 8 nm appeared on the surface, indicating the successful synthesis of Ti-Bi$_2$WO$_6$. In Figure 1c, the Ni-Bi$_2$WO$_6$ image exhibited the appearance of sharp needle-like nanostructures and some agglomerates with the addition of Ni(II), which is consistent with the reports in the literature, indicating the successful loading of Ni(II) [48].

Figure 1. SEM images of (**a**) Bi$_2$WO$_6$, (**b**) Ti-Bi$_2$WO$_6$, and (**c**) Ni-Bi$_2$WO$_6$.

As for the Ni/Ti-Bi$_2$WO$_6$ sample (Figure 2), in addition to the presence of a large number of fine particles on the surface of the material compared to pure Bi$_2$WO$_6$, sharp needle-like nanostructures appeared in the places marked in the figure. To measure the specific composition of 0.4Ni/Ti-Bi$_2$WO$_6$, EDS analysis was used and is shown in the inset of Figure 2. The results show that the signals of Ti(IV) and Ni(II) were clearly visible, where the weight ratio of Ti was 5.07 wt% and that of Ni was 2.42 wt%. The molar ratio of Ni/Ti was 0.377, which is close to the expected value of 0.4Ni/Ti-Bi$_2$WO$_6$, which means that Ti(IV) and Ni(II) were successfully loaded onto the Bi$_2$WO$_6$ surface.

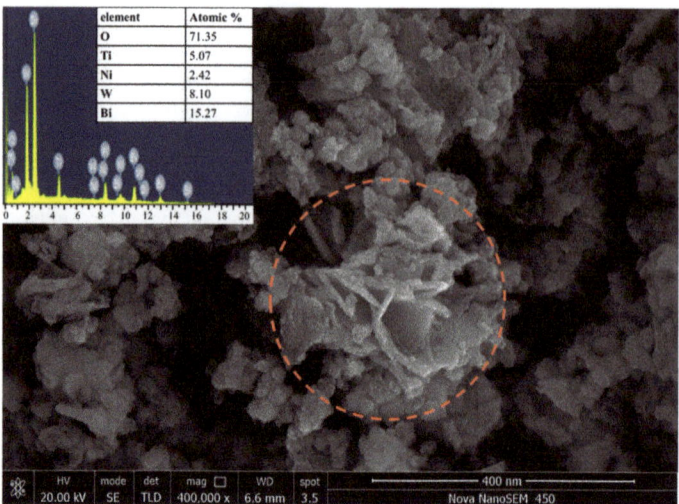

Figure 2. SEM images of Ni/Ti (0.4)-Bi$_2$WO$_6$ and EDX of 0.4Ni/Ti -Bi$_2$WO$_6$.

The detailed morphology of the 0.4Ni/Ti -Bi$_2$WO$_6$ and Bi$_2$WO$_6$ samples was further investigated by TEM. As shown in Figure 3a,b, it was consistent with the results of SEM tests. Observing the TEM of the 0.4Ni/Ti-Bi$_2$WO$_6$ sample, markings of Ni(II) and Ti(IV) nanoparticles modification could be clearly found on the surface of Bi$_2$WO$_6$.

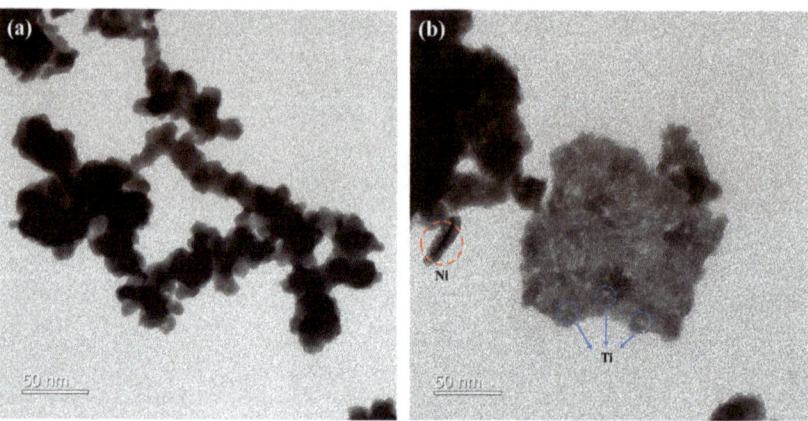

Figure 3. TEM images of (**a**) Bi$_2$WO$_6$ and (**b**) 0.4Ni/Ti-Bi$_2$WO$_6$.

2.1.2. XRD Analysis

To analyze the crystal structures and phase purities of the Bi$_2$WO$_6$, Ti-Bi$_2$WO$_6$, and Ni/Ti-Bi$_2$WO$_6$, XRD was utilized and is shown in Figure 4.

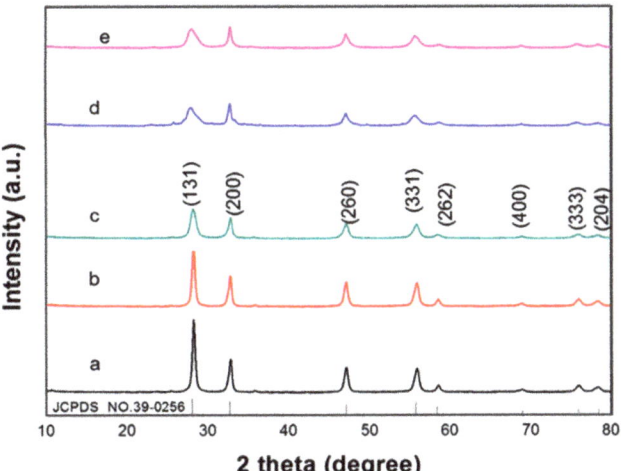

Figure 4. XRD patterns of the samples: (**a**) Bi_2WO_6, (**b**) $Ti\text{-}Bi_2WO_6$, (**c**) $0.1Ni/Ti\text{-}Bi_2WO_6$, (**d**) $0.4Ni/Ti\text{-}Bi_2WO_6$, and (**e**) $0.7Ni/Ti\text{-}Bi_2WO_6$.

As can be seen, the diffraction peaks of all five samples exhibited similar crystal structures as well without any impurity peaks, and all the characteristic peaks could match the pure orthorhombic phase of Bi_2WO_6 (JCPDS Card: 39–0256). The results were assigned to the low contents of Ti and Ni and their good dispersion in the $Ni/Ti\text{-}Bi_2WO_6$ samples. For $Ti\text{-}Bi_2WO_6$ and $Ni/Ti\text{-}Bi_2WO_6$ samples, the positions of diffraction peaks had no noticeable change compared with those of Bi_2WO_6, indicating that the Ti and Ni were only deposited on the surfaces and not incorporated into the lattice of Bi_2WO_6. These results clearly suggested that the loading of Ti and Ni had no impact on the crystal phase of Bi_2WO_6. However, the intensity of the characteristic peak decreased after doping Ni, indicating the crystallite size of Bi_2WO_6 could decrease by doping Ni, in good agreement with the results observed in SEM images. Therefore, it is obvious that the Bi_2WO_6 samples loaded by amorphous Ti(IV) and Ni(II) cocatalysts were well synthesized by the method mentioned above.

2.1.3. XPS Analysis

XPS analysis was employed to demonstrate the surface composition and chemical state of $Ni/Ti\text{-}Bi_2WO_6$ composites. Figure 5 shows the survey scan spectra of pure Bi_2WO_6, $Ti\text{-}Bi_2WO6$, and $0.4Ni/Ti\text{-}Bi_2WO_6$. As can be seen, Bi, W, and O elements were detected in all samples, which can be mainly ascribed to the Bi_2WO_6 phase. Compared with pure Bi_2WO_6, $Ti\text{-}Bi_2WO_6$ and $Ni/Ti\text{-}Bi_2WO_6$ exhibited new XPS peaks of Ti and Ni elements.

To further reveal Bi, W, Ti, and Ni elements and their chemical states, the high-resolution XPS spectra of the above samples were investigated. As shown in Figure 6, the high-resolution spectrum of Bi 4f revealed two typical peaks located at 159 eV ($Bi4f_{7/2}$) and 164.1 ($Bi4f_{5/2}$) eV, which match well with those from Bi_2WO_6 [49]. In Figure 5, the W4f spectrum can be subdivided into two peaks at 35.1 eV and 37.2 eV that were ascribed to the $W4f_{7/2}$ and $W4f_{5/2}$, respectively, indicating that W atoms presented a valence of +6 in the samples [50]. $Ti\text{-}Bi_2WO_6$ and $0.4Ni/Ti\text{-}Bi_2WO_6$ samples showed the obvious Ti2p peaks at about 458.0 eV ($Ti2p_{3/2}$) and 465 eV ($Ti2p_{1/2}$) in Figure 5, implying that the Ti atoms were in the +4 oxidization state in the samples [51]. From Figure 5, the binding energies of Ni 2p were located at 858.2 eV and 873.8 eV, demonstrating that the Ni elements were in +2 states in the samples [52]. In addition, the binding energy of W and Bi in $0.4Ni/Ti\text{-}Bi_2WO_6$ was also slightly shifted to the right by 0.2~0.4 eV compared with that of pure Bi_2WO_6, which may be due to the doping of the co-catalyst producing an electron shielding effect, resulting in a shift in the binding energy to higher energies [53].

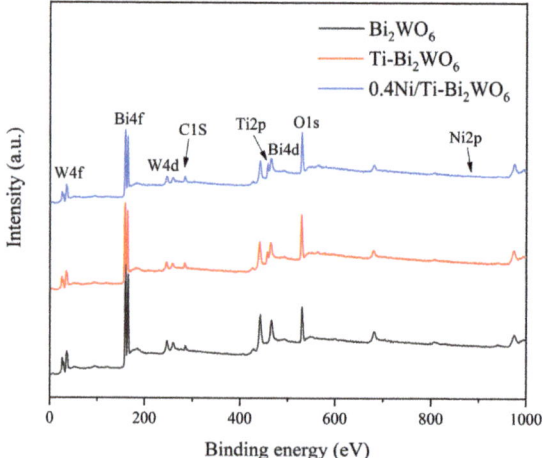

Figure 5. XPS survey spectrum of various samples.

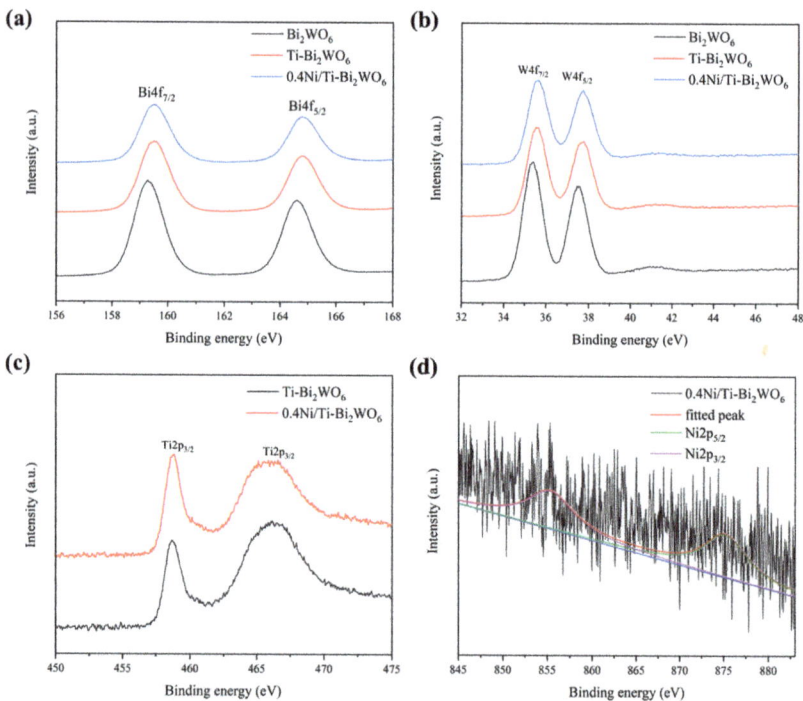

Figure 6. The XPS spectra of various samples: (**a**) Bi4f, (**b**) W4f, (**c**) Ti 2p, and (**d**) Ni 2p.

2.1.4. UV-vis Analysis

The optical properties of all samples were characterized by UV-vis diffuse reflectance spectroscopy in the wavelength range of 200–500 nm. As can be seen in Figure 7a, the absorption edge of pure Bi_2WO_6 was extended up to 430 nm, which presented a wide photo-absorption from UV to visible light, implying its potential photocatalytic activities under visible light. After loading the Ti(IV) cocatalyst onto the Bi_2WO_6, the Ti-Bi_2WO_6 showed a similar absorption curve compared with the pure Bi_2WO_6, owing to the low content of Ti(IV) on the Bi_2WO_6 surface. Compared with the pure Bi_2WO_6, the absorption curves of Ni/Ti-Bi_2WO_6 samples were similar, but there was a small redshift.

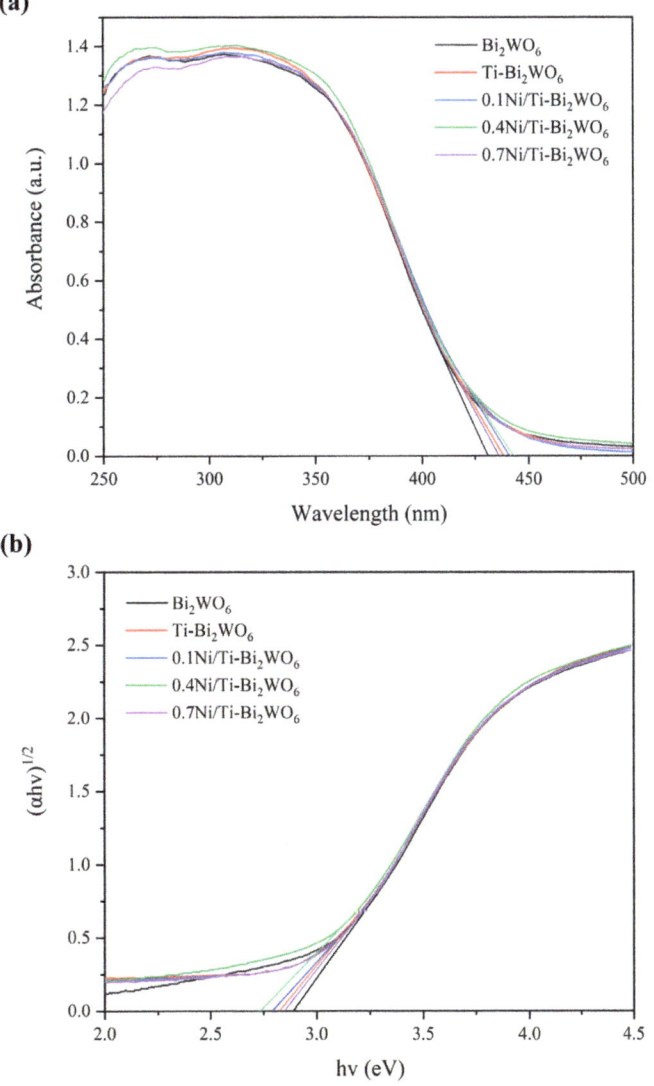

Figure 7. (a) UV-vis absorption spectra of over samples; (b) the corresponding plots of $(\alpha h\nu)^{1/2}$ versus $h\nu$ for the band gap energy over samples.

The approximate bandgap of the catalyst was illustrated from the plot of $(\alpha h\nu)^{1/2}$ versus energy ($h\nu$), as shown in Figure 7b. The bandgaps of Bi_2WO_6, 0.7Ni/Ti-Bi_2WO_6, Ti-Bi_2WO_6, 0.1Ni/Ti-Bi_2WO_6, and 0.4Ni/Ti-Bi_2WO_6 were estimated approximately to be 2.75, 2.80, 2.83, 2.85, and 2.89 eV by extrapolation of the linear part of the dependence. Hence, it is obvious that the doping of Ti(IV) and Ni(II) cocatalysts affected the light absorption capability of Bi_2WO_6. This result may be attributed to the synergistic effect between Ti(IV) as a hole catalyst and Ni(II) as an electron catalyst.

2.1.5. UV-Vis Analysis

The UV-Vis absorption spectra of TC degradation on $0.4Ni/Ti-Bi_2WO_6$ were monitored for the corresponding time. As shown in Figure 8, the characteristic absorption peaks of TC were observed at 275 and 360 nm. With increasing irradiation time, the two typical absorption peaks of TC gradually became smaller, indicating that the structure of TC was disrupted to small molecules.

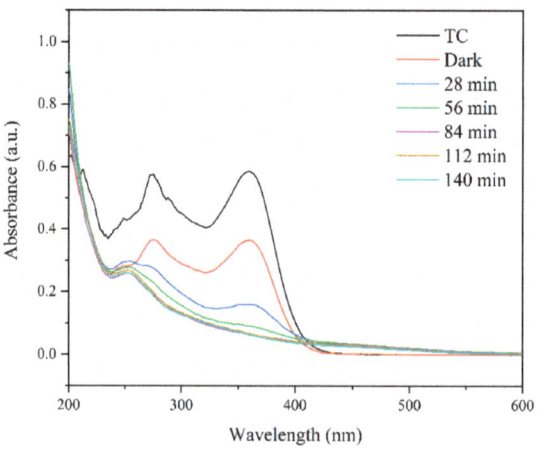

Figure 8. Absorption spectra changes of TC over $0.4Ni/Ti-Bi_2WO_6$.

2.2. Evaluation of Photocatalytic Activity

2.2.1. Photocatalytic Degradation of TC

The photocatalytic performance of the samples was tested mainly by degrading TC under visible-light irradiation. Figure 9 shows the degradation of TC under the conditions of five photocatalysts and no added catalyst. First, a mixture of photocatalyst and tetracycline solution was stirred in the dark for 30 min to exclude the effect of adsorption. After reaching the equilibrium between adsorption and desorption, photoluminescence was started. The adsorption effect of the prepared catalysts showed that Bi_2WO_6 < $Ti-Bi_2WO_6$ < $0.1Ni/Ti-Bi_2WO_6$ < $0.4Ni/Ti-Bi_2WO_6$ < $0.7Ni/Ti-Bi_2WO_6$. This can be attributed to the increase in the specific surface area of the Bi_2WO_6 composite due to the addition of the co-catalyst, which, in turn, led to the increase in the adsorption capacity of the catalysts.

The degradation efficiency of the pure Bi_2WO_6 sample was the worst at 77.8% after 140 min of light exposure, which can be attributed to the rapid recombination of electron and hole pairs generated by light. The degradation rate of $Ti-Bi_2WO_6$ reached 87.2% after the addition of Ti(IV) co-catalyst, which indicated that Ti(IV) as a hole co-catalyst had a good promotion effect on the photocatalytic activity. In addition, the photocatalytic degradation efficiency of the composites was significantly improved when Bi_2WO_6 was modified by Ni(II) and Ti(IV) dual co-catalysts, indicating that Ti and Ni co-catalysts had a synergistic effect on Bi_2WO_6. Among a series of catalysts modified by dual co-catalysts with different Ni/Ti molar ratios, $0.4Ni/Ti-Bi_2WO_6$ exhibited the highest degradation efficiency of about 92.9%. The degradation efficiency of the samples decreased when the Ni/Ti molar ratio was less than 0.4 or more than 0.4. The reason may be that when the molar ratio of Ni/Ti was less than 0.4, the number of photogenerated electrons accepted by the Ni(II) co-catalyst as an electron trap and the number of photogenerated holes captured by the Ti(IV) co-catalyst as a hole trap were reduced, and the separation efficiency of photogenerated electron–hole pairs was not high, so the photocatalytic performance was lower. When the molar ratio of Ni/Ti exceeded 0.4, too many Ti(IV) and Ni(II) co-catalysts covered the active surface sites of Bi_2WO_6, thus leading to the lower photocatalytic activity of Bi_2WO_6.

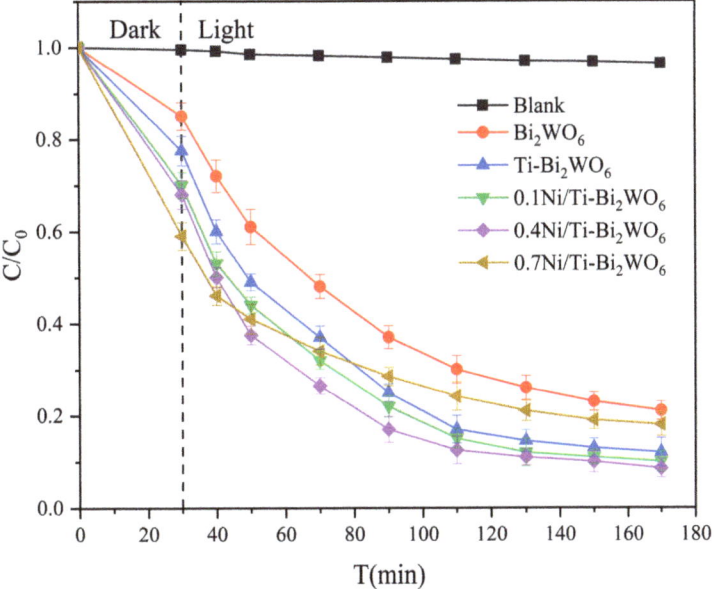

Figure 9. Photocatalytic activities of as-prepared samples for TC degradation under visible-light irradiation (>420 nm).

Ti(IV) has been shown to act as a hole co-catalyst to improve the photocatalytic performance of TiO_2 by effectively trapping photogenerated holes [54]. In the present work, Ti(IV) was relied on as a hole co-catalyst to modify Bi_2WO_6 to improve its photocatalytic ability. The conduction band (CB) and valence band (VB) of Bi_2WO_6 were about +0.3 V and +3.0 V (vs. SHE), respectively [55]. In general, to promote the efficient transfer of electrons from CB to oxygen, the CB potential of the semiconductor should be more damaging than that of the single-electron oxygen reduction reaction (−0.046 V vs. SHE) [56]. However, the CB potential of Bi_2WO_6 was significantly more positive (+0.3 V, vs. SHE) than that of the single-electron oxygen reduction, so it was poorly reduced, thus leading to the poor photocatalytic performance of Bi_2WO_6. $Ni(OH)_2$ and NiO have been widely demonstrated to be effective electron co-catalysts to improve photocatalytic performance by rapidly capturing photogenerated electrons and promoting interfacial H_2 precipitation reactions [57,58]. When the surface of Bi_2WO_6 is modified by the Ni(II) catalyst, the photogenerated electrons of Bi_2WO_6 can be rapidly transferred to the Ni(II) co-catalyst because the potential of Ni(II) is more positive than the CB level of Bi_2WO_6 [59]. When both Ni(II) and Ti(IV) co-catalysts were loaded on the surface of Bi_2WO_6, it is clear that the photocatalytic performance of the synthesized $Ni/Ti-Bi_2WO_6$ photocatalyst could be further improved, which can be well explained by the synergistic effect of Ni(II) and Ti(IV) co-catalysts. The loading of Ni(II) led to the effective transfer of photogenerated electrons in the oxygen reduction reaction, and the loading of the Ti(IV) co-catalyst led to the effective transfer of photogenerated holes in the oxidation reaction of organic matter. This principle is very similar to that reported for co-catalyst-modified photocatalysts such as Ag/AgCl-rGO and Cu(II)/AgCl [60,61]. Table 1 summarizes some of the recently reported degradation capabilities of several bismuth-based photocatalytic materials for different organic compounds.

Table 1. The degradation efficiency of different photocatalytic materials.

Types of Catalyst	Type of Degradate	Degradation Rate	Year	Ref.
Ti-Bi_2WO_6	Ceftriaxone sodium	75%	2021	[62]
0.25% Ni-Bi_2WO_6	Rhodamine B	93%	2022	[63]
30% Bi_2WO_6/$ZnWO_4$	Plasmocorinth B dye	48%	2022	[64]
Ag/WO_3/Bi_2WO_6	chlorobenzene	79%	2019	[49]
$Zn_3In_2S_6$/Bi_2WO_3	metronidazole	98.13%	2022	[65]
$Zn_3In_2S_6$/Bi_2WO_3	Hexavalent chromium	99.67%	2022	[65]
0.4Ni/Ti-Bi_2WO_6	Tetracycline	92.9%	-	-
Bi_2WO_6/C-dots/TiO_2	levofloxacin	99%	2020	[66]
Bi_2WO_6–TiO_2-N	acetone	100%	2022	[67]

2.2.2. Reusability and Stability

Reusability and stability are essential properties for photocatalysts in practical applications. To test the stability of the as-prepared samples, in this section, the 0.4Ni/Ti-Bi_2WO_6 photocatalysts were collected after degrading TC for the recycling experiment. Figure 10 shows the results of the cycling test. It can be clearly noted that the photocatalytic efficiency decreased by only about 6% after five successive cycles for the degradation of TC, due to the inevitable deficiency of the photocatalyst in the recycling process. The results showed that the 0.4Ni/Ti-Bi_2WO_6 photocatalyst had high stability in the photocatalytic reaction. The prepared Ni/Ti-Bi_2WO_6 photocatalyst had good photocatalytic activity and stability, making it an excellent photocatalyst in the treatment of actual pollutants.

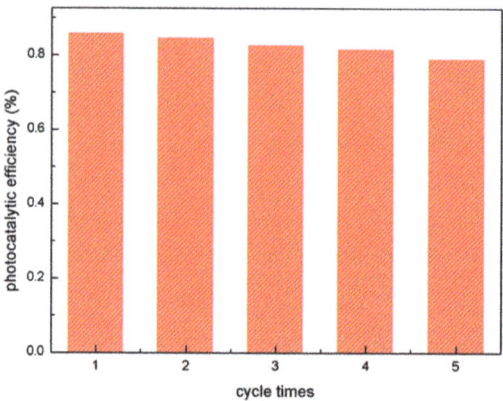

Figure 10. Cycling runs of the photocatalytic activity during the photocatalytic degradation of TC over 0.4Ni/Ti-Bi_2WO_6 photocatalyst under visible-light irradiation.

2.2.3. Roles of Reactive Species

It is vital to explore the predominant reactive species in the photocatalytic degradation of TC to comprehend the photocatalytic mechanism. In this study, the effects of three sacrificial agents on photocatalytic reactions under the same conditions were studied. The three sacrificial agents included tert-Butanol (TBA) for hydroxyl radicals (OH), triethanolamine (TEOA) for holes (h^+), and p-Benzoquinone (p-BQ) for superoxide radicals (O^{2-}). The photocatalytic efficiency would become lower when the corresponding active species was quenched in the photocatalytic degradation of TC. As shown in Figure 11, the photocatalytic performance of 0.4Ni/Ti-Bi_2WO_6 was not obviously inhibited when 1 mmol of TBA was added into the solution, indicating that the ·OH was not involved in the degradation of TC. However, whether the 1 mmol of TEOA or 1 mmol of BQ were added into TC solution, the photocatalytic performance of 0.4Ni/Ti-Bi_2WO_6 could be obviously affected, which indicated that h^+ and O^{2-} radicals were the predominant active species in the reaction system.

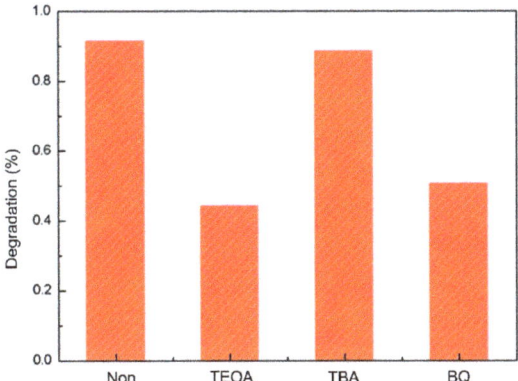

Figure 11. Trapping experiment of active species during the photocatalytic degradation of TC over 0.4 Ni/Ti-Bi$_2$WO$_6$ photocatalyst under visible-light irradiation.

2.3. Possible Photocatalytic Mechanism

To better describe the degradation process of TC, the main intermediates were identified by high-performance liquid chromatography and mass spectrometry in negative ion scan mode, as shown in Figure 12. The main intermediates of tetracycline degradation could be derived with mass-to-charge ratios m/z of 416, 373, and 306, etc. The intermediates of tetracycline degradation in general are mainly formed during the photocatalytic reaction by the removal of functional groups on the ring and ring opening reaction. Therefore, the pathways of tetracycline degradation were inferred, as shown in Figure 13. The m/z = 445 for tetracycline; product 1 (m/z = 416) was probably formed due to the removal of the methyl group from dimethylamine by tetracycline; product 1 was further formed by the removal of the deamidation group to form product 2 (m/z = 373). As the photocatalytic reaction proceeded, the ring opening reaction further occurred and product 2 (m/z = 373) was stripped of hydroxyl, carbonyl, and amino groups to form product 3 (m/z = 306). Eventually, these small molecules were further oxidized to form CO$_2$ and H$_2$O.

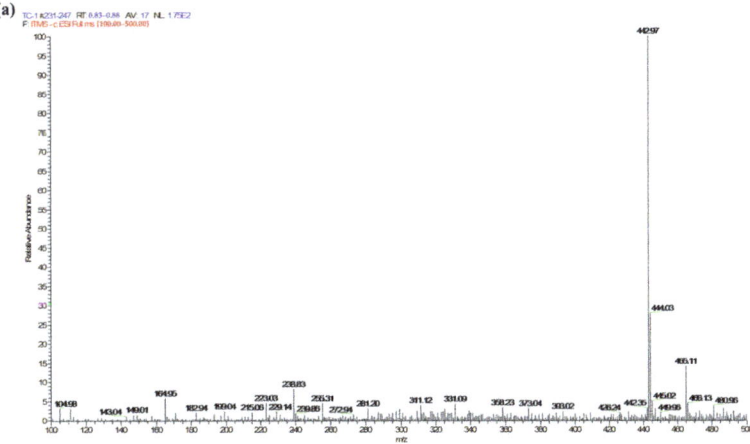

Figure 12. Cont.

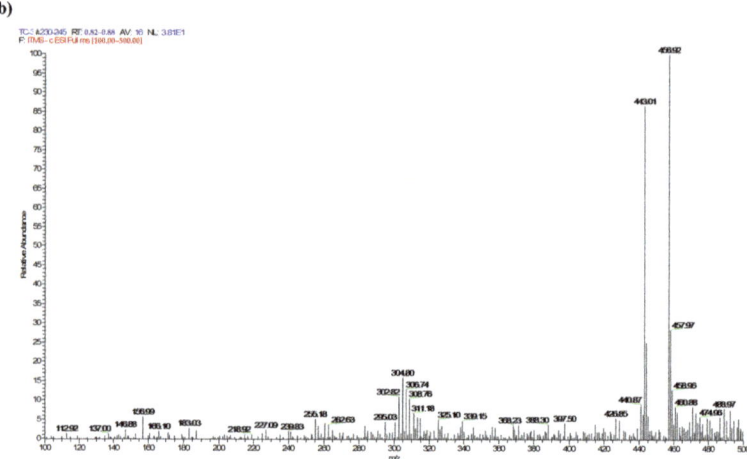

Figure 12. (a) HPLC-MS spectra of 0.4Ni/Ti-Bi$_2$WO$_6$ photocatalytic degradation of tetracycline at 0 min; (b) HPLC-MS spectra of 0.4Ni/Ti-Bi$_2$WO$_6$ photocatalytic degradation of tetracycline at 140 min.

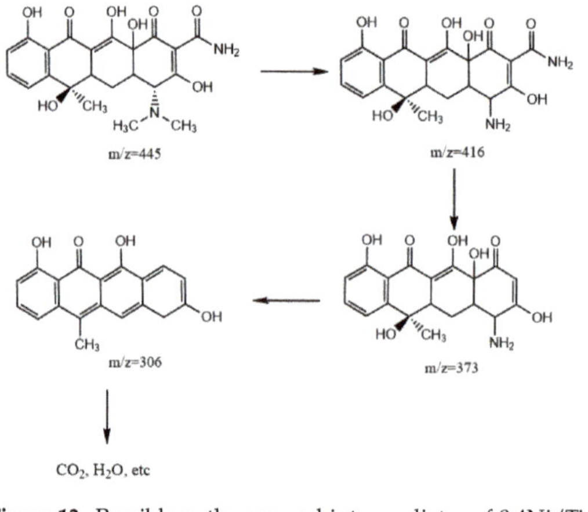

Figure 13. Possible pathways and intermediates of 0.4Ni/Ti-Bi$_2$WO$_6$ photocatalytic degradation of tetracycline.

According to the above results of the radical trapping experiments, the possible photocatalytic mechanism is presented in Figure 14. Under visible-light irradiation, the electrons and holes of Bi$_2$WO$_6$ were generated easily and separated, and the electrons were excited from the valence band (VB) to the conduction band (CB), leaving holes on the VB. However, these photogenerated electrons and holes might recombine and only a small part of electrons and holes could participate in the photocatalytic degradation of TC. Significantly, the Ni(II) cocatalyst that existed in Ni/Ti-Bi$_2$WO$_6$ samples could work as an electron trap to accept the photogenerated electrons. Then, photogenerated electrons reacted with oxygen in solution to form O^{2-} that has a strong oxidation ability to promote the degradation efficiency of TC. The photogenerated holes on the VB of Bi$_2$WO$_6$ could be rapidly transferred to the surface of the Ti(IV) cocatalyst, and directly oxidized the TC under visible-light irradiation. The corresponding reaction process can be expressed as follows:

$$0.4\text{Ni}/\text{Ti-Bi}_2\text{WO}_6 + h\nu \rightarrow h^+ + e^- \quad (1)$$

$$e^- + O_2 \rightarrow \cdot O_2^- \quad (2)$$

$$h^+ + H_2O \rightarrow \cdot OH + H^+ - \quad (3)$$

$$TC + (\cdot O_2^-, h^+) \rightarrow \text{Degradation products} \quad (4)$$

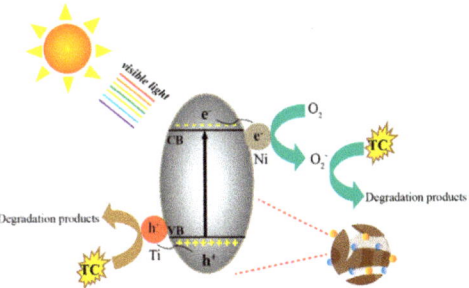

Figure 14. Possible mechanism of the enhanced photocatalytic activity during the photocatalytic degradation of TC over Ni/Ti-Bi$_2$WO$_6$ photocatalyst under visible-light irradiation.

3. Materials and Methods

3.1. Preparation of Photocatalyst

3.1.1. Preparation of Ni-doping Bi$_2$WO$_6$

Ni-doping Bi$_2$WO$_6$ was prepared by the hydrothermal method. The specific preparation process was as follows: Na$_2$WO$_4$·2H$_2$O (0.4948 g) was dissolved in 60 mL of deionized water and sonicated for 5 min. Bi(NO$_3$)$_3$·H$_2$O (1.45521 g) and certain amounts of NiCl$_2$·6H$_2$O were successively added into the solution during stirring and sonicated again for 10 min to obtain a homogeneous solution. The resulting solution was transferred into a 100 mL Teflon-lined stainless-steel autoclave (Autoclave, Zibo Haiyu Chemical Equipment Co., Ltd., Zibo, China) for hydrothermal treatment at 180 °C for 12 h. After the reaction system cooled to room temperature naturally, the products were collected by filtering with a 0.22 μm filter membrane, washed with deionized water and absolute ethanol three times, and dried at 60 °C overnight. According to the molar ratio of Ni to W, the products were referred to as 0.07Ni-Bi$_2$WO$_6$, 0.28Ni-Bi$_2$WO$_6$, and 0.49Ni-Bi$_2$WO$_6$, respectively. At the same time, single Bi$_2$WO$_6$ was also synthesized by the same process without adding NiCl$_2$·6H$_2$O.

3.1.2. Preparation of Ti-doping Bi$_2$WO$_6$

According to the previous studies, it was found that the samples showed the highest photocatalytic activity when the molar ratio of Ti to Bi$_2$WO$_6$ was 0.7. Therefore, the molar ratio of Ti to Bi$_2$WO$_6$ was determined to be 0.7 in this work. The Ti doping Bi$_2$WO$_6$ was prepared by an impregnation method. In a typical preparation, 1.25 g of Bi$_2$WO$_6$ and 0.32984 g of Ti(SO$_4$)$_2$ were dispersed into 200 mL of deionized water, and then stirred at 75 °C for 1 h. The products were collected by filtering with a 0.22 μm filter membrane and washed with deionized water to neutral. Lastly, the sample was dried at 60 °C overnight.

3.1.3. Preparation of Ni/Ti-doping Bi$_2$WO$_6$

The Ni/Ti-doping Bi$_2$WO$_6$ was synthesized by following Sections 3.1.1 and 3.1.2. First, the Ni-doping Bi$_2$WO$_6$ was prepared by the hydrothermal method according to part 2.2.1, and then amorphous Ti was further doped onto the Ni/Bi$_2$WO$_6$ surface to form Ni/Ti-doping Bi$_2$WO$_6$ by the impregnation method according to Section 3.1.2. According to the molar ratio of Ni to Ti, the products were referred to as 0.1Ni/Ti-Bi$_2$WO$_6$, 0.4Ni/Ti-Bi$_2$WO$_6$, and 0.7Ni/Ti-Bi$_2$WO$_6$.

3.2. Characterization

The crystal structure and purity of the prepared photocatalyst were characterized by X-ray diffraction (XRD) patterns, which were collected by a Bruker D8 Advanced (Bruker, Billerica, MA, USA) instrument using Cu-Kα radiation (λ = 0.15405 nm, 40 KV × 60 mA) from 10° to 80° (2θ) with a scanning rate of 15°/min. The morphologies of the samples were performed on a JSM-6700 field-emission scanning electron microscope (FESEM, JEOL Ltd., Tokyo, Japan) equipped with an X-max 50 energy-dispersive X-ray spectroscope (EDS, Oxford Instruments, Abington, UK) using the acceleration voltage of 2 kV. The optical properties of the photocatalyst were investigated by UV-vis diffuse reflectance spectra (DRS), which were monitored by a UV-Vis spectrophotometer (SHIMADZU, UV-2550, Kyoto, Japan), in which $BaSO_4$ served as the reflectance standard. X-ray photoelectron spectroscopy (XPS) measurements were conducted on a Thermo Scientific ESCALAB 250Xi (Thermo Fischer Scientific, Waltham, MA, USA) with a monochromated Al Kα X-ray source.

3.3. Photocatalytic Test

Photocatalytic Degradation

The photocatalytic activity was performed by the degradation of TC under visible-light irradiation using a 300 W Xenon lamp (Xenon lamp, China Education Au-light, Beijing, China) equipped with a 420 nm cut-off filter. The photocatalytic experiments were described as follows: First, 0.05 g of photocatalysts was mixed with 100 mL of 20 mg/L tetracycline solution. Then, the mixed solution was stirred in the dark for 30 min to exclude the effect of the adsorption. The light experiment was started after reaching the equilibrium of adsorption and desorption. During the irradiation process, 4 mL of the suspension was collected and filtered using a 0.22 μm filter membrane to remove the photocatalyst at a certain interval. Subsequently, the absorbance of the solution was measured by a UV-vis spectrometer (UV-vis spectrometer, Hitachi, Tokyo, Japan), where the characteristic absorption wavelengths of TC in solutions was 356 nm. By the standard curve of TC, the degradation rate of prepared samples was calculated. To demonstrate the stability of as-prepared samples, repeated experiments were carried out under the same conditions. The photocatalysts were separated by centrifugation, and washed with distilled water and ethanol three times before being redispersed into the TC solutions.

3.4. Active Species Capturing Experiments

Sacrificial agents, such as tert-Butanol (TBA), triethanolamine (TEOA), and p-Benzoquinone (p-BQ), to quench hydroxyl radicals (·OH), holes (h^+), and superoxide radicals ($·O^{2-}$), respectively, were used to determine the active species in the photocatalytic reaction. Typically, 10 mM scavenger was added into 100 mL of 20 mg/L of TC solution with 0.4Ni/Ti-Bi_2WO_6 as a photocatalyst at room temperature. The other experiment condition was the same as the photocatalytic degradation referred to above, for instance, the 30 min dark reaction process before irradiation. The main active species were decided by the degradation rate of TC.

4. Conclusions

In summary, we successfully designed Ni/Ti-Bi_2WO_6 composites for the degradation of TC under visible-light irradiation by a simple one-step hydrothermal and impregnation method. The photocatalytic efficiency of Ni/Ti-Bi_2WO_6 under visible light was improved compared with that of the pure Bi_2WO_6 photocatalyst. The highest degradation efficiency was achieved when the molar ratio of Ni/Ti in Ni/Ti-Bi_2WO_6 was 0.4. After 140 min of visible-light irradiation, the degradation efficiency of TC could reach 92.9%. This excellent photocatalytic ability of the Ni/Ti-Bi_2WO_6 composite can be attributed to the synergistic effect between Ti(IV) as a hole catalyst and Ni(II) as an electron catalyst, which prevents the recombination of photogenerated electron–hole pairs and increases the amount of active species for photodegradation of TC. The low-cost, non-toxic, and abundant Ti(IV) and Ni(II) co-catalysts can be ideal co-catalysts for potential applications of new photocatalytic materials compared to conventional noble metal co-catalysts such as Pt, Au, and RuO_2. In addition, the synthesis of the dual co-catalyst-modified photocatalysts used in this study can be extended for the synthesis of new dual co-catalyst-modified high-efficiency photocatalytic materials.

Author Contributions: C.S.: Writing—Original Draft; Investigation, Data Curation; K.Z.: Writing—Original Draft; Investigation, Data Curation; C.S. and K.Z. contributed equally to this paper. B.W.: Investigation, Data Curation; R.W.: Writing—Review and Editing, Supervision. All authors have read and agreed to the published version of the manuscript.

Funding: This research was funded by the Key Research and Development Program of Shandong Province, China [2017GSF217006].

Conflicts of Interest: The authors declare no conflict of interest in publishing the result.

References

1. Nie, M.; Yan, C.; Li, M.; Wang, X.; Bi, W.; Dong, W. Degradation of chloramphenicol by persulfate activated by Fe^{2+} and zerovalent iron. *Chem. Eng. J.* **2015**, *279*, 507–515. [CrossRef]
2. Kümmerer, K. Antibiotics in the aquatic environment—A review–part I. *Chemosphere* **2009**, *75*, 417–434. [CrossRef] [PubMed]
3. Chelliapan, S.; Wilby, T.; Sallis, P.J. Performance of an up-flow anaerobic stage reactor (UASR) in the treatment of pharmaceutical wastewater containing macrolide antibiotics. *Water Res.* **2006**, *40*, 507–516. [CrossRef] [PubMed]
4. Trovo, A.G.; Nogueira, R.F.P.; Agüera, A.; Fernandez-Alba, A.R.; Malato, S. Degradation of the antibiotic amoxicillin by photo-Fenton process–chemical and toxicological assessment. *Water Res.* **2011**, *45*, 1394–1402. [CrossRef] [PubMed]
5. Chatzitakis, A.; Berberidou, C.; Paspaltsis, I.; Kyriakou, G.; Sklaviadis, T.; Poulios, I. Photocatalytic degradation and drug activity reduction of chloramphenicol. *Water Res.* **2008**, *42*, 386–394. [CrossRef] [PubMed]
6. Nebel, C.E. A source of energetic electrons. *Nat. Mater.* **2013**, *12*, 780–781. [CrossRef] [PubMed]
7. Romão, J.; Mul, G. Substrate specificity in photocatalytic degradation of mixtures of organic contaminants in water. *ACS Catal.* **2016**, *6*, 1254–1262. [CrossRef]
8. Simsek, E.B. Solvothermal synthesized boron doped TiO_2 catalysts: Photocatalytic degradation of endocrine disrupting compounds and pharmaceuticals under visible light irradiation. *Appl. Catal. B Environ.* **2017**, *200*, 309–322. [CrossRef]
9. Fujishima, A.; Honda, K. Electrochemical photolysis of water at a semiconductor electrode. *Nature* **1972**, *238*, 37–38. [CrossRef]
10. Luo, M.-L.; Zhao, J.-Q.; Tang, W.; Pu, C.-S. Hydrophilic modification of poly (ether sulfone) ultrafiltration membrane surface by self-assembly of TiO_2 nanoparticles. *Appl. Surf. Sci.* **2005**, *249*, 76–84. [CrossRef]
11. Hu, J.; Li, H.; Muhammad, S.; Wu, Q.; Zhao, Y.; Jiao, Q. Surfactant-assisted hydrothermal synthesis of TiO_2/reduced graphene oxide nanocomposites and their photocatalytic performances. *J. Solid State Chem.* **2017**, *253*, 113–120. [CrossRef]
12. Liu, L.; Liu, Z.; Bai, H.; Sun, D.D. Concurrent filtration and solar photocatalytic disinfection/degradation using high-performance Ag/TiO2 nanofiber membrane. *Water Res.* **2012**, *46*, 1101–1112. [CrossRef] [PubMed]
13. Hong, L.-F.; Guo, R.-T.; Yuan, Y.; Ji, X.-Y.; Lin, Z.-D.; Li, Z.-S.; Pan, W.-G. Recent progress of transition metal phosphides for photocatalytic hydrogen evolution. *ChemSusChem* **2021**, *14*, 539–557. [CrossRef] [PubMed]
14. Yan, Y.; Yang, M.; Shi, H.; Wang, C.; Fan, J.; Liu, E.; Hu, X. $CuInS_2$ sensitized TiO_2 for enhanced photodegradation and hydrogen production. *Ceram. Int.* **2019**, *45*, 6093–6101. [CrossRef]
15. Yang, G.; Ding, H.; Chen, D.; Feng, J.; Hao, Q.; Zhu, Y. Construction of urchin-like $ZnIn_2S_4$-Au-TiO_2 heterostructure with enhanced activity for photocatalytic hydrogen evolution. *Appl. Catal. B Environ.* **2018**, *234*, 260–267. [CrossRef]
16. Liu, Y.; Yang, B.; He, H.; Yang, S.; Duan, X.; Wang, S. Bismuth-based complex oxides for photocatalytic applications in environmental remediation and water splitting: A review. *Sci. Total Environ.* **2022**, *804*, 150215. [CrossRef]
17. Zhang, L.; Wang, H.; Chen, Z.; Wong, P.K.; Liu, J. Bi_2WO_6 micro/nano-structures: Synthesis, modifications and visible-light-driven photocatalytic applications. *Appl. Catal. B Environ.* **2011**, *106*, 1–13. [CrossRef]
18. Sun, C.; Wang, R. Enhanced photocatalytic activity of Bi_2WO_6 for the degradation of TC by synergistic effects between amorphous Ti and Ni as hole–electron cocatalysts. *New J. Chem.* **2020**, *44*, 10833–10839. [CrossRef]
19. Yi, H.; Qin, L.; Huang, D.; Zeng, G.; Lai, C.; Liu, X.; Li, B.; Wang, H.; Zhou, C.; Huang, F. Nano-structured bismuth tungstate with controlled morphology: Fabrication, modification, environmental application and mechanism insight. *Chem. Eng. J.* **2019**, *358*, 480–496. [CrossRef]
20. Wang, M.; Han, Q.; Li, L.; Tang, L.; Li, H.; Zhou, Y.; Zou, Z. Construction of an all-solid-state artificial Z-scheme system consisting of Bi_2WO_6/Au/CdS nanostructure for photocatalytic CO_2 reduction into renewable hydrocarbon fuel. *Nanotechnology* **2017**, *28*, 274002. [CrossRef]
21. Li, Z.; Zhu, L.; Wu, W.; Wang, S.; Qiang, L. Highly efficient photocatalysis toward tetracycline under simulated solar-light by Ag+-CDs-Bi_2WO_6: Synergistic effects of silver ions and carbon dots. *Appl. Catal. B Environ.* **2016**, *192*, 277–285. [CrossRef]
22. Wang, Q.; Lu, Q.; Yao, L.; Sun, K.; Wei, M.; Guo, E. Preparation and characterization of ultrathin Pt/CeO_2/Bi_2WO_6 nanobelts with enhanced photoelectrochemical properties. *Dye. Pigm.* **2018**, *149*, 612–619. [CrossRef]

23. Liu, Y.; He, J.; Qi, Y.; Wang, Y.; Long, F.; Wang, M. Preparation of flower-like BiOBr/Bi$_2$WO$_6$ Z-scheme heterojunction through an ion exchange process with enhanced photocatalytic activity. *Mater. Sci. Semicond. Process.* **2022**, *137*, 106195. [CrossRef]
24. Liu, G.; Cui, P.; Liu, X.; Wang, X.; Liu, G.; Zhang, C.; Liu, M.; Chen, Y.; Xu, S. A facile preparation strategy for Bi$_2$O$_4$/Bi$_2$WO$_6$ heterojunction with excellent visible light photocatalytic activity. *J. Solid State Chem.* **2020**, *290*, 121542. [CrossRef]
25. Wang, Y.; Liu, Y.-J.; Li, S.-Y.; Ye, M.; Shao, Y.-H.; Wang, R.; Guo, L.-F.; Zhao, C.-E.; Wei, A. Low temperature synthesis of tungsten trioxide/bismuth tungstate heterojunction with enhanced photocatalytic activity. *J. Nanosci. Nanotechnol.* **2017**, *17*, 5520–5524. [CrossRef]
26. Wang, F.; Gu, Y.; Yang, Z.; Xie, Y.; Zhang, J.; Shang, X.; Zhao, H.; Zhang, Z.; Wang, X. The effect of halogen on BiOX (X = Cl, Br, I)/Bi$_2$WO$_6$ heterojunction for visible-light-driven photocatalytic benzyl alcohol selective oxidation. *Appl. Catal. A Gen.* **2018**, *567*, 65–72. [CrossRef]
27. Ning, J.; Zhang, J.; Dai, R.; Wu, Q.; Zhang, L.; Zhang, W.; Yan, J.; Zhang, F. Experiment and DFT study on the photocatalytic properties of La-doped Bi$_2$WO$_6$ nanoplate-like materials. *Appl. Surf. Sci.* **2022**, *579*, 152219. [CrossRef]
28. Longchin, P.; Sakulsermsuk, S.; Wetchakun, K.; Kidkhunthod, P.; Wetchakun, N. Roles of Mo dopant in Bi$_2$WO$_6$ for enhancing photocatalytic activities. *Dalton Trans.* **2021**, *50*, 12619–12629. [CrossRef]
29. Gu, H.; Yu, L.; Wang, J.; Ni, M.; Liu, T.; Chen, F. Tunable luminescence and enhanced photocatalytic activity for Eu (III) doped Bi$_2$WO$_6$ nanoparticles. *Spectrochim. Acta Part A Mol. Biomol. Spectrosc.* **2017**, *177*, 58–62. [CrossRef]
30. Song, X.C.; Zhou, H.; Zhen Huang, W.; Wang, L.; Zheng, Y.F. Enhanced photocatalytic activity of nano-Bi$_2$WO$_6$ by tin doping. *Curr. Nano.* **2015**, *11*, 627–632. [CrossRef]
31. Ge, L.; Han, C.; Liu, J. Novel visible light-induced g-C$_3$N$_4$/Bi$_2$WO$_6$ composite photocatalysts for efficient degradation of methyl orange. *Appl. Catal. B Environ.* **2011**, *108*, 100–107. [CrossRef]
32. Liu, L.; Ding, L.; Liu, Y.; An, W.; Lin, S.; Liang, Y.; Cui, W. Enhanced visible light photocatalytic activity by Cu$_2$O-coupled flower-like Bi$_2$WO$_6$ structures. *Appl. Surf. Sci.* **2016**, *364*, 505–515. [CrossRef]
33. Ge, M.; Li, Y.; Liu, L.; Zhou, Z.; Chen, W. Bi$_2$O$_3$–Bi$_2$WO$_6$ composite microspheres: Hydrothermal synthesis and photocatalytic performances. *J. Phys. Chem. C* **2011**, *115*, 5220–5225. [CrossRef]
34. Lu, Y.; Xu, Y.; Wu, Q.; Yu, H.; Zhao, Y.; Qu, J.; Huo, M.; Yuan, X. Synthesis of Cu$_2$O nanocrystals/TiO$_2$ photonic crystal composite for efficient p-nitrophenol removal. *Colloids Surf. A Physicochem. Eng. Asp.* **2018**, *539*, 291–300. [CrossRef]
35. Mehraj, O.; Pirzada, B.M.; Mir, N.A.; Sultana, S.; Sabir, S. Ag$_2$S sensitized mesoporous Bi$_2$WO$_6$ architectures with enhanced visible light photocatalytic activity and recycling properties. *RSC Adv.* **2015**, *5*, 42910–42921. [CrossRef]
36. Zhang, N.; Ciriminna, R.; Pagliaro, M.; Xu, Y.-J. Nanochemistry-derived Bi$_2$WO$_6$ nanostructures: Towards production of sustainable chemicals and fuels induced by visible light. *Chem. Soc. Rev.* **2014**, *43*, 5276–5287. [CrossRef]
37. Ran, J.; Zhang, J.; Yu, J.; Jaroniec, M.; Qiao, S.Z. Earth-abundant cocatalysts for semiconductor-based photocatalytic water splitting. *Chem. Soc. Rev.* **2014**, *43*, 7787–7812. [CrossRef]
38. Wen, J.; Li, X.; Li, H.; Ma, S.; He, K.; Xu, Y.; Fang, Y.; Liu, W.; Gao, Q. Enhanced visible-light H$_2$ evolution of g-C$_3$N$_4$ photocatalysts via the synergetic effect of amorphous NiS and cheap metal-free carbon black nanoparticles as co-catalysts. *Appl. Surf. Sci.* **2015**, *358*, 204–212. [CrossRef]
39. Liu, L.; Ji, Z.; Zou, W.; Gu, X.; Deng, Y.; Gao, F.; Tang, C.; Dong, L. In situ loading transition metal oxide clusters on TiO$_2$ nanosheets as co-catalysts for exceptional high photoactivity. *Acs Catal.* **2013**, *3*, 2052–2061. [CrossRef]
40. Yu, H.; Chen, W.; Wang, X.; Xu, Y.; Yu, J. Enhanced photocatalytic activity and photoinduced stability of Ag-based photocatalysts: The synergistic action of amorphous-Ti (IV) and Fe (III) cocatalysts. *Appl. Catal. B Environ.* **2016**, *187*, 163–170. [CrossRef]
41. Ohno, T.; Bai, L.; Hisatomi, T.; Maeda, K.; Domen, K. Photocatalytic water splitting using modified GaN: ZnO solid solution under visible light: Long-time operation and regeneration of activity. *J. Am. Chem. Soc.* **2012**, *134*, 8254–8259. [CrossRef] [PubMed]
42. Yang, J.; Yan, H.; Wang, X.; Wen, F.; Wang, Z.; Fan, D.; Shi, J.; Li, C. Roles of cocatalysts in Pt–PdS/CdS with exceptionally high quantum efficiency for photocatalytic hydrogen production. *J. Catal.* **2012**, *290*, 151–157. [CrossRef]
43. Higashi, M.; Domen, K.; Abe, R. Highly stable water splitting on oxynitride TaON photoanode system under visible light irradiation. *J. Am. Chem. Soc.* **2012**, *134*, 6968–6971. [CrossRef] [PubMed]
44. Xie, Y.P.; Liu, G.; Lu, G.Q.M.; Cheng, H.-M. Boron oxynitride nanoclusters on tungsten trioxide as a metal-free cocatalyst for photocatalytic oxygen evolution from water splitting. *Nanoscale* **2012**, *4*, 1267–1270. [CrossRef] [PubMed]
45. Liang, S.; Xia, Y.; Zhu, S.; Zheng, S.; He, Y.; Bi, J.; Liu, M.; Wu, L. Au and Pt co-loaded g-C$_3$N$_4$ nanosheets for enhanced photocatalytic hydrogen production under visible light irradiation. *Appl. Surf. Sci.* **2015**, *358*, 304–312. [CrossRef]
46. Zhu, X.; Yu, J.; Jiang, C.; Cheng, B. Catalytic decomposition and mechanism of formaldehyde over Pt–Al$_2$O$_3$ molecular sieves at room temperature. *Phys. Chem. Chem. Phys.* **2017**, *19*, 6957–6963. [CrossRef]
47. Yan, H.; Yang, J.; Ma, G.; Wu, G.; Zong, X.; Lei, Z.; Shi, J.; Li, C. Visible-light-driven hydrogen production with extremely high quantum efficiency on Pt–PdS/CdS photocatalyst. *J. Catal.* **2009**, *266*, 165–168. [CrossRef]
48. Keerthana, S.; Rani, B.J.; Ravi, G.; Yuvakkumar, R.; Hong, S.; Velauthapillai, D.; Saravanakumar, B.; Thambidurai, M.; Dang, C. Ni doped Bi$_2$WO$_6$ for electrochemical OER activity. *Int. J. Hydrog. Energy* **2020**, *45*, 18859–18866. [CrossRef]
49. Zhou, H.; Wen, Z.; Liu, J.; Ke, J.; Duan, X.; Wang, S. Z-scheme plasmonic Ag decorated WO$_3$/Bi$_2$WO$_6$ hybrids for enhanced photocatalytic abatement of chlorinated-VOCs under solar light irradiation. *Appl. Catal. B Environ.* **2019**, *242*, 76–84. [CrossRef]
50. Xu, Y.; Song, J.; Chen, F.; Wang, X.; Yu, H.; Yu, J. Amorphous Ti (iv)-modified Bi$_2$WO$_6$ with enhanced photocatalytic performance. *RSC Adv.* **2016**, *6*, 65902–65910. [CrossRef]

51. Singh, J.; Gusain, A.; Saxena, V.; Chauhan, A.; Veerender, P.; Koiry, S.; Jha, P.; Jain, A.; Aswal, D.; Gupta, S. XPS, UV–vis, FTIR, and EXAFS studies to investigate the binding mechanism of N719 dye onto oxalic acid treated TiO_2 and its implication on photovoltaic properties. *J. Phys. Chem. C* **2013**, *117*, 21096–21104. [CrossRef]
52. Wang, Z.; Liu, G.; Ding, C.; Chen, Z.; Zhang, F.; Shi, J.; Li, C. Synergetic effect of conjugated $Ni(OH)_2/IrO_2$ cocatalyst on titanium-doped hematite photoanode for solar water splitting. *J. Phys. Chem. C* **2015**, *119*, 19607–19612. [CrossRef]
53. Li, C.; Chen, G.; Sun, J.; Feng, Y.; Liu, J.; Dong, H. Ultrathin nanoflakes constructed erythrocyte-like Bi_2WO_6 hierarchical architecture via anionic self-regulation strategy for improving photocatalytic activity and gas-sensing property. *Appl. Catal. B Environ.* **2015**, *163*, 415–423. [CrossRef]
54. Liu, M.; Inde, R.; Nishikawa, M.; Qiu, X.; Atarashi, D.; Sakai, E.; Nosaka, Y.; Hashimoto, K.; Miyauchi, M. Enhanced photoactivity with nanocluster-grafted titanium dioxide photocatalysts. *Acs Nano* **2014**, *8*, 7229–7238. [CrossRef] [PubMed]
55. Yu, H.; Liu, R.; Wang, X.; Wang, P.; Yu, J. Enhanced visible-light photocatalytic activity of Bi_2WO_6 nanoparticles by Ag_2O cocatalyst. *Appl. Catal. B Environ.* **2012**, *111*, 326–333. [CrossRef]
56. Bard, A. *Standard Potentials in Aqueous Solution*; Routledge: London, UK, 2017.
57. Yu, J.; Hai, Y.; Cheng, B. Enhanced photocatalytic H_2-production activity of TiO_2 by $Ni(OH)_2$ cluster modification. *J. Phys. Chem. C* **2011**, *115*, 4953–4958. [CrossRef]
58. Xu, Y.; Xu, R. Nickel-based cocatalysts for photocatalytic hydrogen production. *Appl. Surf. Sci.* **2015**, *351*, 779–793. [CrossRef]
59. Ran, J.; Yu, J.; Jaroniec, M. $Ni(OH)_2$ modified CdS nanorods for highly efficient visible-light-driven photocatalytic H_2 generation. *Green Chem.* **2011**, *13*, 2708–2713. [CrossRef]
60. Wang, P.; Ming, T.; Wang, G.; Wang, X.; Yu, H.; Yu, J. Cocatalyst modification and nanonization of Ag/AgCl photocatalyst with enhanced photocatalytic performance. *J. Mol. Catal. A Chem.* **2014**, *381*, 114–119. [CrossRef]
61. Wang, P.; Xia, Y.; Wu, P.; Wang, X.; Yu, H.; Yu, J. Cu (II) as a general cocatalyst for improved visible-light photocatalytic performance of photosensitive Ag-based compounds. *J. Phys. Chem. C* **2014**, *118*, 8891–8898. [CrossRef]
62. Arif, M.; Zhang, M.; Mao, Y.; Bu, Q.; Ali, A.; Qin, Z.; Muhmood, T.; Liu, X.; Zhou, B.; Chen, S.-M. Oxygen vacancy mediated single unit cell Bi_2WO_6 by Ti doping for ameliorated photocatalytic performance. *J. Colloid Interface Sci.* **2021**, *581*, 276–291. [CrossRef] [PubMed]
63. Su, H.; Li, S.; Xu, L.; Liu, C.; Zhang, R.; Tan, W. Hydrothermal preparation of flower-like Ni^{2+} doped Bi_2WO_6 for enhanced photocatalytic degradation. *J. Phys. Chem. Solids* **2022**, *170*, 110954. [CrossRef]
64. Kumar, P.; Verma, S.; Korošin, N.Č.; Žener, B.; Štangar, U.L. Increasing the photocatalytic efficiency of $ZnWO_4$ by synthesizing a $Bi_2WO_6/ZnWO_4$ composite photocatalyst. *Catal. Today* **2022**, *397*, 278–285. [CrossRef]
65. Wang, C.; Liu, H.; Wang, G.; Fang, H.; Yuan, X.; Lu, C. Photocatalytic removal of metronidazole and Cr (VI) by a novel $Zn_3In_2S_6/Bi_2O_3$ S-scheme heterojunction: Performance, mechanism insight and toxicity assessment. *Chem. Eng. J.* **2022**, *450*, 138167. [CrossRef]
66. Sharma, S.; Ibhadon, A.O.; Francesconi, M.G.; Mehta, S.K.; Elumalai, S.; Kansal, S.K.; Umar, A.; Baskoutas, S. Bi_2WO_6/C-Dots/TiO_2: A novel Z-scheme photocatalyst for the degradation of fluoroquinolone levofloxacin from aqueous medium. *Nanomaterials* **2020**, *10*, 910. [CrossRef]
67. Kovalevskiy, N.; Cherepanova, S.; Gerasimov, E.; Lyulyukin, M.; Solovyeva, M.; Prosvirin, I.; Kozlov, D.; Selishchev, D. Enhanced Photocatalytic Activity and Stability of Bi_2WO_6–TiO_2-N Nanocomposites in the Oxidation of Volatile Pollutants. *Nanomaterials* **2022**, *12*, 359. [CrossRef] [PubMed]

Article

Direct Z-Scheme g-C₃N₅/Cu₃TiO₄ Heterojunction Enhanced Photocatalytic Performance of Chromene-3-Carbonitriles Synthesis under Visible Light Irradiation

Murugan Arunachalapandi, Thangapandi Chellapandi , Gunabalan Madhumitha, Ravichandran Manjupriya, Kumar Aravindraj and Selvaraj Mohana Roopan *

Citation: Arunachalapandi, M.; Chellapandi, T.; Madhumitha, G.; Manjupriya, R.; Aravindraj, K.; Roopan, S.M. Direct Z-Scheme g-C₃N₅/Cu₃TiO₄ Heterojunction Enhanced Photocatalytic Performance of Chromene-3-Carbonitriles Synthesis under Visible Light Irradiation. *Catalysts* 2022, *12*, 1593. https:// doi.org/10.3390/catal12121593

Academic Editors: Yongming Fu and Qian Zhang

Received: 1 November 2022
Accepted: 2 December 2022
Published: 6 December 2022

Publisher's Note: MDPI stays neutral with regard to jurisdictional claims in published maps and institutional affiliations.

Copyright: © 2022 by the authors. Licensee MDPI, Basel, Switzerland. This article is an open access article distributed under the terms and conditions of the Creative Commons Attribution (CC BY) license (https:// creativecommons.org/licenses/by/ 4.0/).

Chemistry of Heterocycles & Natural Product Research Laboratory, Department of Chemistry, School of Advanced Sciences, Vellore Institute of Technology, Vellore 632 014, Tamil Nadu, India
* Correspondence: mohanaroopan.s@vit.ac.in; Tel.: +91-0416-220-2313

Abstract: In order to make the synthesis of pharmaceutically active carbonitriles efficient, environmentally friendly, and sustainable, the method is regularly examined. Here, we introduce a brand-new, very effective $Cu_3TiO_4/g-C_3N_5$ photocatalyst for the production of compounds containing chromene-3-carbonitriles. The direct Z-Scheme photo-generated charge transfer mechanism used by the $Cu_3TiO_4/g-C_3N_5$ photocatalyst results in a suppressed rate of electron-hole pair recombination and an increase in photocatalytic activity. Experiments showed that the current method has some advantages, such as using an environmentally friendly and sustainable photocatalyst, having a simple procedure, quick reaction times, a good product yield (82–94%), and being able to reuse the photocatalyst multiple times in a row without noticeably decreasing its photocatalytic performance.

Keywords: chromene-3-carbonitriles; Cu_3TiO_4; g-C₃N₅; photocatalyst; Nanomaterials; Z-Scheme heterojunction

1. Introduction

Multi-component reactions (MCRs) have garnered a lot of interest recently as a method for creating a variety of heterocyclic compounds with applicability in medicine, pharmaceuticals, and agrochemistry [1,2]. One-pot MCRs, which induce organic reactants to react together in a single step, have emerged as an alternative platform for organic chemists due to their simple operation, effective purification techniques, side products, and quick turnaround times [3,4]. In simple terms, a multi-component reaction is one in which three or more reactants mix in a single vessel to create a single output that effectively contains every atom of the starting elements (with the exception of condensation products, such as water molecules, hydrochloric acid, or methanol) [5–7].

A number of naturally occurring O-bearing heterocyclic organic moieties, which are widely present in edible fruits and vegetables, contain the chromene and related structural skeletons as their major structural motifs [8,9]. Numerous industries, like agrochemistry, medicine, as well as material sciences, use carbonitrile-based heterocycles and their derivatives as ligands and catalysts [10]. Carbonitrile compounds have a high level of therapeutic activity and can be employed as antibacterial [11,12], anti-tumor [13,14], anti-inflammatory, anti-cancer [15–17], and anti-convulsant drugs [18]. Compounds with a carbonitrile base are also employed to make electrical and optical materials as well as anion sensors [19,20]. Given the many different uses for chromenes, synthetic chemists have been exploring new and effective catalysts. In the traditional way of making carbonitrile-based chemicals, the product needed to be made at a high temperature and through a long process called "synthesis".

As a result of growing environmental worry, a lot of work has been put into creating innovative procedures that reduce emissions in chemical synthesis. From the perspective of green and sustainable chemistry [21–23], designing effective and affordable chemical processes utilising heterogeneous catalysts to create fine chemicals and pharmaceutical goods through MCRs has recently attracted a lot of attention from academics and businesses. The fundamental drawback of heterogeneous catalysts is, however, their decreased catalytic activity. The best way to deal with this issue is to minimize the size of the particles in the heterogeneous catalyst [24,25]. Metal oxide nanocatalysts combine the benefits of photo and heterogeneous catalysts. Although this type of material has many valuable advantages, including a high surface area, efficient particle distribution, a straightforward recovery procedure, effective light absorption property, good stability, and non-toxic behaviour, their propensity to agglomerate and oxidise is a drawback [26–28]. Immobilizing organic molecules on metal oxide nanoparticles is thought to be a workable way to address this drawback.

Given the significance of metal oxide, nanoparticles (metal oxide NPs) have received a lot of attention and are widely employed in water splitting, catalysis, photocatalysis, energy, and environmental remediation [29–32]. Titanium dioxide (TiO_2) and copper oxide (CuO) nanoparticles have attracted a great deal of interest. It was discovered that CuO and TiO_2 NPs worked effectively as catalysts for the oxidation of organic molecules as well as CO and NO [33,34]. Additionally, in last decade, a lot of attention has been focused on the use of CuO and TiO_2 NPs for chemical reactions [35]. CuO NPs' high levels of electron-hole pair recombination and TiO_2's less obvious active nature are some of its disadvantages. We created the Cu_3TiO_4 NPs for the organic reaction based on the aforementioned investigative information. In the context of heterogeneous nanocatalysts, this class of catalysts seems to be one of the most promising approaches toward effective reactions under mild and ecologically nonthreatening circumstances.

Organic synthesis research in the area of photocatalysis, notably with regard to visible light, is appealing [36]. Heterogeneous photocatalysis has been given a significant role by carbon-based materials. Due to a variety of characteristics, including stability, narrow bandgap, 2D nanomaterial, excellent visible-light active absorption, low density, and low toxicity, graphitic-carbon nitride (g-C_3N_4) is an effective catalyst under visible light [37,38]. Additionally, its layered structure makes the anchor metal an appropriate substrate for its direct use as a heterogeneous photocatalyst. A variety of modified carbon nitrides, including nanosheets, nanorods, metal-doped and non-metal modified carbon nitrides are effective nanomaterials [39–41]. An intriguing method has been recently created to modify the polymeric structure of C_3N_4 in order to create novel carbon nitrides with a wider visible-light sensitive range. For instance, integrating an N-rich 3-amino-1,2,4-triazole unit into the s-heptazine motif led to the highly ordered C_3N_5, which can increase the C:N atom ratio from 3:4–3:5. The lower bandgap energy is a result of the extra N-atoms connected to the terminal NH_2 groups in the form of 1,2,4-triazole, which are different from the usual C_3N_4 in that they are bridged by a N-atom through sp^2 hybridization [42–44]. The construction of heterojunction photocatalysts based on C_3N_4 has received extensive research, due to its viability and efficiency for the spatial separation of photogenerated electron-hole pairs. Appropriate catalyst materials can generate heterojunctions, which can lower the activation energy needed for the reduction reaction. The secondary target is to encourage the division of photogenerated charge carriers while simultaneously providing more active locations for the photocatalytic activity. Meticulous design of a heterojunction is a promising method to increase photocatalytic activity, because of its rapid charge carrier separation caused by the integrated electric field [45–48]. A switch from photocatalysts based on C_3N_5 to those based on the organic transformation reaction is driving recent advancements for cleaner, more sustainable chemistry. The heterojunction is followed by carbon nitride-based catalysts like Type-II and Z-Scheme. A Z-scheme system has also been built with a pertinent shuttle mediator to obtain better light-harvesting potential with high charge separation performance and sturdy redox ability, furthermore to photocatalysts with conventional heterojunctions that can successfully separate photogenerated electron-hole pairs via band alignment [49,50]. Numerous studies on g-C_3N_4 linked to the Z-Scheme electron transfer mechanism have been reported, including those on $TiO_2/g-C_3N_4$ [51], $CuO/g-C_3N_4$ [52], $ZnO/g-C_3N_4$ [53], $WO_3/g-C_3N_4$ [54], V_2O_5/C_3N_4 [55], MnO_2/C_3N_4, [56] and $Bi_2O_3/g-C_3N_4$ [57], among others. In recent years, $g-C_3N_5$-based heterojunction production has become increasingly significant in photocatalysis and other applications. There are currently very few studies on Z-Scheme mechanisms for $g-C_3N_5$-based heterojunctions, including Ag_3PO_4/C_3N_5 [58], $FeOCl/g-C_3N_5$ [59], $Bi_4O_5I_2/g-C_3N_5$ [60], $LaCoO_3/g-C_3N_5$ [61] and $Bi_2WO_6/g-C_3N_5$ [62,63]. Still, there is no report carried out on $Cu_3TiO_4/g-C_3N_5$ heterojunction fabrication and photocatalytic application.

Light-emitting diodes (LEDs) are the best type of light source for photocatalytic applications. BLUE LEDs have benefits over other light sources in terms of efficiency, the qualities of power, compatibility, longevity, and environmental friendliness. Due to their unique characteristics, BLUE LEDs offer an improvement over traditional lighting sources and allow for more innovative scope when designing different photochemical reactions [64].

In this research, we have developed a novel and very effective $Cu_3TiO_4/g-C_3N_5$ photocatalyst for the construction of chromene-3-carbonitrile molecules under BLUE LED light. The direct Z-scheme charge separation mechanism employed by the $Cu_3TiO_4/g-C_3N_5$ hetero-junction photocatalyst reduces the rate of electron-hole (e^-/h^+) pair recombination of Cu_3TiO_4 and increases its photocatalytic activity. The as-prepared $g-C_3N_5/Cu_3TiO_4$ nanocomposites showed better photocatalytic activity and stability when compared to pure Cu_3TiO_4 and pure C_3N_5. The spatial separation of photoinduced charge carriers could be attributed to the improved photocatalytic performance. The characteristics of the obtained catalysts have been studied with the help of XRD, UV-Vis DRS, FT-IR, FE-SEM, TEM, Zetapotential, and XPS. The applications for the catalyst have also been demonstrated and characterized. This work will be opening up a new path in the field of photocatalysts and MCRs and can further be developed for a wide range of applications other than the proposed one.

2. Results and Discussion

2.1. Characterization of Prepared Nanomaterials

The synthesized nanocomposite nanomaterial crystalline phase structure was investigated using the X-ray diffraction technique. In Figure 1, the pristine $g-C_3N_5$ sheets peak at 13.1° was related to (100), and another strong peak at 27.2° was related to (002) for graphitic $g-C_3N_5$ sheets (JCPDS card-87-1526) [65]. Peaks for Cu_3TiO_4 nanoparticles were found at 2theta = 26.4, 34.2, 38.6, 51.5, 59.1, 63.2, and 70.3° in relation to planes (004) (101) (102) (104) (006) (110) (112) (107), which corresponds well to JCPDS card no. 83-1285 [66]. Both are well-reflected in all CN5CT nanocomposites. Figure 1 shows the p-XRD patterns obtained for $g-C_3N_5/Cu_3TiO_4$ with varying wt. % Cu_3TiO_4. The XRD patterns of the 5% CN5CT composite showed no discernible change, while less Cu_3TiO_4 has been deducted because of the low loading percentage. The intensity of Cu_3TiO_4 nanoparticles peaks increased as the Cu_3TiO_4 nanoparticles ratio was increased to 10%, 20%, and 30%. Finally, Cu_3TiO_4 nanoparticles were supported on $g-C_3N_5$ nanosheets. As confirmed, they have shown good crystalline nature.

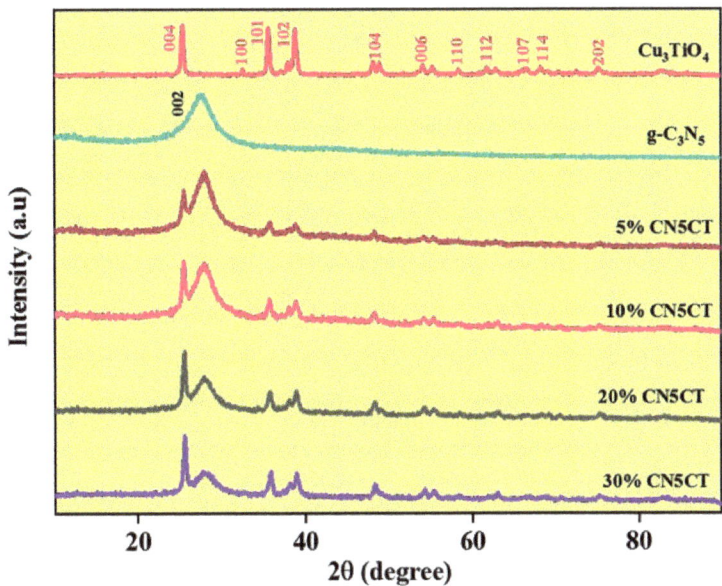

Figure 1. XRD images of prepared CN, CT and various percentages CNCT nanocomposites.

In BET analysis the surface area of the designed nanocomposites was examined, and depicted in Figure 2a. The surface area of Cu_3TiO_4 nanoparticles, g-C_3N_5 and g-C_3N_4 values are 10, 15, and 14 m^2/g respectively. Then 10% CN5CT nanocomposite has 19 m^2/g after supporting Cu_3TiO_4-nanoparticles, which thereby increased the surface area. The Cu_3TiO_4/g-C_3N_4 has less surface area compared to 10% CN5CT, as shown in Figure 2a. From BET analysis, 10%CN5CT exhibited more surface and active sites compared to others. In FT-IR spectroscopy (Figure 2b), g-C_3N_5 sheets showed peaks at 808, 1234, 1405, 1575, and 2160 cm^{-1}, respectively. The signal at 808 cm^{-1} corresponds to the existence of triazine units in g-C_3N_5 nanosheets. NH_2 and NH peaks were obtained at 3500 cm^{-1} and triazine unit stretching bands were found at 1200 cm^{-1}. The g-C_3N_5 nanosheet FT-IR report is consistent with previous literature. [67,68]. The FT-IR spectrum of Cu_3TiO_4 exhibited a range between 400–700 cm^{-1} bands, which is owing to Ti-O-Ti and Cu-O stretching vibration [69]. After that, Cu_3TiO_4 nanoparticles supported the slight shift that happened in 5%, 10%, 20%, and 30% CN5CT nanocomposite.

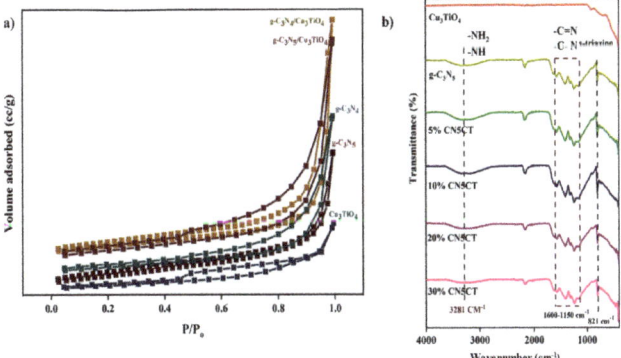

Figure 2. (a) BET isotherm and (b) FT-IR spectra.

In XPS, the chemical state of the prepared 10% CN5CT was exposed in Figure S1. Figure S1a shows the survey spectrum of the synthesized 10% CN5CT nanocomposite and all the signals confirmed the presence of C, N, Cu, Ti, and O. For C 1s, in Figure S1b, the three strong peaks at ~287, 285.8 and 283 eV show the presence of C 1s in nanocomposite, which corresponds to N-C-N, C-N and C-C bonds [70]. Then N 1s displayed two peaks at 399 and 397 eV, providing evidence for the presence of N 1s species (Figure S1c) with different environments of N-C and C-N-C bonds [71]. Cu 2p exposed two major peaks at 953 and 933 eV (Figure S1d) combined with two satellite lines, confirming the presence of Cu 2p species in the CN5CT composite [72]. The two peaks (Figure S1e) at 462 and 456 eV confirm the presence of Ti 2p species in CN5CT nanocomposite [72]. Finally, for O1s species, two strong peaks (Figure S1f) at 529 and 531 eV confirmed the existence of oxygen-metal and oxygen-hydrogen bonds in the composite.

The typical morphology of the g-C_3N_5 and g-C_3N_5/Cu_3TiO_4 nanocomposite is shown in Figure 3a,b, indicating that the prepared g-C_3N_5 has a sheet-like morphology. As seen in Figure 3a, exfoliation caused the evolution of 3-amino-1,2,4-triazole into a typical layered structure, showing the formation of a loosely packed 2D g-C_3N_5 structure. In Figure 3b, 10%CN5CT nanocomposite, and Cu_3TiO_4 nanoparticles are marked. The 10% CN5CT nanocomposites showed a strong connection between g-C_3N_5 and Cu_3TiO_4, indicating that there would be a physical adsorption. It also implied that g-C_3N_5 might prevent the aggregation of Cu_3TiO_4 NPs as compared to Figure 3b. The effectiveness of the electron-hole separation can be increased by closing interfacial contacts between g-C_3N_4 and Cu_3TiO_4 nanoparticles by excellent dispersion. According to the elemental mapping in Figure 3c–f, the elements C, N, Ti, Cu, and O were evenly distributed throughout the 2D layered and 3D nanoparticle structures.

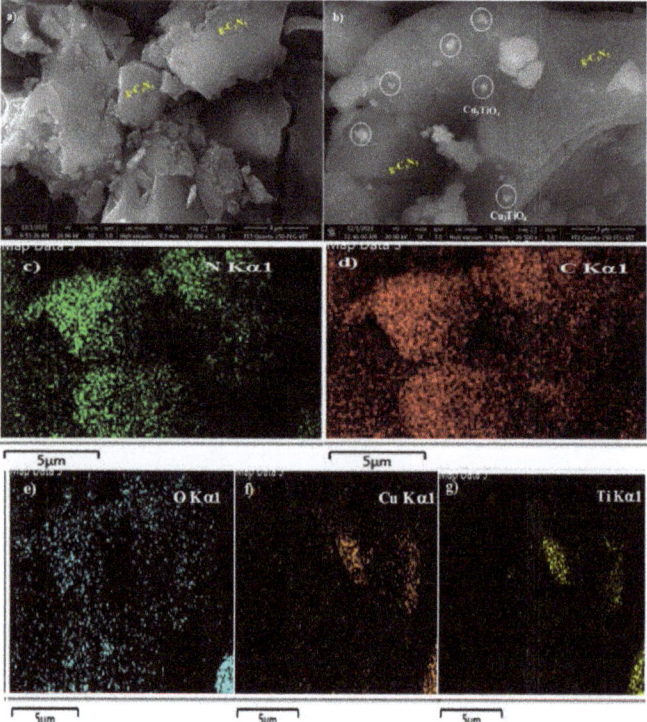

Figure 3. FE-SEM images of (a) g-C_3N_5 (b) 10% CN5CT nanocomposite and (c–g) elemental mapping of 10% CN5CT.

The perfect morphology and structural nature of the g-C_3N_5 and 10% CN5CT were recorded by using TEM (Figure 4a,b). The agglomerated sheet-like structure of g-C_3N_5 was confirmed by a TEM image at 50 nm (Figure 4a). The surface of the g-C_3N_5 was smooth and had high porosity, which was confirmed by the sheet-like structure of the g-C_3N_5. The 3-AT was thermal disintegration, copolymerization, and elimination of CO_2 and NH_3, which may result in the development of g-C_3N_5 nanosheets, which have a smaller and thinner nanosheet like structure than g-C_3N_4. Both g-C_3N_4 and g-C_3N_5 are similar in appearance. The EDS of g-C_3N_5 provides information about the given sample, g-C_3N_5, which has a higher quantity of nitrogen atoms compared to carbon atoms (Figure 4c). On introducing Cu_3TiO_4 into g-C_3N_5 nanosheets, no changes were observed on the surface of g-C_3N_5. The Cu_3TiO_4 NPs are uniformly distributed on the g-C_3N_5 surface and form an effective heterojunction (Figure 4b). In Figure 4d, EDS analysis of 10%CNCT confirms the elements with weight percentage viz., N (52.4%), C (35.1%), O (8.7%), Ti (2%), and Cu (1.8%). Figure 4e depicts the SEAD pattern of the 10%CN5CT, demonstrating that the prepared composite has a good crystalline nature and corresponding planes of (111), (103), (102), and (106) in XRD. The average particle size of Cu_3TiO_4 (~26 nm) in 10%CN5CT was calculated and given in Figure 4f. From the TEM image, the surface area of the compound increased with the decrease in Cu_3TiO_4 particle size. The photocatalyst 10% CN5CT was found to coexist on the surface of g-C_3N_5 sheets. The Cu_3TiO_4 (dark-coloured) nanoparticles are uniformly scattered on the g-C_3N_5 matrix to produce an efficient heterojunction structure, with the average particle size of Cu_3TiO_4 being found to be around 50 nm. The large surface area enhances the photocatalytic activity. Due to their excellent surface area, both g-C_3N_5 and Cu_3TiO_4 have close contact with each other. Due to this nature, the 10%CN5CT nanocomposite has better photocatalytic activity.

The charge separation and transfer of the electron-hole pair were analysed by using photoluminescence (PL) emission spectroscopy. Figure 5a shows the PL results for the prepared CN, CT, and CN5CT at various loading percentages. The g-C_3N_5 nanosheets emission peak was observed at around ~470 nm with high intensity, which confirmed the high recombination rate. On the other hand, the introduction of Cu_3TiO_4 nanoparticles on the surface of g-C_3N_5 nanosheets reduced the intensity of the peak, which confirmed the suppression of recombination rate of the CN5CT nanocomposite. Notably, when compared to other CN5CT nanocomposite, the 10% CN5CT nanocomposite has a lower emission peak. The PL investigation is strong evidence that 10%CN5CT has less electron-hole pair recombination than 5%, 20%, and 30% of CN5CT.

The thermal stability of the prepared nanocomposite was carried out by using thermo gravimetric analysis (TGA). Here, in Figure 5b, is depicted the high stability of Cu_3TiO_4 nanoparticles, because no characteristic decomposition occurred in the Cu_3TiO_4 nanoparticles and up to 98% of Cu_3TiO_4 nanoparticles were getting residue at 800 °C, showing the high stability nature of nanoparticles. At 530 °C, g-C_3N_5 nanosheets began decomposing, and all of the carbon and nitrogen in g-C_3N_5 was easily eliminated above 700 °C (1% residue at 800 °C). In 10% CN5CT nanocomposite, g-C_3N_5 nanosheets got decomposed at 530 °C but the residue obtained was 12%, which shows the presence of Cu_3TiO_4 nanoparticles in g-C_3N_5 nanosheets at a 10% ratio was confirmed. The 10%CN5CT nanocomposite has improved the thermal stability compared to g-C_3N_5.

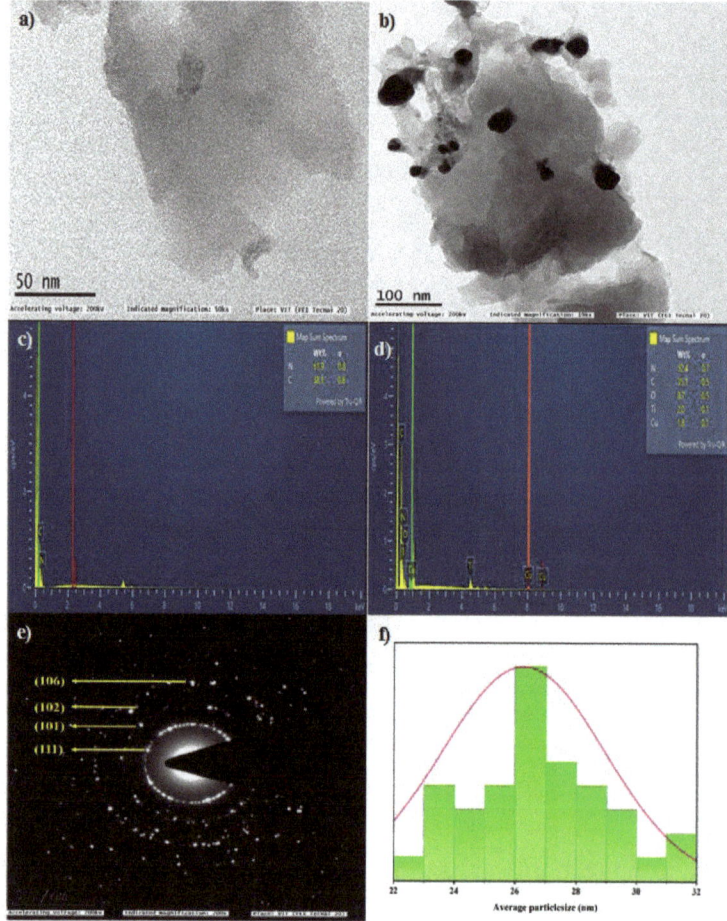

Figure 4. (**a**,**b**) TEM images, (**c**,**d**) EDS spectra of g-C$_3$N$_5$, 10% CN5CT nanocomposite (**e**) SEAD pattern of 10% CN5CT. (**f**) Average particle size of Cu$_3$TiO$_4$ nanoparticles in 10% CN5CT.

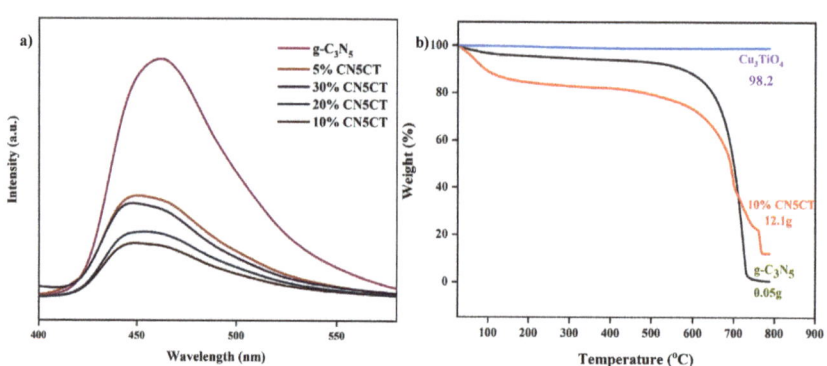

Figure 5. (**a**) PL emission spectra of g-C$_3$N$_5$, 5% CN5CT, 30% CN5CT, 20% CN5CT, 10% CN5CT (**b**) TGA analysis of Cu$_3$TiO$_4$, g-C$_3$N$_5$, and 10% CN5CT nanocomposite.

2.2. Photocatalytic Activity of g-C_3N_5/Cu_3TiO_4 Nanocomposite

In this study, we concentrated on the synthesis of different carbonitriles using a g-C_3N_5/Cu_3TiO_4 (CN5CT) photocatalyst. The catalyst exhibited better efficiency for the synthesis of carbonitriles in the presence of light. Ethanol has been used as solvent, which generated a yield of 94% (Table 1, entry 6). The reaction was carried out in the presence of BLUE LED visible light using a photocatalyst employing diketone, malanonitrile, and aldehyde (Scheme 1). Based on the characterization results, we investigated the CN5CT nanocomposite, which has a strong light absorption capacity and thereby increases the life time of electrons in the conduction band. The high electron-hole recombination rate improves the activity of the catalyst in the production of 1.4a. In this case, 20 mg of the catalyst was combined with the reactants and exposed to Blue LED light. Simultaneously, radicals formed on the catalyst surface, resulting in the formation of activated complexes. These are highly active species that are rapidly degraded to create products on the catalyst surface. In Table 1, the reaction conditions were adjusted to determine the optimal conditions for carbonitrile synthesis. Here, the solvent, catalyst, condition, and time were varied to determine the optimization condition. (Table 1 entries 1 and 3), produced yields (62% and 33%) that were significantly lower than the CN5CT nanocomposite. CN5CT nanocomposite and ethanol solvent worked better in all situations in the presence of light for 3 min. In all situations, the CN5CT nanocomposite and ethanol solvent worked better in all situations in the presence of light for 3 min. The increase in catalyst percentage also did not affect the product yield and for 5%, 20% and 30%, CN5CT nanocomposites give good yields (Table 1, entry 5, 7, 8).

Table 1. Optimization for the synthesis of Carbonitriles in BLUE LED visible light.

S.No	Catalyst	Solvent	Conditions	Time (minutes)	Yield [a] (%)
1.	CuO	EtOH	Light	60	62
2.	CuOAc	EtOH	Light	60	66
3.	TiO_2	EtOH	Light	60	33
4.	Cu_3TiO_4	EtOH	Light	60	75
5.	5% CN5CT	EtOH	Light	3	67
6.	10% CN5CT	EtOH	Light	3	94
7.	20% CN5CT	EtOH	Light	3	81
8.	30% CN5CT	EtOH	Light	3	89
9.	10% CN5CT	H_2O	Light	10	61
10.	10% CN5CT	THF	RT	60	23
11.	10% CN5CT	Acetonitrile	RT	60	21
12	10%CN5CT	EtOH	RT	10	95
13.	Without Catalyst	EtOH	RT	10	Trace
14.	10% CN5CT	Solvent free	Heat	10	78

Diketone (1 mmol), aldehyde (1 mmol), malononitrile (1 mmol), ethanol, photocatalyst (20 mg). [a] Yields—Isolated yields.

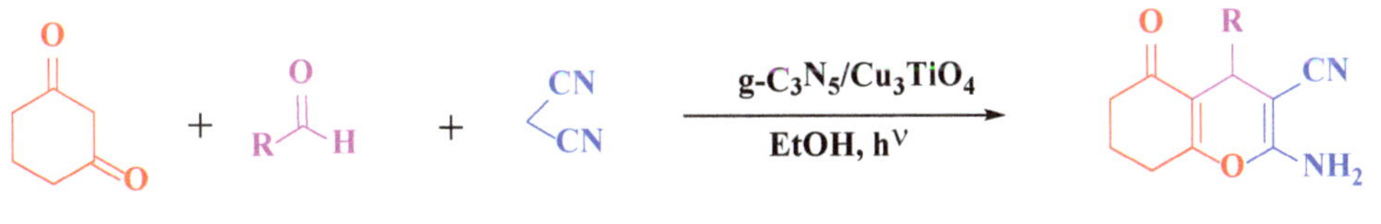

Scheme 1. Synthesis of hexahydroquinoline-2-carbonitrile using CN5CT nanocomposite in Blue LED light.

The optimal condition from Table 1 was utilized to synthesize different carbonitrile derivatives, which is shown in Table 2. The physicochemical properties and evidence (FT-IR, ^{1}H, ^{13}C NMR, and GC-MS) of the synthesized carbonitrile derivatives 1.4a-g are shown in Supporting Information Figures S2–S29. Light excites the electrons on the catalyst surface, causing them to move into the excited conduction band. The radicals ($^{\bullet}$O$_2^-$ and $^{\bullet}$OH) then react with the reactants to generate intermediates, which are then broken down to form the desired products on the catalyst surface. The catalyst, CN5CT nanocomposite, showed exceptional activity in the preparation of compound 1.4a. The recovered catalyst also performed admirably in the subsequent runs, with better yields.

Table 2. Scope of different substrate.

Name	R	Time	TON [a]	TOF [a]	Yield [b]	Ref. [c]
1.4a	-Ph	5	865	3.52	85	[73]
1.4b	4-(OCH$_3$)-Ph	5	832	3.26	82	[73]
1.4c	2-NO$_2$-Ph	3	876	3.78	88	[73]
1.4d	4-OH-Ph	5	831	3.22	81	[73]
1.4e	4-F-Ph	5	878	3.86	90	[73]
1.4f	2-Cl-Ph	3	923	4.63	94	[73]
1.4g	2-Nap	3	902	4.12	92	[73]

Diketone (1 mmol), aldehyde (1 mmol), malononitrile (1 mmol), ethanol, photocatalyst (20 mg). [a] TON—Turn Over Number; TOF—Turn Over Frequency. [b] Yields—Isolated yields. [c]—reference.

In Table 2, we studied how several heterocycles were synthesized. We successfully prepared several substituted aldehydes such as electron donating, electron withdrawing, and halogenated aldehydes here. When donating substitutions were compared to others, they produced moderate yields (Table 2, compound 1.4b). In the proposed procedure, napthaldehyde and benzaldehyde substitutions produced the highest yields of all. The multi-component technique made good yields and cut down on the amount of time needed to make 1.4a.

The reaction was carried out using CN5CT and irradiated for 3 min in an open environment at RT with a BLUE LED light. Remarkably, 94% of the targeted product (1.4a) was achieved. Nevertheless, when ambient O$_2$ was removed from the reaction mixture, the conversion was minimal and only a trace amount of product could be found if the reaction was completed in darkness or without photocatalysts. As a result, CN5CT photocatalyst, atmospheric oxygen, and BLUE LED visible light are all required for this transition.

2.3. Proposed Mechanism of 1.4a Synthesis

Based on the findings, a plausible mechanism was illustrated in Scheme 2. The more stable triplet state CN5CT* of the photoredox catalyst is then activated by the absorption of visible light and performs a single electron transfer (SET). This single electron transfer from 1.1 and 1.3 to the excited state of the CN5CT photocatalyst then lost a hydrogen atom and generated malononitrile radical (1.1a) and 1,3 cyclohexanedione radical (1.3a). The Knoevenagel condensation process between aldehyde (1.2) and activated malononitrile (1.1a) resulted in the synthesis of alkene intermediate (1.2a) [74]. Then, the intermediates 1.2b and 1.3a react in situ via Michael addition to give an adduct 1.3b, followed by intramolecular cyclisation of 1.3c, giving the desired product 1.4.

2.4. Transport Process of Photo-Excited Charge Carriers

The prepared samples' light absorption in the 200–800 nm region was examined in order to obtain an appearance in the optical harvest property. The absorption edge of g-C$_3$N$_5$ is located at approximately 680 nm, as shown in Figure 6a. In contrast, the 10% CN5CT nanocomposite clearly showed a shift, which may have been caused by the intense interface interaction between Cu$_3$TiO$_4$ and g-C$_3$N$_5$. As a result, the representative

10 weight percentage (10%) CN5CT nanocomposite photocatalyst exhibits a greater capacity to capture visible light than 2D g-C_3N_5.

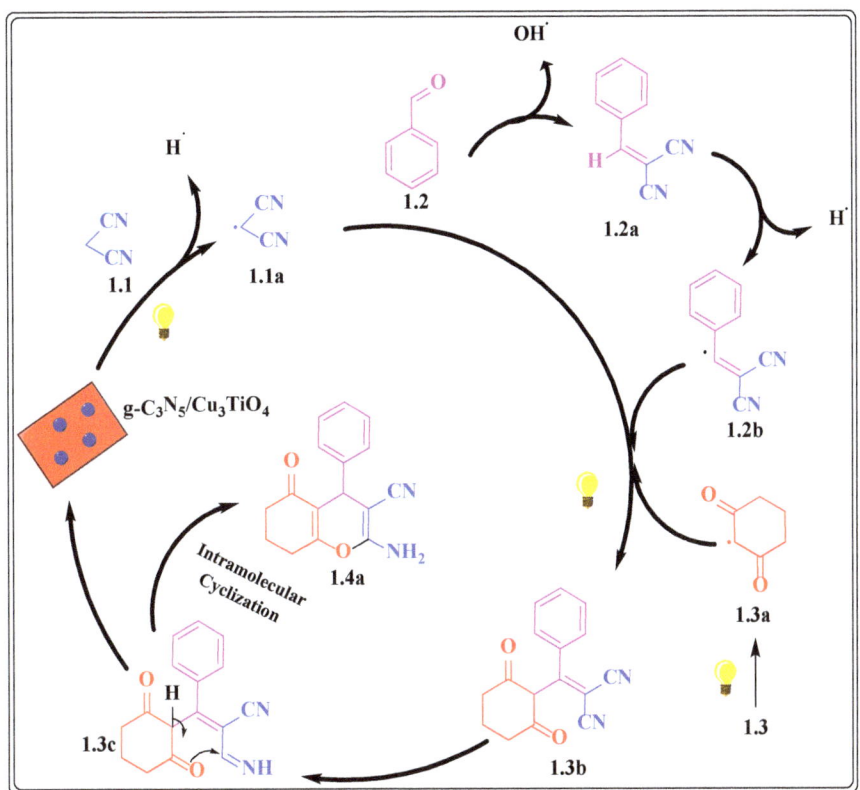

Scheme 2. Mechanism of hexahydroquinoline-2-carbonitrile using g-C_3N_5/Cu_3TiO_4.

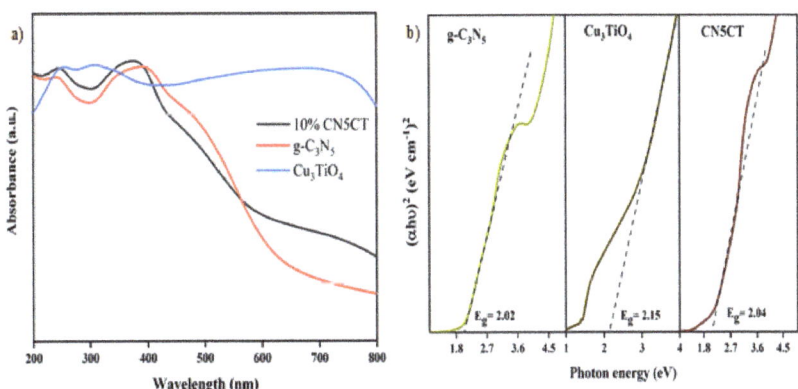

Figure 6. (a) UV-Vis DRS spectra and (b) bandgap plot.

As seen in Figure 6b, the band-gap values were also calculated from the intercept of tangents to plots of $(Ah\nu)^{1/2}$ versus photon energy. The optical band gap of g-C_3N_5,

Cu$_3$TiO$_4$ and 10% CN5CT was calculated and found to be 2.02, 2.15, and 2.04 eV respectively. In comparison to pure g-C$_3$N$_5$ and Cu$_3$TiO$_4$, the 10% CN5CT nanocomposite exhibits a narrower band gap. According to the findings, the photocatalyst for the 10% CN5CT heterojunction complex has good reactivity to visible light, which may increase its photocatalytic activity. The results show that photo-generated electron-hole charge carriers are enhanced, which will boost photocatalytic activity compared to CN5 and CT.

To obtain the band levels of the g-C$_3$N$_5$/Cu$_3$TiO$_4$, the valance band (CB) and conduction band (CB) potentials for g-C$_3$N$_5$ & Cu$_3$TiO$_4$ were calculated with the aid of Equations (1) and (2) [75].

$$E_{VB} = \chi - E_e + ((1/2) * E_g) \qquad (1)$$

$$E_{CB} = E_{VB} - E_g \qquad (2)$$

where, Eg-band gap calculated from Figure 6b (2.02 eV for g-C$_3$N$_5$, and 2.15 eV for Cu$_3$TiO$_4$); E$_{CB}$-CB potential; E$_{VB}$-VB potential; Ee-energy of free electrons evaluated on the H scale (4.5 eV); χ–electro negativity (6.89 eV for g-C$_3$N$_5$ and 5.63 for Cu$_3$TiO$_4$) calculated from literature [64]. The calculated E$_{VB}$ and E$_{CB}$ were +1.38 and −0.64 eV for g-C$_3$N$_5$, +0.055 and −2.095 eV for Cu$_3$TiO$_4$, respectively.

It is well known that the heterojunction photocatalyst g-C$_3$N$_5$/Cu$_3$TiO$_4$ will be produced when g-C$_3$N$_5$ and Cu$_3$TiO$_4$ are coupled. According to the band gap structures of g-C$_3$N$_5$ & Cu$_3$TiO$_4$, the separation processes of photo-induced electron–hole (e$^-$/h$^+$) is conveyed in Figure 7a,b, respectively.

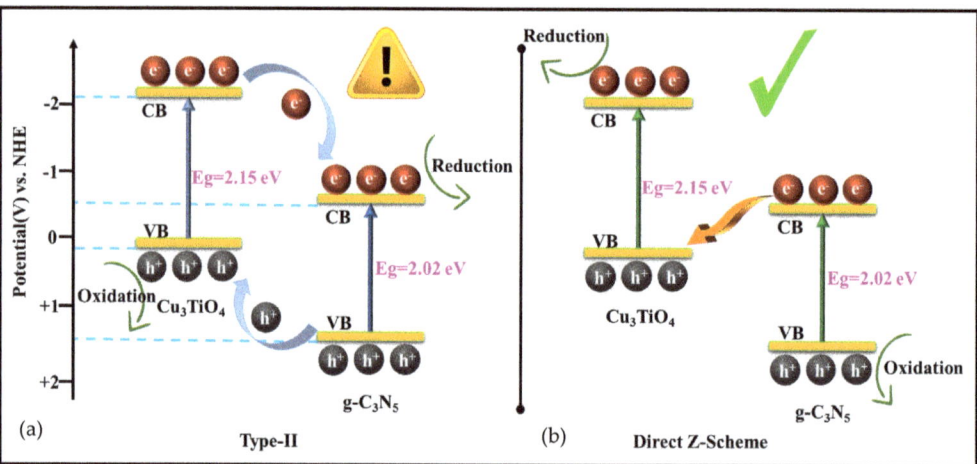

Figure 7. Proposed electron-hole pair transfer mechanism of g-C$_3$N$_5$/Cu$_3$TiO$_4$ nanocomposite (**a**) type-II, (**b**) Direct Z-Scheme.

According to the type-II heterojunction (Figure 7a), the holes of g-C$_3$N$_5$ formed by visible-light will transfer to the VB of Cu$_3$TiO$_4$ (0.055 eV). Altogether, e$^-$s instantaneously migrate from the CB of Cu$_3$TiO$_4$ (−2.09 eV) to the CB of g-C$_3$N$_5$, (−0.64 eV). However, the potential energy (V vs. NHE) of e$^-$s from g-C$_3$N$_5$ is much lower than O$_2$/O$_2^-$ (−0.33 V) [76], which designates that these electrons are unable to reducing O$_2$ to ·O$_2^-$. Correspondingly, holes of Cu$_3$TiO$_4$ have inadequate potential energy to oxidize H$_2$O to ·OH (·OH/H$_2$O = 2.34 V vs. NHE) [77]. Consequently, the type-II mechanism does not well define the photocatalytic activity by the g-C$_3$N$_5$/Cu$_3$TiO$_4$ heterojunction.

Our photocatalyst works in total tandem with the visible light wavelength from BLUE LED light sources. Both g-C$_3$N$_5$ and Cu$_3$TiO$_4$ can be stimulated to produce electrons and holes when exposed to light. The electrons in CB of g-C$_3$N$_5$ tend to migrate to VB of Cu$_3$TiO$_4$ and recombine with the holes there under the influence of the electron transport system, following the Z scheme charge transfer path. As a result, the highly redox active photogenerated electrons in CB of Cu_3TiO_4 and holes in VB of $g-C_3N_5$ are retained to participate in the photocatalytic activity. If the photo-induced charge transfer process follows the direct Z-Scheme mechanism as illustrated in Figure 7b, the photon-induced excited electrons would thermodynamically transfer from the CB of $g-C_3N_5$ to VB of Cu_3TiO_4. Meanwhile, the O_2 could be reduced into $·O_2^-$ (reduction reaction) by the electrons generated by CB of Cu_3TiO_4 because of the more negative potential (−2.09 eV). Further, the holes of $g-C_3N_5$ are arrested by OH to produce the ·OH, and the ·OH are moved to the superficial of the nanocomposite to oxidize molecules, because of the more positive potential (+1.38 eV). This charge transfer process in the direct Z-scheme greatly improves the efficiency of photo-generated electron-hole separation, resulting in the enhancement of the photocatalytic activity [78–80].

The nanocomposite exhibited excellent stability, visible absorption nature, high yields, a shortened reaction time, and reusability (Figure S30a). The photocatalyst 10%CN5CT does not affected by the reaction mixture and giving higher yields for the next few runs. Then after recycled catalyst was showing good crystalline nature in XRD analysis shown in Figure S30b. In the presence of visible light, the narrow bandgap feature demonstrates a visible active photocatalyst. In visible light, green synthesis of different heterocycles using CN5CT nanocomposite produced good yields in a short period. The photocatalyst can be used repeatedly without affecting the activity of the catalyst.

3. Materials and Methods

3.1. Materials Information

All of the chemicals and reagents were obtained from Avra Chemicals Private Limited, which is located in Andhra Pradesh, India. The column solvents and other chemicals were brought from Avra chemicals and used without further purification or processing. We purchased the lemons from the Vellore local market in Katpadi, Tamil Nadu, India. The NMR solvents came from Merck and was said to be 99.9% pure by the manufacturer.

3.2. Preparation of Citrus Limon Extract

Fresh lemons were split into half and squeezed to gather the pulp. There is some excess waste in the gathered juice (seeds and peel). To eliminate the additional trash, the pulp was centrifuged at 10,000 rpm. The clear liquid from the top layer was collected and it was stored in refrigerator.

3.3. Preparation of Cu_3TiO_4 NPs

Copper acetate (1 M) and titanium tert-butoxide (1 M) were taken in two separate beakers with 50mL of deionized water. A titanium tert-butoxide solution was added into the copper acetate solution with constant stirring. Then 5 mL of citrus extract obtained was added and the mixture was kept stirring overnight. Bluish-white coloured residue was separated and washed with a water-ethanol solution to remove the other impurities. The precipitate was heated at 500 °C for 4 h in a silica crucible. Finally, a black precipitate was obtained and subjected to characterization.

3.4. Preparation of $g-C_3N_5$ Nanosheets and Composite

In a beaker, 1 mmol of 3-amino-1,2,4-triazole was stirred well with 20 pellets of NaOH for 20 min. The colour of the mixture was altered to light brown after it was ground. For 4 h, the mixture was heated at 550 °C at a rate of 20 °C/min. Finally, grey $g-C_3N_5$ was collected and rinsed 10 times with distilled water to eliminate unreacted NaOH before being kept for characterization. The 5%, 10%, 20%, and 30% CN5CT were produced after processing $g-C_3N_5$ with Cu_3TiO_4 at various concentrations.

3.5. General Synthesis of Carbonitriles (1.4a-g) under Visible Light

1,3-cyclohexadione (1 mmol), corresponding aldehydes (1 mmol), malononitrile (1 mmol), CN5CT photocatalyst (20 mg), and 5 mL of ethanol were placed into a 25 mL round-bottom flask in the BLUE LED (12 W, 450 to 495 wavelength) light chamber. The reaction mixture was illuminated, and TLC was tested every minute to identify the result. After reaction completion, the reaction mixture was poured into crushed ice and precipitated out. The solids were removed from the ethanol and recrystallized.

4. Conclusions

In conclusion, we described the fabrication of direct Z-scheme Cu_3TiO_4/g-C_3N_5 heterojunction and their characterization. The Cu_3TiO_4/g-C_3N_5 photocatalyst has expressed excellent activity for the synthesis of carbonitriles derivatives. In an as-prepared heterojunction, 10% Cu_3TiO_4 is dispersed on surface the surface of g-C_3N_5 exhibited better photocatalytic activity. When compared to pure Cu_3TiO_4 and g-C_3N_5, Cu_3TiO_4/g-C_3N_5 nanocomposites perform better at photocatalysis due to their distinctive nanostructure and direct Z-scheme heterojunction. Low-energy consumption visible light BLUE LED lights were used for the irradiation. The reaction has been carried out immediately and the photocatalyst was showing excellent activity, reusability, easily recoverable, less recombination rate in visible light and high surface area.

Supplementary Materials: The following supporting information can be downloaded at: https://www.mdpi.com/article/10.3390/catal12121593/s1, Figure S1: XPS of Cu_3TiO_4/g-C_3N_5 nanocomposites; Figures S2–S29: FT-IR, ^{1}H, ^{13}C NMR and GC-MS of carbonitrile derivatives 1.4a–g; Figure S30: Reusability and reused XRD.

Author Contributions: Original draft preparation, data collection and writing, M.A. and T.C.; Writing and editing, R.M.; Data analysis, K.A.; Writing-review and editing, G.M. and S.M.R.; Supervision, S.M.R. All authors discussed the results and contributed to the final manuscript. All authors have read and agreed to the published version of the manuscript.

Funding: This research received no external funding.

Data Availability Statement: Not applicable.

Acknowledgments: Our profound thanks go to the management of the Vellore Institute of Technology for providing the Sophisticated Instrument Facility (SIF) to aid in the research endeavour.

Conflicts of Interest: The authors declared no potential conflicts of interest.

References

1. Nazeri, M.T.; Shaabani, A. Synthesis of Polysubstituted Pyrroles via Isocyanide-Based Multicomponent Reactions as an Efficient Synthesis Tool. *New J. Chem.* **2021**, *45*, 21967–22011. [CrossRef]
2. Chopra, P.K.P.G.; Lambat, T.L.; Mahmood, S.H.; Chaudhary, R.G.; Banerjee, S. Sulfamic Acid as Versatile Green Catalyst Used for Synthetic Organic Chemistry: A Comprehensive Update. *Chem. Sel.* **2021**, *6*, 6867–6889. [CrossRef]
3. Dekamin, M.G.; Eslami, M. Highly Efficient Organocatalytic Synthesis of Diverse and Densely Functionalized 2-Amino-3-Cyano-4H-Pyrans under Mechanochemical Ball Milling. *Green Chem.* **2014**, *16*, 4914–4921. [CrossRef]
4. Wang, X.; Yin, F.; Bi, Y.; Cheng, G.; Li, J.; Hou, L.; Li, Y.; Yang, B.; Liu, W.; Yang, L. Rapid and Sensitive Detection of Zika Virus by Reverse Transcription Loop-Mediated Isothermal Amplification. *J. Virol. Methods* **2016**, *238*, 86–93. [CrossRef]
5. Singh, S.B. Copper Nanocatalysis in Multi-Component Reactions: A Green to Greener Approach. *Curr. Catal.* **2018**, *7*, 80–88. [CrossRef]
6. Ghamari Kargar, P.; Bagherzade, G. Robust, Highly Active, and Stable Supported Co(ii) Nanoparticles on Magnetic Cellulose Nanofiber-Functionalized for the Multi-Component Reactions of Piperidines and Alcohol Oxidation. *RSC Adv.* **2021**, *11*, 23192–23206. [CrossRef]
7. Kamanna, K.; Khatavi, S.Y. Microwave-Accelerated Carbon-Carbon and Carbon-Heteroatom Bond Formation via Multi-Component Reactions: A Brief Overview. *Curr. Microw. Chem.* **2020**, *7*, 23–39. [CrossRef]
8. Ramesh, R.; Maheswari, S.; Malecki, J.G.; Lalitha, A. NaN$_3$ Catalyzed Highly Convenient Access to Functionalized 4 H - Chromenes: A Green One-Pot Approach for Diversity Amplification. *Polycycl. Aromat. Compd.* **2020**, *40*, 1581–1594. [CrossRef]
9. Majumdar, N.; Paul, N.D.; Mandal, S.; de Bruin, B.; Wulff, W.D. Catalytic Synthesis of 2 H -Chromenes. *ACS Catal.* **2015**, *5*, 2329–2366. [CrossRef]

10. Hauguel, C.; Ducellier, S.; Provot, O.; Ibrahim, N.; Lamaa, D.; Balcerowiak, C.; Letribot, B.; Nascimento, M.; Blanchard, V.; Askenatzis, L.; et al. Design, Synthesis and Biological Evaluation of Quinoline-2-Carbonitrile-Based Hydroxamic Acids as Dual Tubulin Polymerization and Histone Deacetylases Inhibitors. *Eur. J. Med. Chem.* **2022**, *240*, 114573. [CrossRef]
11. Khan, S.A.; Asiri, A.M.; Basisi, H.M.; Asad, M.; Zayed, M.E.M.; Sharma, K.; Wani, M.Y. Synthesis and Evaluation of Quinoline-3-Carbonitrile Derivatives as Potential Antibacterial Agents. *Bioorg. Chem.* **2019**, *88*, 102968. [CrossRef] [PubMed]
12. Patel, K.B.; Kumari, P. A Review: Structure-Activity Relationship and Antibacterial Activities of Quinoline Based Hybrids. *J. Mol. Struct.* **2022**, *1268*, 133634. [CrossRef]
13. Liu, T.; Wu, Z.; He, Y.; Xiao, Y.; Xia, C. Single and Dual Target Inhibitors Based on Bcl-2: Promising Anti-Tumor Agents for Cancer Therapy. *Eur. J. Med. Chem.* **2020**, *201*, 112446. [CrossRef]
14. Sharma, D.; Kumar, M.; Das, P. Application of Cyclohexane-1,3-Diones for Six-Membered Oxygen-Containing Heterocycles Synthesis. *Bioorg. Chem.* **2021**, *107*, 104559. [CrossRef] [PubMed]
15. Oliveira-Pinto, S.; Pontes, O.; Baltazar, F.; Costa, M. In Vivo Efficacy Studies of Chromene-Based Compounds in Triple-Negative Breast Cancer—A Systematic Review. *Eur. J. Pharmacol.* **2020**, *887*, 173452. [CrossRef]
16. Costa, M.; Dias, T.A.; Brito, A.; Proença, F. Biological Importance of Structurally Diversified Chromenes. *Eur. J. Med. Chem.* **2016**, *123*, 487–507. [CrossRef]
17. Sadek, K.U.; Mekheimer, R.A.H.; Abd-Elmonem, M.; Abdel-Hameed, A.; Elnagdi, M.H. Recent Developments in the Enantioselective Synthesis of Polyfunctionalized Pyran and Chromene Derivatives. *Tetrahedron Asymmetry* **2017**, *28*, 1462–1485. [CrossRef]
18. Chaudhary, A.; Singh, K.; Verma, N.; Kumar, S.; Kumar, D.; Sharma, P.P. Chromenes—A Novel Class of Heterocyclic Compounds: Recent Advancements and Future Directions. *Mini.-Rev. Med. Chem.* **2022**, *22*, 2736–2751. [CrossRef]
19. Son, Y.-A.; Gwon, S.-Y.; Kim, S.-H. Chromene and Imidazole Based D-π-A Chemosensor Preparation and Its Anion Responsive Effects. *Mol. Cryst. Liq. Cryst.* **2014**, *599*, 16–22. [CrossRef]
20. Dagilienė, M.; Markuckaitė, G.; Krikštolaitytė, S.; Šačkus, A.; Martynaitis, V. Cyanide Anion Determination Based on Nucleophilic Addition to 6-[(E)-(4-Nitrophenyl)Diazenyl]-1′,3,3′,4-Tetrahydrospiro[Chromene-2,2′-Indole] Derivatives. *Chem. Sens.* **2022**, *10*, 185. [CrossRef]
21. Kar, S.; Sanderson, H.; Roy, K.; Benfenati, E.; Leszczynski, J. Green Chemistry in the Synthesis of Pharmaceuticals. *Chem. Rev.* **2022**, *122*, 3637–3710. [CrossRef] [PubMed]
22. Clarke, C.J.; Tu, W.-C.; Levers, O.; Bröhl, A.; Hallett, J.P. Green and Sustainable Solvents in Chemical Processes. *Chem. Rev.* **2018**, *118*, 747–800. [CrossRef]
23. Kamanna, K.; Amaregouda, Y. Synthesis of Bioactive Scaffolds Catalyzed by Agro-Waste-Based Solvent Medium. *Phys. Sci. Rev.* **2022**. [CrossRef]
24. Védrine, J. Heterogeneous Catalysis on Metal Oxides. *Catalysts* **2017**, *7*, 341. [CrossRef]
25. Chellapandi, T.; Madhumitha, G. Montmorillonite Clay-Based Heterogenous Catalyst for the Synthesis of Nitrogen Heterocycle Organic Moieties: A Review. *Mol. Divers.* 2021. [CrossRef] [PubMed]
26. Singh, B.K.; Lee, S.; Na, K. An Overview on Metal-Related Catalysts: Metal Oxides, Nanoporous Metals and Supported Metal Nanoparticles on Metal Organic Frameworks and Zeolites. *Rare Met.* **2020**, *39*, 751–766. [CrossRef]
27. Shifrina, Z.B.; Matveeva, V.G.; Bronstein, L.M. Role of Polymer Structures in Catalysis by Transition Metal and Metal Oxide Nanoparticle Composites. *Chem. Rev.* **2020**, *120*, 1350–1396. [CrossRef]
28. Terna, A.D.; Elemike, E.E.; Mbonu, J.I.; Osafile, O.E.; Ezeani, R.O. The Future of Semiconductors Nanoparticles: Synthesis, Properties and Applications. *Mater. Sci. Eng. B* **2021**, *272*, 115363. [CrossRef]
29. Ge, M.Z.; Cao, C.Y.; Li, S.H.; Tang, Y.X.; Wang, L.N.; Qi, N.; Lai, Y.K. In situ plasmonic Ag nanoparticle anchored TiO_2 nanotube arrays as visible-light-driven photocatalysts for enhanced water splitting. *Nanoscale* **2016**, *8*, 5226–5234. [CrossRef]
30. Ge, M.; Cai, J.; Iocozzia, J.; Cao, C.; Huang, J.; Zhang, X.; Lin, Z. A review of TiO_2 nanostructured catalysts for sustainable H_2 generation. *Int. J. Hydrogren Energy* **2017**, *42*, 8418–8449. [CrossRef]
31. Ge, M.; Li, Q.; Cao, C.; Huang, J.; Li, S.; Zhang, S.; Lai, Y. One-dimensional TiO_2 nanotube photocatalysts for solar water splitting. *Adv. Sci.* **2017**, *4*, 1600152. [CrossRef] [PubMed]
32. Hu, X.; Luo, T.; Lin, Y.; Yang, M. Construction of Novel Z-Scheme g-C_3N_4/AgBr-Ag Composite for Efficient Photocatalytic Degradation of Organic Pollutants under Visible Light. *Catalysts* **2022**, *12*, 1309. [CrossRef]
33. Cuong, H.N.; Pansambal, S.; Ghotekar, S.; Oza, R.; Thanh Hai, N.T.; Viet, N.M.; Nguyen, V.-H. New Frontiers in the Plant Extract Mediated Biosynthesis of Copper Oxide (CuO) Nanoparticles and Their Potential Applications: A Review. *Environ. Res.* **2022**, *203*, 111858. [CrossRef] [PubMed]
34. Kumar, A.; Choudhary, P.; Kumar, A.; Camargo, P.H.C.; Krishnan, V. Recent Advances in Plasmonic Photocatalysis Based on TiO_2 and Noble Metal Nanoparticles for Energy Conversion, Environmental Remediation, and Organic Synthesis. *Small* **2022**, *18*, 2101638. [CrossRef]
35. Gnanasekaran, L.; Pachaiappan, R.; Kumar, P.S.; Hoang, T.K.A.; Rajendran, S.; Durgalakshmi, D.; Soto-Moscoso, M.; Cornejo-Ponce, L.; Gracia, F. Visible Light Driven Exotic p (CuO)–n (TiO_2) Heterojunction for the Photodegradation of 4-Chlorophenol and Antibacterial Activity. *Environ. Pollut.* **2021**, *287*, 117304. [CrossRef]
36. Marzo, L.; Pagire, S.K.; Reiser, O.; König, B. Visible-Light Photocatalysis: Does It Make a Difference in Organic Synthesis? *Angew. Chem. Int. Ed.* **2018**, *57*, 10034–10072. [CrossRef]

37. Alaghmandfard, A.; Ghandi, K. A Comprehensive Review of Graphitic Carbon Nitride (g-C_3N_4)–Metal Oxide-Based Nanocomposites: Potential for Photocatalysis and Sensing. *Nanomaterials* **2022**, *12*, 294. [CrossRef]
38. Arunachalapandi, M.; Roopan, S.M. Environment Friendly G-C_3N_4-Based Catalysts and Their Recent Strategy in Organic Transformations. *High Energy Chem.* **2022**, *56*, 73–90. [CrossRef]
39. Mun, S.J.; Park, S.-J. Graphitic Carbon Nitride Materials for Photocatalytic Hydrogen Production via Water Splitting: A Short Review. *Catalysts* **2019**, *9*, 805. [CrossRef]
40. Chen, J.; Fang, S.; Shen, Q.; Fan, J.; Li, Q.; Lv, K. Recent Advances of Doping and Surface Modifying Carbon Nitride with Characterization Techniques. *Catalysts* **2022**, *12*, 962. [CrossRef]
41. Wang, J.; Wang, S. A Critical Review on Graphitic Carbon Nitride (g-C_3N_4)-Based Materials: Preparation, Modification and Environmental Application. *Coord. Chem. Rev.* **2022**, *453*, 214338. [CrossRef]
42. Mane, G.P.; Talapaneni, S.N.; Lakhi, K.S.; Ilbeygi, H.; Ravon, U.; Al-Bahily, K.; Mori, T.; Park, D.H.; Vinu, A. Highly ordered nitrogen-rich mesoporous carbon nitrides and their superior performance for sensing and photocatalytic hydrogen generation. *Angew. Chem. Int. Ed.* **2017**, *56*, 8481–8485. [CrossRef] [PubMed]
43. Liu, D.; Yao, J.; Chen, S.; Zhang, J.; Li, R.; Peng, T. Construction of rGO-coupled C_3N_4/C_3N_5 2D/2D Z-scheme heterojunction to accelerate charge separation for efficient visible light H_2 evolution. *Appl. Catal. B.* **2022**, *318*, 121822. [CrossRef]
44. Luo, T.; Hu, X.; She, Z.; Wei, J.; Feng, X.; Chang, F. Synergistic effects of Ag-doped and morphology regulation of graphitic carbon nitride nanosheets for enhanced photocatalytic performance. *J. Mol. Liq.* **2021**, *324*, 114772. [CrossRef]
45. Li, S.; Cai, M.; Liu, Y.; Wang, C.; Yan, R.; Chen, X. Constructing $Cd_{0.5}Zn_{0.5}S/Bi_2WO_6$ S-scheme heterojunction for boosted photocatalytic antibiotic oxidation and Cr (VI) reduction. *Adv. Powder Technol.* **2023**, *2*, 100073. [CrossRef]
46. Li, Z.; Li, H.; Wang, S.; Yang, F.; Zhou, W. Mesoporous black $TiO_2/MoS_2/Cu_2S$ hierarchical tandem heterojunctions toward optimized photothermal-photocatalytic fuel production. *J. Chem. Eng.* **2022**, *427*, 131830. [CrossRef]
47. Wang, S.; Guan, B.Y.; Lou, X.W.D. Construction of $ZnIn_2S_4$–In_2O_3 hierarchical tubular heterostructures for efficient CO_2 photoreduction. *J. Am. Chem. Soc.* **2018**, *140*, 5037–5040. [CrossRef]
48. Xiao, Y.; Guo, S.; Tian, G.; Jiang, B.; Ren, Z.; Tian, C.; Li, W.; Fu, H. Synergetic enhancement of surface reactions and charge separation over holey C3N4/TiO2 2D heterojunctions. *Sci. Bull.* **2021**, *66*, 275–283. [CrossRef]
49. Humayun, M.; Ullah, H.; Tahir, A.A.; bin Mohd Yusoff, A.R.; Mat Teridi, M.A.; Nazeeruddin, M.K.; Luo, W. An Overview of the Recent Progress in Polymeric Carbon Nitride Based Photocatalysis. *Chem. Rec.* **2021**, *21*, 1811–1844. [CrossRef]
50. Palanivel, B.; Mani, A. Conversion of a Type-II to a Z-Scheme Heterojunction by Intercalation of a 0D Electron Mediator between the Integrative $NiFe_2O_4$/g-C_3N_4 Composite Nanoparticles: Boosting the Radical Production for Photo-Fenton Degradation. *ACS Omega* **2020**, *5*, 19747–19759. [CrossRef]
51. Lu, L.; Wang, G.; Zou, M.; Wang, J.; Li, J. Effects of Calcining Temperature on Formation of Hierarchical TiO_2/g-C_3N_4 Hybrids as an Effective Z-Scheme Heterojunction Photocatalyst. *Appl. Surf. Sci.* **2018**, *441*, 1012–1023. [CrossRef]
52. Wang, M.; Jin, C.; Kang, J.; Liu, J.; Tang, Y.; Li, Z.; Li, S. CuO/g-C_3N_4 2D/2D Heterojunction Photocatalysts as Efficient Peroxymonosulfate Activators under Visible Light for Oxytetracycline Degradation: Characterization, Efficiency and Mechanism. *Chem. Eng. J.* **2021**, *416*, 128118. [CrossRef]
53. Shen, D.; Li, X.; Ma, C.; Zhou, Y.; Sun, L.; Yin, S.; Huo, P.; Wang, H. Synthesized Z-Scheme Photocatalyst ZnO/g-C_3N_4 for Enhanced Photocatalytic Reduction of CO_2. *New J. Chem.* **2020**, *44*, 16390–16399. [CrossRef]
54. Meng, J.; Wang, X.; Liu, Y.; Ren, M.; Zhang, X.; Ding, X.; Guo, Y.; Yang, Y. Acid-Induced Molecule Self-Assembly Synthesis of Z-Scheme WO_3/g-C_3N_4 Heterojunctions for Robust Photocatalysis against Phenolic Pollutants. *Chem. Eng. J.* **2021**, *403*, 126354. [CrossRef]
55. Dadigala, R.; Bandi, R.; Gangapuram, B.R.; Dasari, A.; Belay, H.H.; Guttena, V. Fabrication of Novel 1D/2D V_2O_5/g-C_3N_4 Composites as Z-Scheme Photocatalysts for CR Degradation and Cr (VI) Reduction under Sunlight Irradiation. *J. Environ. Chem. Eng.* **2019**, *7*, 102822. [CrossRef]
56. Wu, B.; Li, Y.; Su, K.; Tan, L.; Liu, X.; Cui, Z.; Yang, X.; Liang, Y.; Li, Z.; Zhu, S.; et al. The Enhanced Photocatalytic Properties of MnO_2/g-C_3N_4 Heterostructure for Rapid Sterilization under Visible Light. *J. Hazard. Mater.* **2019**, *377*, 227–236. [CrossRef]
57. He, R.; Zhou, J.; Fu, H.; Zhang, S.; Jiang, C. Room-Temperature in Situ Fabrication of Bi_2O_3/g-C_3N_4 Direct Z-Scheme Photocatalyst with Enhanced Photocatalytic Activity. *Appl. Surf. Sci.* **2018**, *430*, 273–282. [CrossRef]
58. Yin, H.; Cao, Y.; Fan, T.; Zhang, M.; Yao, J.; Li, P.; Chen, S.; Liu, X. In Situ Synthesis of Ag_3PO_4/C_3N_5 Z-Scheme Heterojunctions with Enhanced Visible-Light-Responsive Photocatalytic Performance for Antibiotics Removal. *Sci. Total Environ.* **2021**, *754*, 141926. [CrossRef]
59. Vadivel, S.; Fujii, M.; Rajendran, S. Facile Synthesis of Broom Stick like FeOCl/g-C_3N_4 Nanocomposite as Novel Z-Scheme Photocatalysts for Rapid Degradation of Pollutants. *Chemosphere* **2022**, *307*, 135716. [CrossRef]
60. Chu, W.; Li, S.; Zhou, H.; Shi, M.; Zhu, J.; He, P.; Liu, W.; Wu, M.; Wu, J.; Yan, X. A Novel Bionic Flower-like Z-Scheme $Bi_4O_5I_2$/g-C_3N_5 Heterojunction with I^-/I_3^- active sites and π-conjugated structure for increasing photocatalytic oxidation of elemental mercury. *J. Environ. Chem. Eng.* **2022**, *10*, 108623. [CrossRef]
61. Wang, R.; Zhang, K.; Zhong, X.; Jiang, F. Z-Scheme $LaCoO_3/C_3N_5$ for Efficient Full-Spectrum Light-Simulated Solar Photocatalytic Hydrogen Generation. *RSC Adv.* **2022**, *12*, 24026–24036. [CrossRef]

62. Rajendran, S.; Chellapandi, T.; UshaVipinachandran, V.; Venkata Ramanaiah, D.; Dalal, C.; Sonkar, S.K.; Madhumitha, G.; Bhunia, S.K. Sustainable 2D Bi_2WO_6/g-C_3N_5 Heterostructure as Visible Light-Triggered Abatement of Colorless Endocrine Disruptors in Wastewater. *Appl. Surf. Sci.* **2022**, *577*, 151809. [CrossRef]
63. Deng, S.; Li, Z.; Zhao, T.; Huang, G.; Wang, J.; Bi, J. Direct Z-Scheme Covalent Triazine-Based Framework/Bi_2WO_6 Heterostructure for Efficient Photocatalytic Degradation of Tetracycline: Kinetics, Mechanism and Toxicity. *J. Water Process Eng.* **2022**, *49*, 103021. [CrossRef]
64. Das, A. LED Light Sources in Organic Synthesis: An Entry to a Novel Approach. *Lett. Org. Chem.* **2022**, *19*, 283–292. [CrossRef]
65. Cai, Z.; Huang, Y.; Ji, H.; Liu, W.; Fu, J.; Sun, X. Type-II Surface Heterojunction of Bismuth-Rich $Bi_4O_5Br_2$ on Nitrogen-Rich g-C_3N_5 Nanosheets for Efficient Photocatalytic Degradation of Antibiotics. *Sep. Purif. Technol.* **2022**, *280*, 119772. [CrossRef]
66. Arunachalapandi, M.; Roopan, S.M. Ultrasound/Visible Light-Mediated Synthesis of N-Heterocycles Using g-C_3N_4/Cu_3TiO_4 as Sonophotocatalyst. *Res. Chem. Intermed.* **2021**, *47*, 3363–3378. [CrossRef]
67. Wei, W.; Gong, H.; Sheng, L.; Wu, W.; Hu, S.; Feng, L.; Li, X.; You, W. Highly Efficient Photocatalytic Activity and Mechanism of Novel Er^{3+} and Tb^{3+} Co-Doped BiOBr/g-C_3N_5 towards Sulfamethoxazole Degradation. *Ceram. Int.* **2021**, *47*, 24062–24072. [CrossRef]
68. Vadivel, S.; Fujii, M.; Rajendran, S. Novel S-Scheme 2D/2D $Bi_4O_5Br_2$ Nanoplatelets/g-C_3N_5 Heterojunctions with Enhanced Photocatalytic Activity towards Organic Pollutants Removal. *Environ. Res.* **2022**, *213*, 113736. [CrossRef]
69. Ramezanalizadeh, H.; Rafiee, E. Design, Fabrication, Electro- and Photoelectrochemical Investigations of Novel $CoTiO_3$/$CuBi_2O_4$ Heterojunction Semiconductor: An Efficient Photocatalyst for the Degradation of DR16 Dye. *Mater. Sci. Semicond. Process.* **2020**, *113*, 105055. [CrossRef]
70. Li, M.; Lu, Q.; Liu, M.; Yin, P.; Wu, C.; Li, H.; Zhang, Y.; Yao, S. Photsoinduced Charge Separation via the Double-Electron Transfer Mechanism in Nitrogen Vacancies g-C_3N_5/BiOBr for the Photoelectrochemical Nitrogen Reduction. *ACS Appl. Mater. Interfaces* **2020**, *12*, 38266–38274. [CrossRef]
71. Rajendran, R.; Vignesh, S.; Suganthi, S.; Raj, V.; Kavitha, G.; Palanivel, B.; Shkir, M.; Algarni, H. G-C_3N_4/TiO_2/CuO S-Scheme Heterostructure Photocatalysts for Enhancing Organic Pollutant Degradation. *J. Phys. Chem. Solids* **2022**, *161*, 110391. [CrossRef]
72. Saravanakumar, K.; Maheskumar, V.; Yea, Y.; Yoon, Y.; Muthuraj, V.; Park, C.M. 2D/2D nitrogen-rich graphitic carbon nitride coupled Bi_2WO_6 S-scheme heterojunction for boosting photodegradation of tetracycline: Influencing factors, intermediates, and insights into the mechanism. *Compos. Part B Eng.* **2022**, *234*, 109726. [CrossRef]
73. Chen, L.; Bao, S.; Yang, L.; Zhang, X.; Li, B.; Li, Y. Cheap Thiamine Hydrochloride as Efficient Catalyst for Synthesis of 4H-Benzo[b]Pyrans in Aqueous Ethanol. *Res. Chem. Intermed.* **2017**, *43*, 3883–3891. [CrossRef]
74. Tabassum, S.; Govindaraju, S.; Kendrekar, P. (Mes-Acr-Me)$^+ClO_4^-$ Catalyzed Visible Light-Supported, One-Pot Green Synthesis of 1,8-Naphthyridine-3-Carbonitriles. *Top. Catal.* **2022**, 1–9. [CrossRef]
75. Aditya, M.N.; Chellapandi, T.; Prasad, G.K.; Venkatesh, M.J.P.; Khan, M.M.R.; Madhumitha, G.; Roopan, S.M. Biosynthesis of Rod Shaped Gd_2O_3 on G-C_3N_4 as Nanocomposite for Visible Light Mediated Photocatalytic Degradation of Pollutants and RSM Optimization. *Diam. Relat. Mater.* **2022**, *121*, 108790. [CrossRef]
76. Mondal, M.; Banerjee, S.; Mal, S.; Das, S.; Pradhan, S.K. Nanocomposites of $GaBr_3$ and $BiBr_3$ Nanocrystals on BiOBr for the Photocatalytic Degradation of Dyes and Tetracycline. *ACS Appl. Nano. Mater.* **2022**, *5*, 15676–15691. [CrossRef]
77. Deng, S.; Yang, Z.; Lv, G.; Zhu, Y.; Li, H.; Wang, F.; Zhang, X. WO_3 Nanosheets/g-C_3N_4 Nanosheets' Nanocomposite as an Effective Photocatalyst for Degradation of Rhodamine B. *Appl. Phys. A* **2019**, *125*, 44. [CrossRef]
78. Miao, Z.; Zhang, Y.; Wang, N.; Xu, P.; Wang, X. $BiOBr$/Bi_2S_3 Heterojunction with S-Scheme Structureand Oxygen Defects: In-Situ Construction and Photocatalytic Behavior for Reduction of CO_2 with H2O. *J. Colloid Interface Sci.* **2022**, *620*, 407–418. [CrossRef]
79. Xing, P.; Zhou, F.; Li, Z. Preparation of WO_3/g-C_3N_4 Composites with Enhanced Photocatalytic Hydrogen Production Performance. *Appl. Phys. A* **2019**, *125*, 788. [CrossRef]
80. Ye, R.; Fang, H.; Zheng, Y.-Z.; Li, N.; Wang, Y.; Tao, X. Fabrication of $CoTiO_3$/g-C_3N_4 Hybrid Photocatalysts with Enhanced H_2 Evolution: Z-Scheme Photocatalytic Mechanism Insight. *ACS Appl. Mater. Interfaces* **2016**, *8*, 13879–13889. [CrossRef]

Article

Efficient Visible-Light Driven Photocatalytic Hydrogen Production by Z-Scheme ZnWO$_4$/Mn$_{0.5}$Cd$_{0.5}$S Nanocomposite without Precious Metal Cocatalyst

Tingting Ma [1], Zhen Li [2], Gan Wang [1], Jinfeng Zhang [3],* and Zhenghua Wang [1],*

[1] Key Laboratory of Functional Molecular Solids, Ministry of Education, College of Chemistry and Materials Science, Anhui Normal University, Wuhu 241000, China
[2] School of Food Engineering, Anhui Science and Technology University, Fengyang 233100, China
[3] Anhui Province Key Laboratory of Pollutant Sensitive Materials and Environmental Remediation, School of Physics and Electronic Information, Huaibei Normal University, Huaibei 235000, China
* Correspondence: jfzhang@chnu.edu.cn (J.Z.); zhwang@ahnu.edu.cn (Z.W.)

Abstract: How to restrain the recombination of photogenerated electrons and holes is still very important for photocatalytic hydrogen production. Herein, Z-scheme ZnWO$_4$/Mn$_{0.5}$Cd$_{0.5}$S (ZWMCS) nanocomposites are prepared and are applied as visible-light driven precious metal cocatalyst free photocatalyst for hydrogen generation. The ZnWO$_4$/Mn$_{0.5}$Cd$_{0.5}$S nanocomposites with 30 wt% ZnWO$_4$ (ZWMCS-2) can reach a photocatalytic hydrogen evolution rate of 3.36 mmol g^{-1} h^{-1}, which is much higher than that of single ZnWO$_4$ (trace) and Mn$_{0.5}$Cd$_{0.5}$S (1.96 mmol g^{-1} h^{-1}). Cycling test reveals that the ZMWCS-2 nanocomposite can maintain stable photocatalytic hydrogen evolution for seven cycles (21 h). The type of heterojunction in the ZWMCS-2 nanocomposite can be identified as Z-scheme heterojunction. The Z-scheme heterojunction can effectively separate the electrons and holes, so that the hydrogen generation activity and stability of the ZWMCS-2 nanocomposite can be enhanced. This work provides a highly efficient and stable Z-scheme heterojunction photocatalyst for hydrogen generation.

Keywords: hydrogen energy; Z-scheme; nanocomposite; heterojunction

Citation: Ma, T.; Li, Z.; Wang, G.; Zhang, J.; Wang, Z. Efficient Visible-Light Driven Photocatalytic Hydrogen Production by Z-Scheme ZnWO$_4$/Mn$_{0.5}$Cd$_{0.5}$S Nanocomposite without Precious Metal Cocatalyst. *Catalysts* **2022**, *12*, 1527. https://doi.org/10.3390/catal12121527

Academic Editors: Yongming Fu and Qian Zhang

Received: 9 November 2022
Accepted: 24 November 2022
Published: 27 November 2022

Publisher's Note: MDPI stays neutral with regard to jurisdictional claims in published maps and institutional affiliations.

Copyright: © 2022 by the authors. Licensee MDPI, Basel, Switzerland. This article is an open access article distributed under the terms and conditions of the Creative Commons Attribution (CC BY) license (https://creativecommons.org/licenses/by/4.0/).

1. Introduction

In order to alleviate the environmental pollution and energy shortage originated from the excessive use of fossil fuels, renewable and clean energy sources are urgently needed to replace the fossil fuels [1,2]. Hydrogen, as an ideal renewable and clean fuel, has attracted extensive attentions [3–5]. Visible light driven photocatalytic hydrogen evolution from water splitting is an efficient and environmentally friendly method to harvest the endless solar energy and turn it into hydrogen energy [6–9]. This method is very helpful to alleviate the environmental and energy issues originated from fossil fuels.

Currently, many photocatalysts are continuously discovered and intensively studied [10,11]. Among them, CdS, as a kind of classical semiconductor material, is widely used in solar energy generation, paints, photocatalysis, as well as other fields. Since CdS has strong absorption of visible light, and its conduction band position is lower than hydrogen evolution potential, it is expected to be an excellent photocatalytic hydrogen evolution photocatalyst [12–15]. However, numerous studies have shown that the actual photocatalytic hydrogen generation performance of CdS is not high, and the main problems are its low separation efficiency of electrons and holes and serious photo-corrosion [16,17]. Fortunately, an increasing number of reports have proposed methods to improve the weakness of CdS, such as modification of nanostructures [18], non-noble metal ion doping [19], construction of heterojunctions [20,21], and introducing surface defects [22].

Recent studies have revealed that introducing other metal ions into CdS can form solid solutions such as Mn$_x$Cd$_{1-x}$S [23], Zn$_x$Cd$_{1-x}$S [24], Cd$_x$In$_{1-x}$S [25], and (Zn$_{0.95}$Cu$_{0.05}$)$_{(1-x)}$Cd$_x$S [26].

$Mn_xCd_{1-x}S$ solid solution can be formed by substitute part of Cd^{2+} in CdS by Mn^{2+}. The substitution of Cd^{2+} by Mn^{2+} can effectively enlarge the interlayer spacing of CdS, thus promoting the ionic diffusion kinetic properties [27,28]. By changing the molar ratio of Mn^{2+}/Cd^{2+}, the band edge position and the band gap width of $Mn_xCd_{1-x}S$ can be tuned [29]. So the photocatalytic hydrogen evolution activity of $Mn_xCd_{1-x}S$ can be optimized. However, some of the drawbacks of CdS are still retained in $Mn_xCd_{1-x}S$, such as low separation efficiency of carriers [30].

As a typical metal tungstate semiconductor, $ZnWO_4$ has been widely used in the fields of photocatalytic hydrogen generation and organic pollutant degradation due to its non-toxicity, low cost, ultra-wide band gap (3.5 eV) as well as excellent physicochemical stability [31]. However, the problems of rapid recombination of photogenerated electron-hole and weak visible light absorption still hinder the practical usage of $ZnWO_4$ in the field of photocatalysis [32]. Therefore, the study on modification of $ZnWO_4$ is of great significance.

In recent years, more and more studies tend to construct Z-scheme heterojunction to further promote the photocatalytic hydrogen evolution rate and stability of the photocatalyst [33,34]. For instance, Zuo et al. constructed TiO_2-$ZnIn_2S_4$ nanoflower with Z-scheme heterostructure, which effectively suppressed the recombination of photogenerated electrons–holes, and improved the photocatalytic H_2 generation efficiency and stability [35]. Li and co-workers successfully synthesized a Z-scheme heterostructured $NiTiO_3/Cd_{0.5}Zn_{0.5}S$ photocatalyst with high photocatalytic hydrogen evolution activity and stability relative to constituent materials [36].

In this work, $ZnWO_4/Mn_{0.5}Cd_{0.5}S$ (ZMWCS) nanocomposites with varied $Mn_{0.5}Cd_{0.5}S$ contents are successfully prepared through a two-step hydrothermal method. their photocatalytic hydrogen evolution performances are measured under visible light and free of any noble-metal cocatalysts. Among them, the ZMWCS-2 nanocomposite with $ZnWO_4$ mass ratio of 30 % shows the optimal photocatalytic hydrogen evolution activity, and it still maintains stable photocatalytic hydrogen evolution activity after 21 h cycling test. Through experiments as well as theoretical calculations, the separation method of photogenerated carriers through the heterojunction in ZMWCS-2 nanocomposite is confirmed as Z-scheme. Through the Z-scheme heterojunction, the photogenerated electrons and holes can be effectively separated, thus the hydrogen evolution reaction can be accelerated, and the self-corrosion of $Mn_{0.5}Cd_{0.5}S$ can be alleviated. The existence of Z-scheme heterojunction lead to the excellent photocatalytic hydrogen evolution activity and stability of the ZMWCS nanocomposites.

2. Results

2.1. Process of Materials Synthesis

As shown in Scheme 1, $ZnWO_4/Mn_{0.5}Cd_{0.5}S$ (ZWMCS) nanocomposites are synthesized via two-step hydrothermal method. First, $NaWO_4$ solution was slowly dropped into $Zn(NO_3)_3$ solution, and then the mixed solution was heated at 180 °C for 8 h to obtain $ZnWO_4$ nanoparticles. Then, $ZnWO_4$ nanoparticles, $Cd(CH_3COO)_2 \cdot 2H_2O$, $Mn(CH_3COO)_2 \cdot 4H_2O$, and TAA were together added in water, fully dissolved, and heated at 160 °C for 24 h. Through these two steps, ZWMCS nanocomposites were obtained.

2.2. Phase and Morphology Analysis

Figure 1 shows the X-ray powder diffraction (XRD) patterns of $Mn_{0.5}Cd_{0.5}S$, $ZnWO_4$ and ZWMCS nanocomposites. The XRD pattern of $Mn_{0.5}Cd_{0.5}S$ is very similar to that of hexagonal phase CdS (JCPDS 80-0006), except a slight shift in peak positions (Figure S1). The peak shift can be attributed to the partial substitution of Cd^{2+} by Mn^{2+} [37]. The XRD pattern of $ZnWO_4$ is in good agreement with that of monoclinic $ZnWO_4$ (Figure S2, JCPDS No.73-0554). These results confirm the successful preparation of $Mn_{0.5}Cd_{0.5}S$ and $ZnWO_4$. The ZWMCS nanocomposites (ZWMCS-1, ZWMCS-2, ZWMCS-3, and ZWMCS-4) contain all the characteristic peaks of $ZnWO_4$ and $Mn_{0.5}Cd_{0.5}S$, which confirms that both $ZnWO_4$ and $Mn_{0.5}Cd_{0.5}S$ are contained in ZWMCS nanocomposites. In addition, with the increase of ZnWO$_4$ content, the diffraction peak intensity of ZnWO$_4$ in ZWMCS nanocomposites gradually enhances.

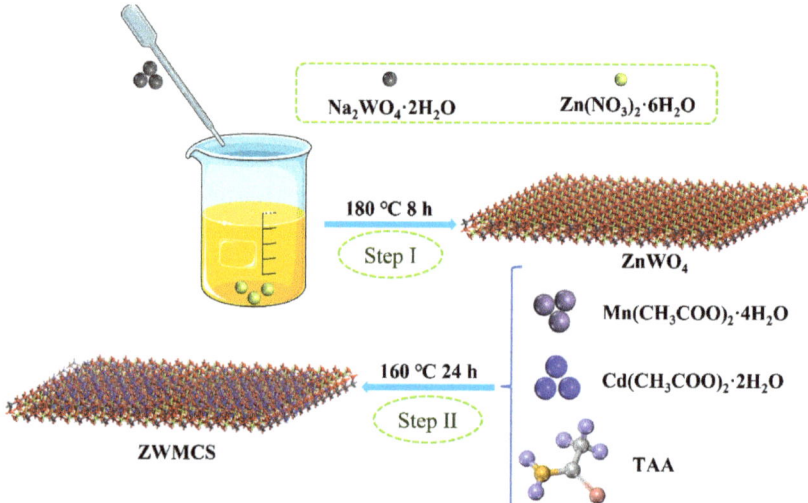

Scheme 1. Schematic diagram for the synthesis of ZnWO$_4$/Mn$_{0.5}$Cd$_{0.5}$S (ZWMCS) nanocomposites.

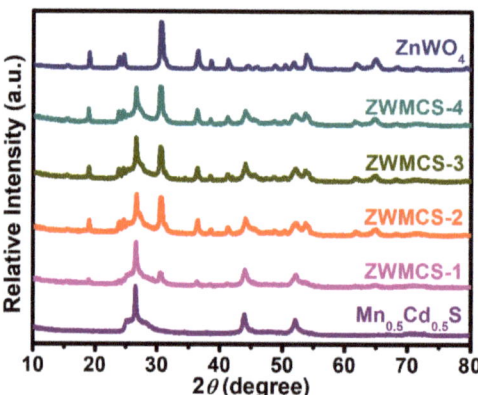

Figure 1. XRD patterns of ZWMCS nanocomposites.

Figure 2 shows the scanning electron microscopy (SEM) images of Mn$_{0.5}$Cd$_{0.5}$S, ZnWO$_4$ and ZWMCS-2 nanocomposite. In Figure 2a, Mn$_{0.5}$Cd$_{0.5}$S presented nanoparticle morphology (~50 nm). In Figure 2b, ZnWO$_4$ exhibits rod-like morphology. In Figure 2c, numerous Mn$_{0.5}$Cd$_{0.5}$S and ZnWO$_4$ are attached to each other to form the ZWMCS-2 nanocomposite. Figure 2d–f shows the energy dispersive spectroscopy (EDS) spectra of Mn$_{0.5}$Cd$_{0.5}$S, ZnWO$_4$ and ZWMCS-2 nanocomposite. It reveals that Mn$_{0.5}$Cd$_{0.5}$S, ZnWO$_4$, and ZWMCS-2 nanocomposite only contain the respective constituent prime peaks and no impurity peaks, which indicates the purity of the synthesized samples. The EDS mapping images in Figure 2g confirm that the constituent elements of ZWMCS-2 nanocomposite are well distributed, which further indicates the composition of the ZWMCS-2 nanocomposite.

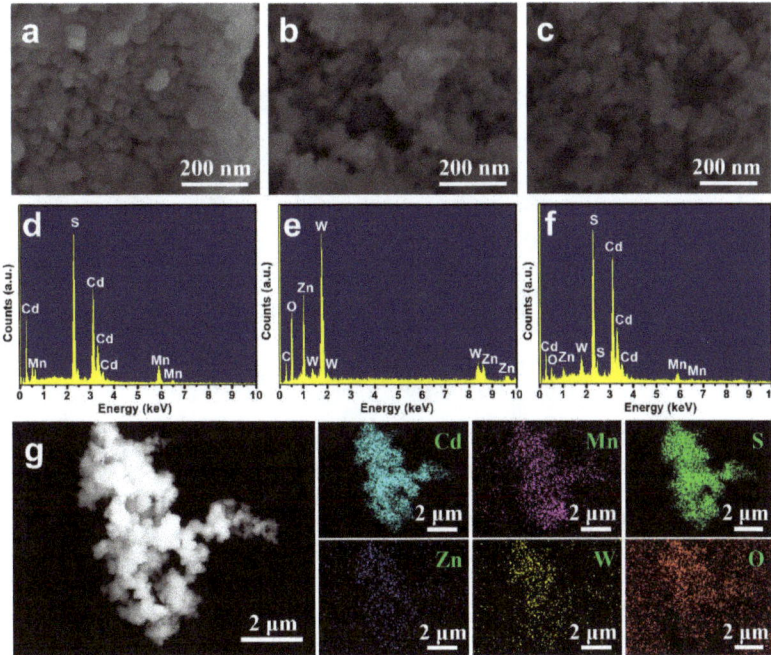

Figure 2. SEM images of (**a**) Mn$_{0.5}$Cd$_{0.5}$S, (**b**) ZnWO$_4$, and (**c**) ZWMCS-2 nanocomposite; EDS spectra of (**d**) Mn$_{0.5}$Cd$_{0.5}$S, (**e**) ZnWO$_4$, and (**f**) ZWMCS-2 nanocomposite; (**g**) SEM image with corresponding elemental mapping images for the EDS mapping images of Cd, Mn, S, Zn, W, and O elements of ZWMCS-2 nanocomposite.

The morphology of the material can be further observed by transmission electron microscopy (TEM). As shown in Figure 3a, Mn$_{0.5}$Cd$_{0.5}$S presents nanoparticles with diameter of about 50 nm. In Figure 3b, ZnWO$_4$ takes on short nanorods morphology with a diameter in the range of 5–50 nm. Figure 3c shows the morphology of ZWMCS-2 nanocomposite. It can be seen that Mn$_{0.5}$Cd$_{0.5}$S and ZnWO$_4$ cluster together to form the nanocomposite. Figure 3d shows a HRTEM image of ZWMCS-2 nanocomposite, the lattice fringe spacing of ZnWO$_4$ is 0.293 nm, corresponding to the (111) crystal plane of ZnWO$_4$, and the lattice fringe spacing of Mn$_{0.5}$Cd$_{0.5}$S is 0.173 nm, corresponding to the (112) crystal plane of Mn$_{0.5}$Cd$_{0.5}$S. The lattice distortion appears at the junction between them, which indicates the formation of heterojunction. The presence of heterojunctions indicates the successful formation of composites between ZnWO$_4$ and Mn$_{0.5}$Cd$_{0.5}$S.

2.3. X-ray Photoelectron Spectroscopy (XPS) and Elemental Analysis

The chemical state and elemental composition of ZWMCS-2 nanocomposite are analyzed by X-ray photoelectron spectroscopy (XPS). As shown in Figure S3, the characteristic peaks of Cd, Mn, S, Zn, W, and O elements can be clearly seen in the XPS survey spectrum of ZWMCS-2 nanocomposite. Figure 4a–f shows the high-resolution XPS spectra of Cd 3d, Mn 2p, S 2p, Zn 2p, W 4f, and O 1s elements, respectively. In Figure 4a, the two peaks at 411.07 and 404.33 eV belong to Cd 3d$_{3/2}$ and Cd 3d$_{5/2}$, respectively [23,30]. The two peaks at 651.56 and 639.96 eV (Figure 4b) correspond to Mn 2p$_{1/2}$ and Mn 2p$_{3/2}$, respectively [23,30]. In Figure 4c, the peaks at 161.95 and 160.69 eV correspond to S 2p$_{3/2}$ and S 2p$_{1/2}$, respectively [23,30]. The peaks at 1044.08 and 1021.06 eV (Figure 4d) represent Zn 2p$_{3/2}$ and Zn 2p$_{1/2}$, respectively [31,38]. As shown in Figure 4e, the two peaks at 36.98 and 34.84 eV belong to W f$_{5/2}$ and W f$_{7/2}$, respectively [31,38]. In the high-resolution

XPS spectra of O 1s (Figure 4f), the peaks located at 530.72 and 529.48 eV represent the characteristic peaks of lattice oxygen in ZnWO$_4$ [31,38].

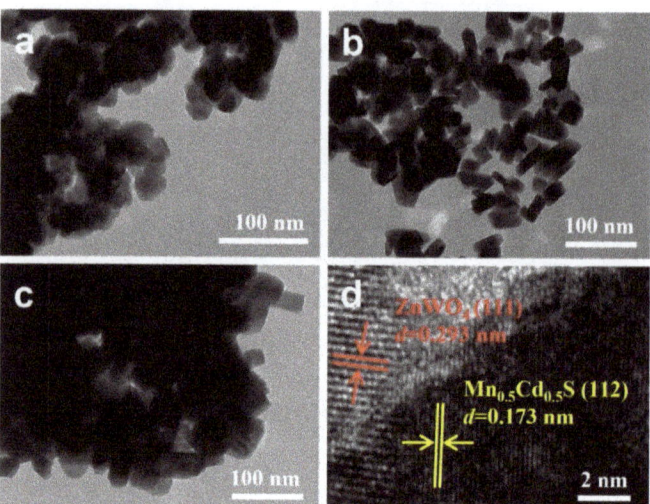

Figure 3. TEM images of (**a**) Mn$_{0.5}$Cd$_{0.5}$S, (**b**) ZnWO$_4$, and (**c**) ZWMCS-2 nanocomposite; (**d**) HRTEM images of the ZWMCS-2 nanocomposite.

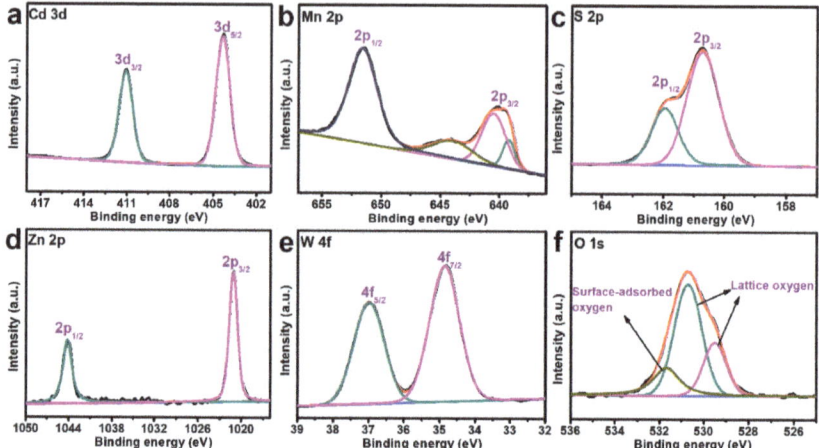

Figure 4. XPS spectra of the as-prepared samples: (**a**) Cd 3d; (**b**) Mn 2p; (**c**) S 3d; (**d**) Zn 2p; (**e**) W 4f; (**f**) O 1s.

2.4. BET Surface Area Analysis

Figure 5a,b show the N$_2$ adsorption–desorption isotherms of Mn$_{0.5}$Cd$_{0.5}$S, ZnWO$_4$ and ZWMCS nanocomposites. It can be seen from the figure that all samples have hysteresis loops, corresponding to Type IV isotherms [35,39]. Figure 5c shows the BET surface areas of each sample (specific values are shown in Table 1), and the specific surface area of ZWMCS nanocomposites are gradually elevated with the increase of the content of ZnWO$_4$.

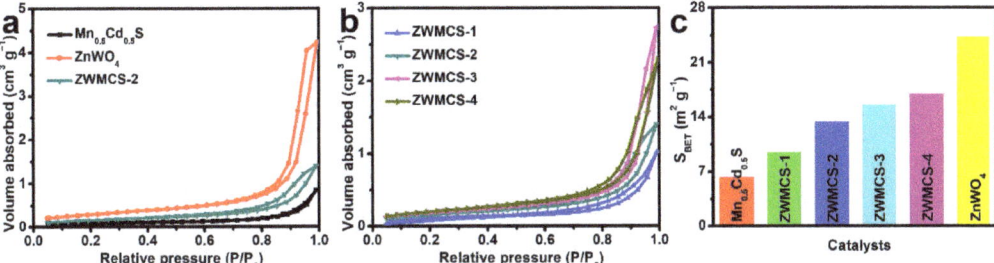

Figure 5. N$_2$ adsorption–desorption isotherms of (**a**) Mn$_{0.5}$Cd$_{0.5}$S, ZnWO$_4$, ZWMCS-2 nanocomposite and (**b**) ZWMCS nanocomposites; (**c**) the BET surface area for Mn$_{0.5}$Cd$_{0.5}$S, ZnWO$_4$, and ZWMCS nanocomposites.

Table 1. The BET surface areas and H$_2$ evolution rate of the Mn$_{0.5}$Cd$_{0.5}$S, ZnWO$_4$, and ZWMCS nanocomposites.

Samples	BET Surface Area (m^2 g^{-1})	H$_2$ Evolution Rate (mmol g^{-1} h^{-1})
Mn$_{0.5}$Cd$_{0.5}$S	6.29	1.96
ZWMCS-1	9.48	1.20
ZWMCS-2	13.34	3.36
ZWMCS-3	15.52	2.28
ZWMCS-4	16.91	1.89
ZnWO$_4$	24.24	0.06

2.5. UV-Vis Diffuse Reflectance Spectroscopy and Band Gap Analysis

Figure 6a shows the UV-*vis* diffuse reflectance spectroscopy (DRS) of Mn$_{0.5}$Cd$_{0.5}$S, ZnWO$_4$ and ZWMCS-2 nanocomposite. The absorption edges of Mn$_{0.5}$Cd$_{0.5}$S and ZnWO$_4$ are at 578.15 and 358.76 nm, respectively. This result indicates that ZnWO$_4$ is a typical invisible light excitation semiconductor material, while Mn$_{0.5}$Cd$_{0.5}$S has good visible light absorption ability. Figure 6a shows that ZWMCS-2 nanocomposite also has good light absorption ability. These results indicate that the ZWMCS-2 nanocomposite formed by Mn$_{0.5}$Cd$_{0.5}$S and ZnWO$_4$ retains the good visible light absorption ability of Mn$_{0.5}$Cd$_{0.5}$S. The linear conversion of absorption curves of Mn$_{0.5}$Cd$_{0.5}$S and ZnWO$_4$ (Figure 6b) shows that the band gap (E_g) of Mn$_{0.5}$Cd$_{0.5}$S and ZnWO$_4$ are 2.24 and 3.68 eV, respectively. According to (Supporting Information Equations (S2) and (S3)) [39,40], the E_{VB} of Mn$_{0.5}$Cd$_{0.5}$S and ZnWO$_4$ are 1.63 and 3.65 eV, respectively, and the E_{CB} of Mn$_{0.5}$Cd$_{0.5}$S and ZnWO$_4$ are −0.61 and −0.03 eV, respectively. Therefore, the conduction and valence band positions of Mn$_{0.5}$Cd$_{0.5}$S are interlaced with those of ZnWO$_4$, the type of heterojunction formed between Mn$_{0.5}$Cd$_{0.5}$S and ZnWO$_4$ in ZWMCS-2 nanocomposite can be Type-II or Z-scheme.

2.6. Photocatalytic H$_2$ Evolution Performance and Electrochemical Analysis

Figure 7a shows the photocatalytic hydrogen evolution rate of each photocatalyst under visible light and without cocatalyst conditions. Among them, the photocatalytic hydrogen evolution rate of Mn$_{0.5}$Cd$_{0.5}$S is 1.96 mmol g^{-1} h^{-1}, which is significantly higher than that of CdS (trace). This indicates that Mn^{2+} doping can significantly increase the photocatalytic hydrogen evolution activity of CdS. The ZWMCS nanocomposites synthesized by combining ZnWO$_4$ and Mn$_{0.5}$Cd$_{0.5}$S together have improved photocatalytic hydrogen evolution activity than the single material. Among them, the ZWMCS-2 nanocomposite has the highest photocatalytic hydrogen evolution rate of 3.32 mmol g^{-1} h^{-1}, which is much higher than that of Mn$_{0.5}$Cd$_{0.5}$S. Cycling test shows that the ZWMCS-2 nanocomposite still maintains good photocatalytic activity after seven cycles (21 h) of photocatalytic hydrogen evolution test (Figure 7b).

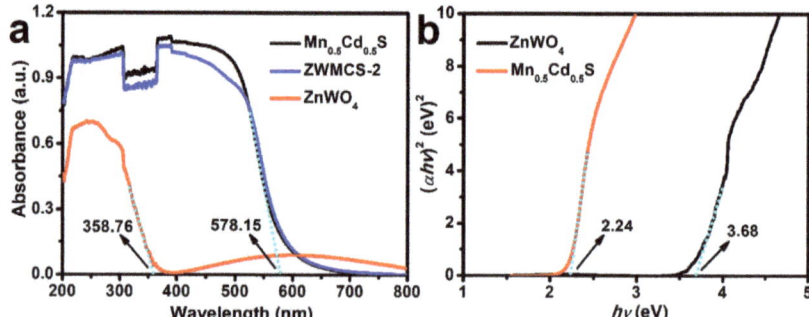

Figure 6. (**a**) UV-vis DRS of $Mn_{0.5}Cd_{0.5}S$, $ZnWO_4$ and ZWMCS-2 nanocomposite; (**b**) plots of $(\alpha h\nu)^2$ versus energy ($h\nu$) for $Mn_{0.5}Cd_{0.5}S$ and $ZnWO_4$.

2.7. Photocatalytic H_2 Evolution Performance and Electrochemical Analysis

Figure 7a shows the photocatalytic hydrogen evolution rate of each photocatalyst under visible light and without cocatalyst conditions. Among them, the photocatalytic hydrogen evolution rate of $Mn_{0.5}Cd_{0.5}S$ is 1.96 mmol g^{-1} h^{-1}, which is significantly higher than that of CdS (trace). This indicates that Mn^{2+} doping can significantly increase the photocatalytic hydrogen evolution activity of CdS. The ZWMCS nanocomposites synthesized by combining $ZnWO_4$ and $Mn_{0.5}Cd_{0.5}S$ together have improved photocatalytic hydrogen evolution activity than the single material. Among them, the ZWMCS-2 nanocomposite has the highest photocatalytic hydrogen evolution rate of 3.32 mmol g^{-1} h^{-1}, which is much higher than that of $Mn_{0.5}Cd_{0.5}S$. Cycling test shows that the ZWMCS-2 nanocomposite still maintains good photocatalytic activity after seven cycles (21 h) of photocatalytic hydrogen evolution test (Figure 7b).

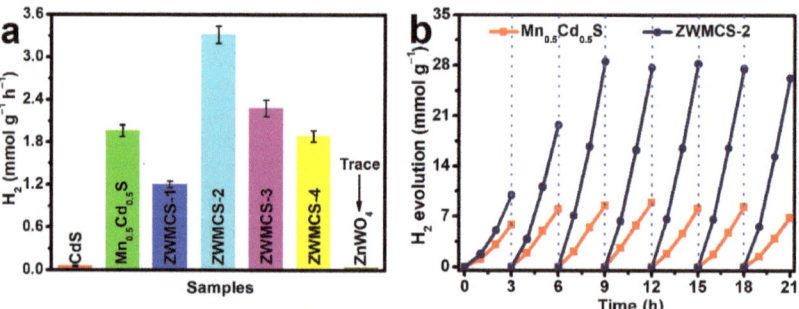

Figure 7. (**a**) Photocatalytic H_2 production of $Mn_{0.5}Cd_{0.5}S$, $ZnWO_4$ and ZWMCS nanocomposites; (**b**) cycling stability for $Mn_{0.5}Cd_{0.5}S$ and ZWMCS-2 nanocomposite.

In order to further explore the photogenerated electron hole separation efficiency of each photocatalyst, their photocurrent responses are tested, and the results are shown in Figure 8. The curve of $ZnWO_4$ is almost a straight line, which indicates that the photocurrent response of $ZnWO_4$ under visible light is very weak. Unlike $ZnWO_4$, $Mn_{0.5}Cd_{0.5}S$ show obvious photocurrent response under visible light. Compared with $Mn_{0.5}Cd_{0.5}S$, the ZWMCS-2 nanocomposite has obvious higher photocurrent. This indicates that the ZWMCS-2 nanocomposite has higher photogenerated electron-hole separation efficiency. The excellent electron-hole separation efficiency of the ZWMCS-2 nanocomposite is conducive to the photocatalytic hydrogen evolution reaction.

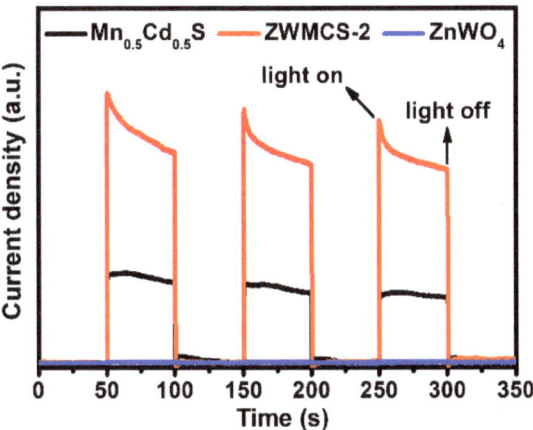

Figure 8. Photocurrent density–time curves of $Mn_{0.5}Cd_{0.5}S$, $ZnWO_4$, and ZWMCS-2 nanocomposite.

2.8. Photocatalytic H_2 Evolution Mechanism

The type of heterojunction is studied by theoretical calculation and EPR. Figure 9a and b show the electrostatic potential diagrams of $Mn_{0.5}Cd_{0.5}S$ and $ZnWO_4$, respectively. The insets in the figure show the optimization models of $Mn_{0.5}Cd_{0.5}S$ (112) and $ZnWO_4$ (111) crystal planes, respectively. The calculation results reveal that the surface work functions of $Mn_{0.5}Cd_{0.5}S$ (112) and $ZnWO_4$ (111) are 4.29 and 5.07 eV, respectively. So, the fermi level of $ZnWO_4$ (111) is significantly lower than that of $Mn_{0.5}Cd_{0.5}S$ (112). When $ZnWO_4$ and $Mn_{0.5}Cd_{0.5}S$ are combined together, a built-in electric field is formed between $ZnWO_4$ and $Mn_{0.5}Cd_{0.5}S$, and the electron movement direction is from $Mn_{0.5}Cd_{0.5}S$ to $ZnWO_4$. The continuous movement of electrons in the built-in electric field and the continuous accumulation of holes make the bands of $Mn_{0.5}Cd_{0.5}S$ and $ZnWO_4$ gradually bend, forming Z-scheme heterojunction, as shown in Figure 9c [40]. It should be noted that the heterojunction herein cannot be Type-II. According to Type-II mechanism (Figure S4), the effective valence band of the ZWMCS-2 nanocomposite is only 1.63 eV, which is much lower than the potential of H_2O/OH (2.27 eV) [39], so the ·OH radical cannot be generated. However, the ZWMCS-2 nanocomposite is found to have a significant ·OH signal by EPR test (Figure 9d). Therefore, the separation mechanism of electrons–holes through the heterojunction in ZWMCS-2 nanocomposite does not belong to Type-II. Following the Z-scheme mechanism (Figure 9e), the effective valence band of the ZWMCS-2 nanocomposite is 3.68 eV, which is much higher than the H_2O/OH potential (2.27 eV), and ·OH can be generated. As revealed by Figure 9d, obvious signals of ·OH can be detected in ZWMCS-2 nanocomposite by EPR. Therefore, the type of heterojunction in ZWMCS-2 nanocomposite is determined to be Z-scheme.

The photocatalytic hydrogen evolution mechanism of the ZWMCS-2 nanocomposite can be described as following. After irradiated by visible light, ZWMCS-2 nanocomposite generates electron-hole pairs. Due to the existence of Z-scheme heterojunction, the photo-generated electrons in the conduction band of $ZnWO_4$ are transferred to the valence band of $Mn_{0.5}Cd_{0.5}S$, and then recombines with the photogenerated holes there. Through this process, the holes in the valence band of $Mn_{0.5}Cd_{0.5}S$ are consumed, avoiding the recombination with electrons, so that a large number of photogenerated electrons are collected in the conduction band of $Mn_{0.5}Cd_{0.5}S$. The effective conduction band of the ZWMCS-2 nanocomposite is −0.61 eV, which is higher than the H^+/H_2 potential (0 eV) [33,34]. Therefore, the photogenerated electrons in the conduction band of $Mn_{0.5}Cd_{0.5}S$ can reduce hydrogen ions in water to hydrogen. At the same time, the holes in the valence band of $ZnWO_4$ are consumed by the sacrificial agents (S^{2-} and SO_3^{2-}). The existence of Z-scheme heterojunction promotes the effective separation of photogenerated electrons and holes, and promotes the performance of photocatalytic hydrogen evolution. Therefore, the ZWMCS-2 nanocomposite has excellent photocatalytic activity and stability for hydrogen production.

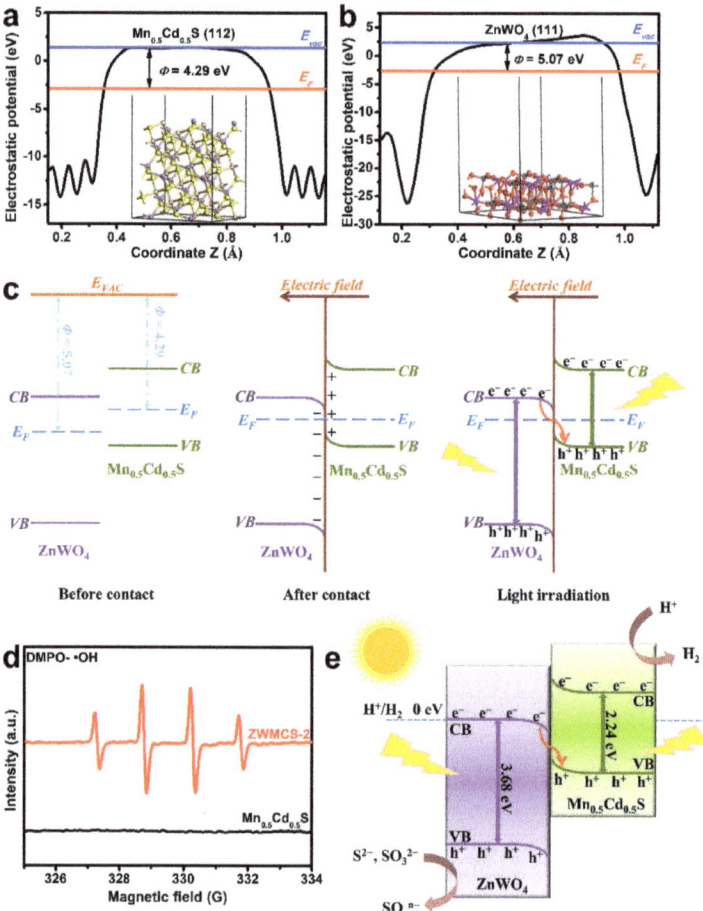

Figure 9. Electrostatic potentials and optimized models of (**a**) $Mn_{0.5}Cd_{0.5}S$ (112) facet and (**b**) $ZnWO_4$ (111) facet. (**c**) The Z-scheme formation process of $Mn_{0.5}Cd_{0.5}S$ (112) and $ZnWO_4$ (111) facet. (**d**) DMPO spin trapping EPR spectra of $Mn_{0.5}Cd_{0.5}S$, and ZWMCS-2 nanocomposite. (**e**) The photocatalysis mechanism of ZWMCS-2 nanocomposite under visible light illumination.

3. Materials and Methods

3.1. Synthesis of $ZnWO_4$ Nanoparticles

Solution A was obtained by dissolving 1.65 g of $NaWO_4 \cdot 2H_2O$ in 15 mL of distilled water. Solution B was obtained by dissolving 1.49 g of $Zn(NO_3)_3 \cdot 6H_2O$ in 15 mL of distilled water. Under continuous stirring, solution A was slowly dropped into solution B. The mixed solution was stirred for 1 h, then it was transferred to a 50 mL reaction kettle, heated at 180 °C for 8 h. After cooling to room temperature, the sample was collected by centrifugation, washed repeatedly with distilled water, and freeze-dried for 18 h.

3.2. Synthesis of $ZnWO_4/Mn_{0.5}Cd_{0.5}S$ Nanocomposites

First, 0.0992 g of $ZnWO_4$, 0.2451 g of $Mn(CH_3COO)_2 \cdot 4H_2O$, 0.2665 g of $Cd(CH_3COO)_2 \cdot 2H_2O$, and 0.1503 g of TAA were dissolved in 35 mL distilled water. After continuous agitation for 1 h, the mixed suspension was transferred to a 50 mL reaction kettle and heated at 160 °C for 48 h. After cooling to room temperature, the sample was collected by centrifugation, washed repeatedly by distilled water, and freeze-dried for 18 h. This sample was named as ZWMCS-2. Other $ZnWO_4/Mn_{0.5}Cd_{0.5}S$ nanocomposites and $Mn_{0.5}Cd_{0.5}S$ nanoparticles were synthesized by similar method with different mass contents of $ZnWO_4$. Table 2 lists the abbreviations of the as-synthesized samples.

Table 2. The amounts of raw materials for preparing the $ZnWO_4/Mn_{0.5}Cd_{0.5}S$ nanocomposites.

Samples	$ZnWO_4$ (g)	$Mn(CH_3COO)_2 \cdot 4H_2O$ (g)	$Cd(CH_3COO)_2 \cdot 2H_2O$ (g)	TAA (g)	Content of $ZnWO_4$ (%)
$Mn_{0.5}Cd_{0.5}S$	0	0.2451	0.2665	0.1503	0
ZWMCS-1	0.0534	0.2451	0.2665	0.1503	20
ZWMCS-2	0.0992	0.2451	0.2665	0.1503	30
ZWMCS-3	0.1542	0.2451	0.2665	0.1503	40
ZWMCS-4	0.2313	0.2451	0.2665	0.1503	50

3.3. Photocatalytic Measurements

Photocatalytic measurements were performed in a 250 mL three-necked flask sealed by rubber stopper. The visible light source is a 300 W Xenon lamp with a $\lambda \geq 420$ nm filter. In a typical photocatalytic hydrogen production process, 0.05 g of catalyst, 8.40 g of $Na_2S \cdot 9H_2O$, and 3.15 g of Na_2SO_3 were dissolved in 100 mL distilled water in the flask, and then N_2 gas was passed into the flask for 30 min to remove air. The flask was illuminated by the Xenon lamp. At an interval of one hour, the gases in the flask was collected by a 1 mL syringe, and was measured by gas chromatography.

3.4. Characterizations

All the characterization equipment and their working parameters are given in the supporting information.

4. Conclusions

$ZnWO_4/Mn_{0.5}Cd_{0.5}S$ nanocomposites are prepared through a two-step hydrothermal method. Under visible light irradiation, and without the help of any noble metal cocatalyst, the ZWMCS-2 nanocomposite exhibits a high hydrogen evolution rate of 3.36 mmol g^{-1} h^{-1}, and does not show obvious deterioration after 21 h cycling test. Moreover, the photocatalytic hydrogen evolution activity of the ZWMCS-2 nanocomposite is significantly higher than that of the constituent materials. Through experimental analysis and theoretical calculation, it is confirmed that the type of the heterojunction in the ZWMCS-2 nanocomposite is Z-scheme. The Z-scheme heterojunction can effectively separate the photogenerated electrons and holes, reduce the photo-corrosion of the single material, and maximize the photocatalytic hydrogen evolution.

Supplementary Materials: The following supporting information can be downloaded at: https://www.mdpi.com/article/10.3390/catal12121527/s1. Figure S1: (a) XRD patterns of as-prepared CdS, MnS, and $Mn_{0.5}Cd_{0.5}S$ (S1. (b) is an enlarged view of the dotted box area.). Figure S2: XRD patterns of as-prepared $ZnWO_4$. Figure S3: XPS survey spectra of ZWMCS-2 nanocomposite. Figure S4: The schematic diagrams of charge transfer in supposed Type-II heterojunction.

Author Contributions: T.M.: conceptualization, data curation, formal analysis, investigation, methodology, writing—original draft preparation; Z.L.: investigation, formal analysis; G.W.: writing—review and editing; J.Z.: funding acquisition, resources; Z.W.: conceptualization, funding acquisition, project administration, resources, writing—review and editing. All authors have read and agreed to the published version of the manuscript.

Funding: This work was financial supported by the National Natural Science Foundation of China (no. 51973078) and Natural Science Foundation of Anhui Province (2108085MB48).

Data Availability Statement: The data presented in this study can be obtained from the first author.

Conflicts of Interest: The authors declare no conflict of interest.

References

1. Song, H.; Luo, S.; Huang, H.; Deng, B.; Ye, J. Solar-Driven Hydrogen Production: Recent Advances, Challenges, and Future Perspectives. *ACS Energy Lett.* **2022**, *7*, 1043–1065. [CrossRef]
2. Feng, C.; Wu, Z.P.; Huang, K.W.; Ye, J.; Zhang, H. Surface Modification of 2D Photocatalysts for Solar Energy Conversion. *Adv. Mater.* **2022**, *34*, 2200180. [CrossRef]
3. Yang, Y.; Zhou, C.; Wang, W.; Xiong, W.; Zeng, G.; Huang, D.; Zhang, C.; Song, B.; Xue, W.; Li, X.; et al. Recent Advances in Application of Transition Metal Phosphides for Photocatalytic Hydrogen Production. *Chem. Eng. J.* **2021**, *405*, 126547. [CrossRef]
4. Tiwari, J.N.; Singh, A.N.; Sultan, S.; Kim, K.S. Recent Advancement of P- and D-Block Elements, Single Atoms, and Graphene-Based Photoelectrochemical Electrodes for Water Splitting. *Adv. Energy Mater.* **2020**, *10*, 2000280. [CrossRef]
5. Lakhera, S.K.; Rajan, A.; Rugma, T.P.; Bernaurdshaw, N. A Review on Particulate Photocatalytic Hydrogen Production System: Progress Made in Achieving High Energy Conversion Efficiency and Key Challenges Ahead. *Renew. Sustain. Energy Rev.* **2021**, *152*, 111694. [CrossRef]
6. Bie, C.; Wang, L.; Yu, J. Challenges for Photocatalytic Overall Water Splitting. *Chem* **2022**, *8*, 1567–1574. [CrossRef]
7. Abdul Nasir, J.; Munir, A.; Ahmad, N.; ul Haq, T.; Khan, Z.; Rehman, Z. Photocatalytic Z-Scheme Overall Water Splitting: Recent Advances in Theory and Experiments. *Adv. Mater.* **2021**, *33*, 2105195. [CrossRef]
8. Yu, J.M.; Lee, J.; Kim, Y.S.; Song, J.; Oh, J.; Lee, S.M.; Jeong, M.; Kim, Y.; Kwak, J.H.; Cho, S.; et al. High-Performance and Stable Photoelectrochemical Water Splitting Cell with Organic-Photoactive-Layer-Based Photoanode. *Nat. Commun.* **2020**, *11*, 5509. [CrossRef]
9. Sharma, C.; Pooja, D.; Thakur, A.; Negi, Y.S. Review—Combining Experimental and Engineering Aspects of Catalyst Design for Photoelectrochemical Water Splitting. *ECS Adv.* **2022**, *1*, 030501. [CrossRef]
10. Shen, R.; Ren, D.; Ding, Y.; Guan, Y.; Ng, Y.H.; Zhang, P.; Li, X. Nanostructured CdS for Efficient Photocatalytic H_2 Evolution: A Review. *Sci. China Mater.* **2020**, *63*, 2153–2188. [CrossRef]
11. Chiarello, G.L.; Dozzi, M.V.; Selli, E. TiO_2-based materials for photocatalytic hydrogen production. *J. Energy Chem.* **2017**, *26*, 250–258. [CrossRef]
12. Nasir, J.A.; Rehman, Z.U.; Shah, S.N.A.; Khan, A.; Butler, I.S.; Catlow, C.R.A. Recent Developments and Perspectives in CdS-Based Photocatalysts for Water Splitting. *J. Mater. Chem. A* **2020**, *8*, 20752–20780. [CrossRef]
13. Ning, X.; Lu, G. Photocorrosion Inhibition of CdS-Based Catalysts for Photocatalytic Overall Water Splitting. *Nanoscale* **2020**, *12*, 1213–1223. [CrossRef]
14. Cheng, L.; Xiang, Q.; Liao, Y.; Zhang, H. CdS-Based Photocatalysts. *Energy Environ. Sci.* **2018**, *11*, 1362–1391. [CrossRef]
15. Mamiyev, Z.; Balayeva, N.O. Metal Sulfide Photocatalysts for Hydrogen Generation: A Review of Recent Advances. *Catalysts* **2022**, *12*, 1316. [CrossRef]
16. Wang, P.; Li, H.; Sheng, Y.; Chen, F. Inhibited Photocorrosion and Improved Photocatalytic H_2-Evolution Activity of CdS Photocatalyst by Molybdate Ions. *Appl. Surf. Sci.* **2019**, *463*, 27–33. [CrossRef]
17. Chen, Y.; Zhong, W.; Chen, F.; Wang, P.; Fan, J.; Yu, H. Photoinduced Self-Stability Mechanism of CdS Photocatalyst: The Dependence of Photocorrosion and H_2-Evolution Performance. *J. Mater. Sci. Technol.* **2022**, *121*, 19–27. [CrossRef]
18. Wang, B.; Chen, C.; Jiang, Y.; Ni, P.; Zhang, C.; Yang, Y.; Lu, Y.; Liu, P. Rational Designing 0D/1D Z-Scheme Heterojunction on CdS Nanorods for Efficient Visible-Light-Driven Photocatalytic H_2 Evolution. *Chem. Eng. J.* **2021**, *412*, 128690. [CrossRef]
19. She, H.; Sun, Y.; Li, S.; Huang, J.; Wang, L.; Zhu, G.; Wang, Q. Synthesis of Non-Noble Metal Nickel Doped Sulfide Solid Solution for Improved Photocatalytic Performance. *Appl. Catal. B Environ.* **2019**, *245*, 439–447. [CrossRef]
20. Iqbal, S. Spatial Charge Separation and Transfer in L-Cysteine Capped NiCoP/CdS Nano-Heterojunction Activated with Intimate Covalent Bonding for High-Quantum-Yield Photocatalytic Hydrogen Evolution. *Appl. Catal. B Environ.* **2020**, *274*, 119097. [CrossRef]
21. Ren, D.; Shen, R.; Jiang, Z.; Lu, X.; Li, X. Highly Efficient Visible-Light Photocatalytic H_2 Evolution over 2D-2D CdS/Cu_7S_4 Layered Heterojunctions. *Chin. J. Catal.* **2020**, *41*, 31–40. [CrossRef]
22. Kumar, A.; Krishnan, V. Vacancy Engineering in Semiconductor Photocatalysts: Implications in Hydrogen Evolution and Nitrogen Fixation Applications. *Adv. Funct. Mater.* **2021**, *31*, 2009807. [CrossRef]
23. Li, L.; Liu, G.; Qi, S.; Liu, X.; Gu, L.; Lou, Y.; Chen, J.; Zhao, Y. Highly Efficient Colloidal $Mn_xCd_{1-x}S$ Nanorod Solid Solution for Photocatalytic Hydrogen Generation. *J. Mater. Chem. A* **2018**, *6*, 23683–23689. [CrossRef]
24. Huang, D.; Wen, M.; Zhou, C.; Li, Z.; Cheng, M.; Chen, S.; Xue, W.; Lei, L.; Yang, Y.; Xiong, W.; et al. $Zn_xCd_{1-x}S$ Based Materials for Photocatalytic Hydrogen Evolution, Pollutants Degradation and Carbon Dioxide Reduction. *Appl. Catal. B Environ.* **2020**, *267*, 118651. [CrossRef]
25. Dan, M.; Prakash, A.; Cai, Q.; Xiang, J.; Ye, Y.; Li, Y.; Yu, S.; Lin, Y.; Zhou, Y. Energy-Band-Controlling Strategy to Construct Novel $Cd_xIn_{1-x}S$ Solid Solution for Durable Visible Light Photocatalytic Hydrogen Sulfide Splitting. *Solar RRL* **2019**, *3*, 1800237. [CrossRef]
26. Liu, J.; Feng, J.; Lu, L.; Wu, B.; Ren, P.; Shi, W.; Cheng, P. A Metal-Organic-Framework-Derived $(Zn_{0.95}Cu_{0.05})_{0.6}Cd_{0.4}S$ Solid Solution as Efficient Photocatalyst for Hydrogen Evolution Reaction. *ACS Appl. Mater. Interfaces* **2020**, *12*, 10261–10267. [CrossRef]

27. Huang, H.; Wang, Z.; Luo, B.; Chen, P.; Lin, T.; Xiao, M.; Wang, S.; Dai, B.; Wang, W.; Kou, J.; et al. Design of Twin Junction with Solid Solution Interface for Efficient Photocatalytic H_2 Production. *Nano Energy* **2020**, *69*, 104410. [CrossRef]
28. Deng, W.; Chen, J.; Yang, L.; Liang, X.; Yin, S.; Deng, X.; Zou, G.; Hou, H.; Ji, X. Solid Solution Metal Chalcogenides for Sodium-Ion Batteries: The Recent Advances as Anodes. *Small* **2021**, *17*, 2101058. [CrossRef]
29. Li, H.; Wang, Z.; He, Y.; Meng, S.; Xu, Y.; Chen, S.; Fu, X. Rational Synthesis of $Mn_xCd_{1-x}S$ for Enhanced Photocatalytic H_2 Evolution: Effects of S Precursors and the Feed Ratio of Mn/Cd on Its Structure and Performance. *J. Colloid Interface Sci.* **2019**, *535*, 469–480. [CrossRef]
30. Xiong, M.; Qin, Y.; Chai, B.; Yan, J.; Fan, G.; Xu, F.; Wang, C.; Song, G. Unveiling the Role of Mn-Cd-S Solid Solution and MnS in $Mn_xCd_{1-x}S$ Photocatalysts and Decorating with CoP Nanoplates for Enhanced Photocatalytic H_2 Evolution. *Chem. Eng. J.* **2022**, *428*, 131069. [CrossRef]
31. Kumar, G.M.; Lee, D.J.; Jeon, H.C.; Ilanchezhiyan, P.; Young, K.D.; Won, K.T. One Dimensional $ZnWO_4$ Nanorods Coupled with WO_3 Nanoplates Heterojunction Composite for Efficient Photocatalytic and Photoelectrochemical Activity. *Ceram. Int.* **2022**, *48*, 4332–4340. [CrossRef]
32. Zhang, N.; Chen, D.; Cai, B.; Wang, S.; Niu, F.; Qin, L.; Huang, Y. Facile Synthesis of $CdS-ZnWO_4$ Composite Photocatalysts for Efficient Visible Light Driven Hydrogen Evolution. *Int. J. Hydrogen Energy* **2017**, *42*, 1962–1969. [CrossRef]
33. Ng, B.J.; Putri, L.K.; Kong, X.Y.; Teh, Y.W.; Pasbakhsh, P.; Chai, S.P. Z-Scheme Photocatalytic Systems for Solar Water Splitting. *Adv. Sci.* **2020**, *7*, 1903171. [CrossRef]
34. Liao, G.; Li, C.; Liu, S.Y.; Fang, B.; Yang, H. Emerging Frontiers of Z-Scheme Photocatalyric Systems. *Trends Chem.* **2022**, *4*, 111–127. [CrossRef]
35. Zuo, G.; Wang, Y.; Teo, W.L.; Xian, Q.; Zhao, Y. Direct Z-Scheme TiO_2-$ZnIn_2S_4$ Nanoflowers for Cocatalyst-Free Photocatalytic Water Splitting. *Appl. Catal. B Environ.* **2021**, *291*, 120126. [CrossRef]
36. Li, B.; Wang, W.; Zhao, J.; Wang, Z.; Su, B.; Hou, Y.; Ding, Z.; Ong, W.J.; Wang, S. All-Solid-State Direct Z-Scheme $NiTiO_3/Cd_{0.5}Zn_{0.5}S$ Heterostructures for Photocatalytic Hydrogen Evolution with Visible Light. *J. Mater. Chem. A* **2021**, *9*, 10270–10276. [CrossRef]
37. Jiang, X.; Gong, H.; Liu, Q.; Song, M.; Huang, C. In Situ Construction of $NiSe/Mn_{0.5}Cd_{0.5}S$ Composites for Enhanced Photocatalytic Hydrogen Production under Visible Light. *Appl. Catal. B Environ.* **2020**, *268*, 118439. [CrossRef]
38. Cui, H.; Li, B.; Zheng, X.; Zhang, Y.; Yao, C.; Li, Z.; Sun, D.; Xu, S. Efficient Activity and Stability of $ZnWO_4/CdS$ Composite Towards Visible Light Photocatalytic H_2 Evolution. *J. Photochem. Photobiol. A Chem.* **2019**, *384*, 112046. [CrossRef]
39. Li, Z.; Ma, T.; Zhang, J.; Wang, Z. Converting the Charge Transfer in $ZnO/Zn_xCd_{1-x}S$-DETA Nanocomposite from Type-I to S-Scheme for Efficient Photocatalytic Hydrogen Production. *Adv. Mater. Interfaces* **2022**, *9*, 2102497. [CrossRef]
40. Shen, C.H.; Wen, X.J.; Fei, Z.H.; Liu, Z.T.; Mu, Q.M. Novel Z-Scheme $W_{18}O_{49}/CeO_2$ Heterojunction for Improved Photocatalytic Hydrogen Evolution. *J. Colloid Interface Sci.* **2020**, *579*, 297–306. [CrossRef]

Article

Efficient and Stable Catalytic Hydrogen Evolution of ZrO₂/CdSe-DETA Nanocomposites under Visible Light

Zhen Li [1], Ligong Zhai [1], Tingting Ma [2], Jinfeng Zhang [3,*] and Zhenghua Wang [2,*]

[1] School of Food Engineering, Anhui Science and Technology University, Fengyang 233100, China
[2] Key Laboratory of Functional Molecular Solids, Ministry of Education, College of Chemistry and Materials Science, Anhui Normal University, Wuhu 241000, China
[3] Anhui Province Key Laboratory of Pollutant Sensitive Materials and Environmental Remediation, School of Physics and Electronic Information, Huaibei Normal University, Huaibei 235000, China
* Correspondence: jfzhang@chnu.edu.cn (J.Z.); zhwang@ahnu.edu.cn (Z.W.)

Citation: Li, Z.; Zhai, L.; Ma, T.; Zhang, J.; Wang, Z. Efficient and Stable Catalytic Hydrogen Evolution of ZrO₂/CdSe-DETA Nanocomposites under Visible Light. *Catalysts* 2022, 12, 1385. https://doi.org/10.3390/catal12111385

Academic Editors: Yongming Fu and Qian Zhang

Received: 9 October 2022
Accepted: 4 November 2022
Published: 8 November 2022

Publisher's Note: MDPI stays neutral with regard to jurisdictional claims in published maps and institutional affiliations.

Copyright: © 2022 by the authors. Licensee MDPI, Basel, Switzerland. This article is an open access article distributed under the terms and conditions of the Creative Commons Attribution (CC BY) license (https://creativecommons.org/licenses/by/4.0/).

Abstract: Composite photocatalysts are crucial for photocatalytic hydrogen evolution. In this work, ZrO₂/CdSe-diethylenetriamine (ZrO₂/CdSe-DETA) heterojunction nanocomposites are synthesized, and efficiently and stably catalyzed hydrogen evolution under visible light. X-ray photoelectron spectroscopy (XPS) and high resolution transmission electron microscope (HRTEM) confirm the formation of heterojunctions between ZrO₂ (ZO) and CdSe-DETA (CS). Ultraviolet–visible spectroscopy diffuse reflectance spectra (UV-vis DRS), Mott–Schottky, and theoretical calculations confirm that the mechanism at the heterojunction of the ZrO₂/CdSe-DETA (ZO/CS) nanocomposites is Type-I. Among the ZO/CS nanocomposites (ZO/CS-0.4, ZO/CS-0.6, and ZO/CS-0.8; in the nanocomposites, the mass ratio of ZO to CS is 0.1:0.0765, 0.1:0.1148, and 0.1:0.1531, respectively). ZO/CS-0.6 nanocomposite has the best photocatalytic hydrogen evolution activity (4.27 mmol g^{-1} h^{-1}), which is significantly higher than ZO (trace) and CS (1.75 mmol g^{-1} h^{-1}). Within four cycles, the ZO/CS-0.6 nanocomposite maintains an efficient catalytic hydrogen evolution rate. Due to the existence of the heterojunction of the composites, the photogenerated electron-hole pairs can be effectively separated, which accelerates the photocatalytic hydrogen evolution reaction and reduces the progress of photocorrosion. This work reveals the feasibility of ZO/CS nanocomposite photocatalysts for hydrogen evolution.

Keywords: photocatalysts; stably; nanocomposites; photogenerated electron-hole pairs; hydrogen

1. Introduction

The increasing consumption of nonrenewable energy has caused many environmental problems, and the exploration of clean energy is gradually increasing [1–4]. Hydrogen energy is considered as one of the ideal green energy sources because of its high heat, it only leaves water behind after combustion, it is completely pollution-free, and it is recyclable [5–9]. Photocatalytic technology can effectively solve the hydrogen acquisition problem by utilizing the continuous production of hydrogen from a wide range of solar energy photocatalytic semiconductor material sources [10–14]. As a classical semiconductor material, cadmium selenide (CdSe) has been widely used in photocatalytic hydrogen evolution experiments due to its suitable band gap, visible light absorption, and better hydrogen evolution activity [15–18]. However, the inherent drawbacks of a single photocatalyst still limit the development of CdSe. Therefore, research is needed to explore ways to effectively solve the above problems.

The traditional way to change the photocatalytic properties of a single semiconductor is to modify its morphology and thus improve its intrinsic properties. Therefore, we synthesized CdSe-diethylenetriamine (CdSe-DETA) with a large Brunauer–Emmett–Teller (BET) surface area by the method of hybridizing diethylenetriamine with CdSe, obtaining better photocatalytic hydrogen evolution activity and stability, which alleviated the inherent drawbacks of CdSe [11]. Yet, there were still no substantial changes in the band structure, light absorption characteristics, or the charge mobility of a single photocatalyst. In recent years, more and more researchers have found that the mutual recombination between semiconductor materials can effectively solve the shortcomings of a single photocatalyst [19–26]. For example, Ma et al. combined CdSe with $WO_3(H_2O)_{0.333}$ to significantly enhance the photocatalytic hydrogen evolution activity of single materials [24]. Wang et al. combined CoO_x with Pt to effectively enhance the hydrogen evolution activity and stability of single materials [25]. These studies are based on the combination of photocatalysts that have a broad forbidden band width, good stability, and are non-toxic and inexpensive. Among numerous semiconductor photocatalysts, ZrO_2 (ZO) perfectly meets the above conditions. However, the study of the composite of (ZO) and CdSe-DETA (CS) in the photocatalytic system has not been reported. Therefore, we combined ZO with CS to address the inherent drawbacks of a single semiconductor photocatalyst.

After the successful composite of ZO and CS, the nanocomposites showed excellent photocatalytic hydrogen evolution activity and good stability. This is attributed to the fact that the composites can effectively separate the photogenerated electrons and holes, resulting in continuous photocatalytic hydrogen evolution. We confirmed the existence of the composites by HRTEM, XPS, and photocatalytic hydrogen evolution activity experiments [27–35]. The accuracy of the mechanism at the heterojunctions derived from the experiments is confirmed by theoretical calculations and Mott–Schottky analysis deriving the band gap and conduction band positions of the single materials, respectively. The reasons for the excellent photocatalytic hydrogen evolution activity and stability of the ZO/CS nanocomposites are explored in detail by combining theory and experiment. This work provides a feasible way to explore the way in which composite semiconductor materials can effectively solve the inherent defects of a single semiconductor material.

2. Results

2.1. Flow Chart of Materials Synthesis

The synthesis process of ZO/CS nanocomposites is plotted by constructing a model, as shown in Scheme 1. First, ZO is added to the reactor as a substrate material. After that, Cd^{2+} (derived from $CdCl_2·2.5H_2O$), Se^{2-} (derived from selenium powder), DETA, and $N_2H_4·H_2O$ are added to the reaction solution and stirred for 1 h at room temperature. Then, the above solution is transferred to the reaction kettle and heated in the oven (100 °C for 2 h). Finally, ZO/CS nanocomposites are obtained by repeated centrifugation (five times) and freeze-drying ($\geq$18 h, $\leq$-45 °C).

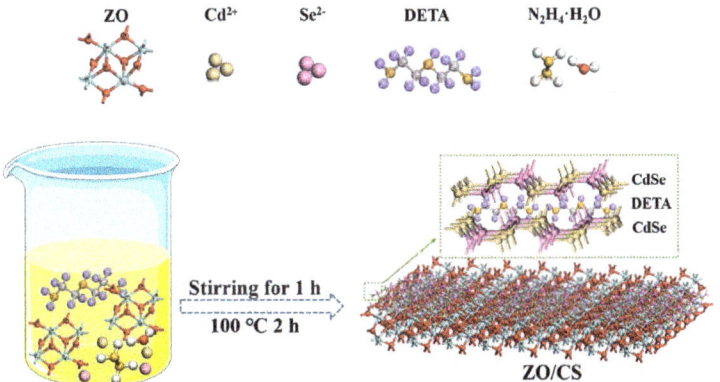

Scheme 1. Schematic diagram for the synthesis of ZO/CS nanocomposites.

2.2. Phase and Microscopic Morphology Analysis

Figure 1 shows the X-ray diffraction (XRD) patterns of ZO, CS, and ZO/CS nanocomposites. ZO has a good crystallinity, and its XRD pattern is consistent with ZrO_2 of a monoclinic phase (JCPDS No. 65-1025) [23]. The main peaks at 24.05°, 28.18°, 31.47°, 34.16°, 35.31°, 40.73°, 49.26°, 50.11°, 54.10°, 55.28°, 60.05° and 62.84° can be indexed to the (011), (-111), (111), (002), (200), (-211), (022), (220), (202), (221), (-302), and (311) crystal planes. The crystallinity of CS is low, which is due to its special synthesis conditions. When the reaction time of CS is prolonged, it will crystallize completely (shown in Figure 1b). In this work, the XRD of CS is in agreement with CdSe of the hexagonal phase (JCPDS No. 77-2307), which has been demonstrated in previous work [27]. The main peaks at 23.88°, 25.39°, 27.10°, 42.00°, 45.81°, and 49.72° can be indexed to the (100), (002), (101), (110), (103), and (112) crystal planes. Here, the crystallite sizes of ZO and CS are calculated from Scherrer's formula [21] to be 28.04 and 20.04 nm, respectively. XRD patterns of ZO/CS nanocomposites contain peaks of ZO and CS, and peaks of CS are gradually highlighted with increasing CS content in ZO/CS (as shown by dashed boxes in Figure 1a). This indicates that ZO and CS are contained in the ZO/CS nanocomposites. In addition, there are no other miscellaneous peaks in the synthesized samples, which indicates the purity of the synthesized samples.

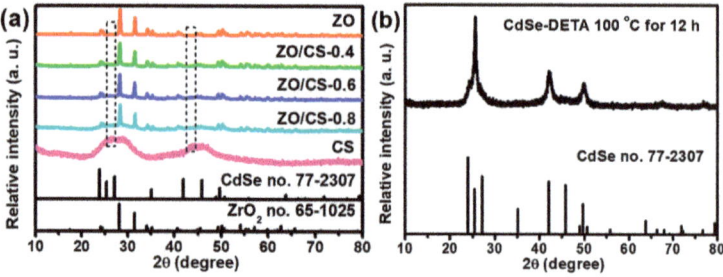

Figure 1. XRD patterns of (**a**) ZO, ZO/CS nanocomposites, and CS; (**b**) CdSe-DETA in 100 °C for 12 h.

Scanning electron microscope (SEM) and transmission electron microscope (TEM) were used in order to better observe the micro morphology and element composition of the synthesized samples, as shown in Figure 2. It can be easily observed from the SEM images in Figure 2a–c that CS (Figure 2a) has the morphology of a flower bud, with a diameter of about 20 nm; ZO (Figure 2b) shows a nanospherical appearance, and its size is in the range of 20–100 nm; ZO/CS-0.6 (Figure 2c) nanocomposite shows the morphology of mutual wrapping of CS and ZO. Figure 2d–f is the EDS spectrum of the characterized samples, showing the elemental composition of the photographed samples. Among them, Figure 2d shows the elemental composition of CS. The results showed that only Cd, Se, C, and N elements are contained in CS, and there are no impurity peaks [11,27]. The atomic contents of Se, Cd, N, and C are 32.23%, 12.41%, 16.48%, and 38.88%, respectively. Figure 2e shows the elemental composition of ZO. ZO contained only O and Zr elements with no impurity peaks [31]. The atomic contents of O and Zr are 59.53% and 40.47%, respectively. Figure 2f shows the elemental composition of ZO/CS-0.6 nanocomposite. ZO/CS-0.6 nanocomposite contains all the pure material elements and has no impurity peaks. The atomic contents of Se, Cd, N, C, O, and Zr are 4.87%, 0.34%, 10.75%, 28.34%, 11.67%, and 6.62%, respectively. Here, the C element comes from DETA or conducting resin. The N element comes from DETA or N_2H_4. The difference in the content of each element is due to the fact that different elements have different energy, and the stacking of samples leads to the deviation of the test results. The results show that the tested samples contain all the constituent elements and there are no impurity peaks, which indicates the purity of the synthesized samples, in agreement with the result in Figure 1.

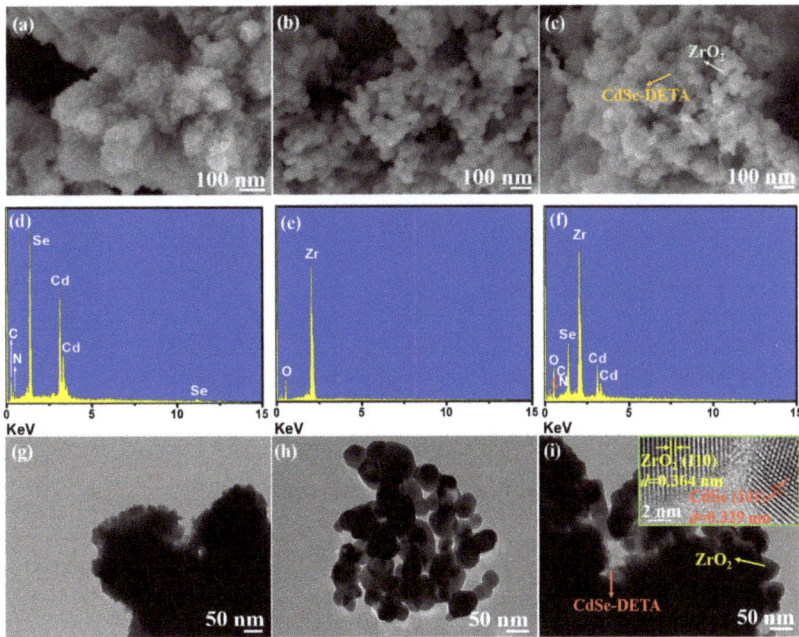

Figure 2. SEM images of (**a**) CS, (**b**) ZO, and (**c**) ZO/CS-0.6; EDS spectra of (**d**) CS, (**e**) ZO, and (**f**) ZO/CS-0.6. TEM images of (**g**) CS, (**h**) ZO, and (**i**) ZO/CS-0.6 (inset shows the HRTEM images of the ZO/CS-0.6).

Figure 2g–i is the clearer TEM images of CS, ZO, and ZO/CS-0.6, which is consistent with the morphology of Figure 2a–c. Among them, the illustration in Figure 2i is the high-resolution TEM (HRTEM) image of the ZO/CS-0.6 nanocomposite. The lattice fringes of ZrO$_2$ (110) and CdSe (101) can be clearly seen from this figure, while the appearance of a fuzzy interface for lattice fringes between them confirms the existence of heterojunctions. The presence of heterojunctions confirms the successful preparation of the ZO/CS-0.6 nanocomposite. Furthermore, Figure S1 presents the high angle annular dark field (HAADF) and elemental mapping images of ZO/CS-0.6 nanocomposite. As can be seen in Figure S1, the elements consisted of ZO/CS-0.6 nanocomposite are well distributed, indicating the homogeneous texture of the synthesized samples.

2.3. X-ray Photoelectron Spectroscopy (XPS) and Elemental Analysis

Figure 3 shows the XPS spectra of ZO, CS, and ZO/CS-0.6 nanocomposite. In Figure 3a, ZO/CS-0.6 nanocomposite contains all the elements in ZO (O, C, and Zr) and CS (Cd, N, and Se). Except for the elements of the respective samples themselves, the N element is from DETA or N$_2$H$_4$·H$_2$O, and the C element is from DETA or surface adsorbed carbon dioxide. Figure 3b–f shows the high-resolution XPS spectra of some core elements of ZO, CS, and ZO/CS-0.6 nanocomposite. In Figure 3b, the two peaks (Zr 3d$_{3/2}$ and Zr 3d$_{5/2}$) of Zr 3d of ZO are located at 184.28 and 181.92 eV, respectively. The two peaks (Zr 3d$_{3/2}$ and Zr 3d$_{5/2}$) of Zr 3d of ZO/CS-0.6 nanocomposite are located at 183.50 and 181.14 eV, respectively [30]. In Figure 3c, O 1s of ZO is divided into lattice oxygen and surface-adsorbed oxygen, and the positions are located at 531.45 and 529.74 eV, respectively [30]. Similarly, the two peaks of the O 1s of ZO/CS-0.6 nanocomposite are located at 530.51 and 528.95 eV, respectively. In Figure 3d, the Cd 3d$_{3/2}$ and Cd 3d$_{5/2}$ peaks of the Cd 3d of CS are located at 410.90 and 404.14 eV, respectively [11]. Similarly, the two peaks of the Cd 3d of ZO/CS-0.6 nanocomposite are located at 410.93 and 404.17 eV, respectively. In

Figure 3e, the Se 3d of CS can be divided into two peaks: Se $3d_{3/2}$ and Se $3d_{5/2}$, which are located at 53.60 and 52.74 eV, respectively [11]. The two peaks of the Se 3d of ZO/CS-0.6 nanocomposite are located at 52.64 and 52.78 eV, respectively. In Figure 3f, the N 1s peak of CS is located at 398.76 eV [11], which is shifted to the right by 0.11 eV relative to ZO/CS-0.6 nanocomposites. The above results show that the peak positions of each element of ZO/CS-0.6 nanocomposite are shifted relative to ZO or CS, which confirms the heterojunctions between ZO and CS. This result is consistent with that of Figure 2i.

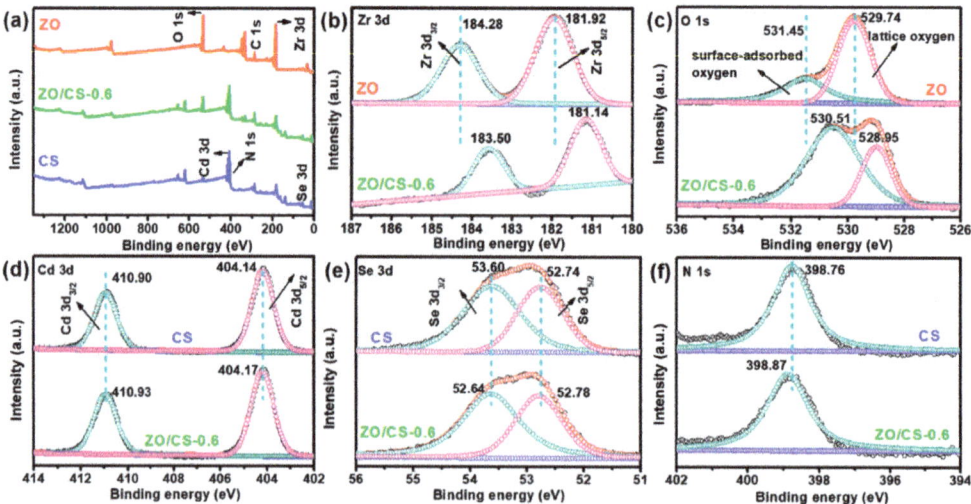

Figure 3. XPS spectra: (**a**) survey scan; (**b**) Zr 3d; (**c**) O 1s; (**d**) Cd 3d; (**e**) Se 3d and (**f**) N 1s.

2.4. Optical Property and Band Gap Analysis

In order to explore the optical absorption properties and band gap of materials, ZO, CS, and ZO/CS nanocomposites are characterized by ultraviolet–visible spectroscopy diffuse reflectance spectra (UV-vis DRS), as shown in Figure 4. In Figure 4a, CS behaves as a visible light absorbing material with excellent visible light absorption ability, which tends to represent its smaller band gap. In contrast, ZO shows strong UV light absorption ability and is not excited in the visible range, while usually represents a large band gap. In the ZO/CS nanocomposites, the light absorption range of the nanocomposites increased gradually with increasing CS content. Furthermore, the color of each sample was uniform and without variegation, which illustrated the purity of the samples. After that, according to Figure 4a, the linear transformation plots of CS and ZO absorption curves are drawn (Figure 4b). As can be seen from the figure, the band gap of CS is 2.36 eV. The band gap of ZO is 5.17 eV, which is significantly larger than that of CS. In addition, the theoretical band gaps of ZO and CS are obtained by theoretical calculation (Figure S2). The results show that the band gap of ZO is significantly larger than that of CS, confirming the accuracy of the above results.

2.5. Fourier Transform Infrared Spectoscopy (FT-IR) Analysis

FT-IR is used to explore the functional groups of each sample in order to further explore its elemental composition. As shown in Figure 5, CS contains strong vibration bands of N-H (about 3090–3500 and 1000–1320 cm^{-1}), -CH$_2$- (approximately 2750–3000 cm^{-1}), C-N (around 1468 cm^{-1}) and C-H (roughly 550–850 and 1590 cm^{-1}) [24]. Wherein, C-H, C-N, and -CH$_2$- are from DETA. N-H comes from DETA or $N_2H_4 \cdot H_2O$. ZO does not contain the above functional groups. Nevertheless, the ZO/CS nanocomposites formed by the composite of ZO and CS contain the vibration bands mentioned above, and the frequency is strong. This indicates that both CS and ZO/CS nanocomposites contain DETA.

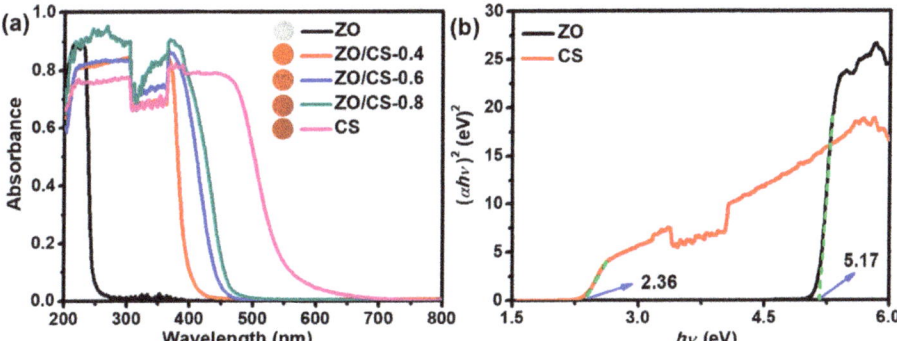

Figure 4. (a) UV-vis DRS of ZO, CS, and ZO/CS nanocomposites; (b) plots of $(\alpha h\nu)^2$ versus energy ($h\nu$) for ZO and CS.

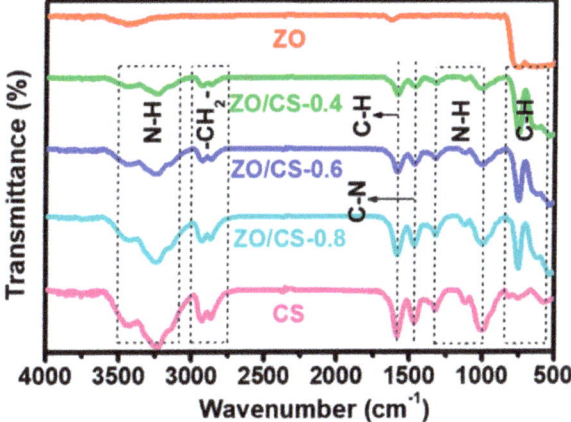

Figure 5. FT-IR spectra of ZO, ZO/CS nanocomposites, and CS.

2.6. BET Surface Area

Figure 6 shows the specific surface analysis of ZO, CS, and ZO/CS nanocomposites. In Figure 6a, all samples show Type IV isotherms and H3 hysteresis loops [35]. As can be seen from the illustrations in Figure 6a, all the samples characterized are mesoporous materials. Among them, most of the pore sizes of the tested materials are distributed in the range of 2–50 nm, conforming to the characteristics of mesoporous materials. Figure 6b shows the BET surface areas of the characterized materials. In Table S1, the BET surface areas of ZO, CS, ZO/CS-0.4, ZO/CS-0.6, and ZO/CS-0.8 are 16.41, 14.87, 13.73, 18.64, and 14.97 m^2 g^{-1}, respectively. The results show that the BET surface areas of CS and ZO are smaller, and the ZO/CS nanocomposites also exhibit smaller BET surface areas. However, ZO/CS-0.6 nanocomposite exhibits the optimal BET surface area. A larger BET surface area will provide more active sites for the reaction, which is helpful to the photocatalytic hydrogen evolution reaction. In addition, the details of the average pore size and total pore volume of the tested samples are shown in Table S1.

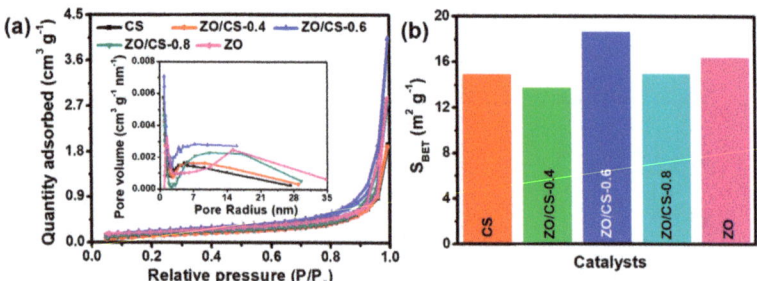

Figure 6. (**a**) N$_2$ adsorption–desorption isotherms of ZO, ZO/CS nanocomposites, and CS, inserts are the pore size distribution curves; (**b**) BET surface area for the above samples.

2.7. Photocatalytic H$_2$ Evolution Performance and Electrochemical Analysis

Figure 7 shows the photocatalytic hydrogen evolution rate of ZO, CS, and ZO/CS nanocomposites, and the photocatalytic hydrogen evolution stability of ZO/CS-0.6. In Figure 7a, CS has a relatively excellent photocatalytic hydrogen evolution rate (1.75 mmol g^{-1} h^{-1}). Yet, ZO has no photocatalytic hydrogen evolution activity, which is represented here by trace. When the two pure materials are compounded, ZO/CS nanocomposites show excellent photocatalytic hydrogen evolution rates, which are much higher than that of CS and ZO alone. Among them, ZO/CS-0.6 shows the best photocatalytic hydrogen evolution rate, reaching 4.27 mmol g^{-1} h^{-1}. Moreover, the present work still possesses excellent photocatalytic hydrogen evolution activity compared with the photocatalysts in other literatures (Table S2). Figure 7b shows the photocatalytic hydrogen evolution stability test of ZO/CS-0.6 nanocomposite. The results showed that ZO/CS-0.6 nanocomposite showes excellent photocatalytic activity for hydrogen evolution in four cycles. In Figure S3a, ZO/CS-0.6 nanocomposite is recrystallized after cycling, showing a slightly sharp XRD peak, while the other peaks remained almost unchanged. In Figure S3b,c, the recrystallized ZO/CS-0.6 nanocomposites are recrystallized after cycling, but the overall morphology did not change obviously, which verified the results of Figure S3a. This fully shows the excellent stability of ZO/CS-0.6 nanocomposite.

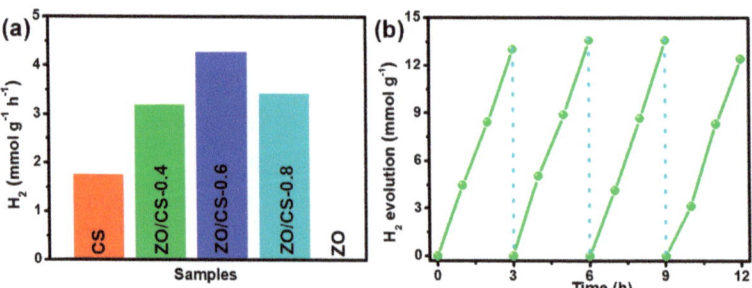

Figure 7. (**a**) Photocatalytic H$_2$ production rates of as-prepared photocatalysts. (**b**) Cycling stability for the ZO/CS-0.6.

The photocurrent response can effectively explore the photoexcitation ability of photocatalysts. Figure 8 shows the photocurrent density-time curves of ZO, CS, and ZO/CS-0.6 nanocomposite. It can be seen from the figure that ZO shows a very weak photocurrent response curve, which is approximately a straight line. However, CS shows a higher photocurrent response curve, which is obviously better than that of ZO. After forming the ZO/CS-0.6 nanocomposite with the composite of ZO and CS, it exhibits an excellent photocurrent response curve, which is much higher than that of ZO and CS alone. This shows that the composite of ZO and CS can effectively improve the light excitation ability of the materials and contribute to the photocatalytic hydrogen evolution reaction. This is consistent with the results of Figure 7.

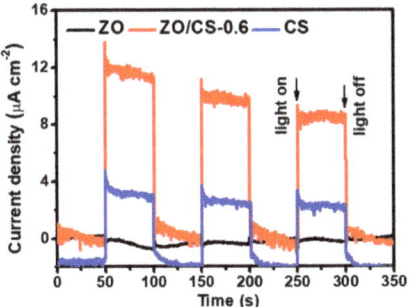

Figure 8. Photocurrent density–time curves of CS, ZO/CS-0.6, and ZO.

2.8. Photocatalytic Mechanism

The photocatalytic hydrogen evolution mechanism at the heterojunction of ZO/CS nanocomposites is shown in Figure 9, which is the classical Type-I model [36–38]. According to the formula (S1)–(S3) (Supporting information) [11] and the results of Figure 4, the conduction band of ZO is at -1.17 eV and the valence band is at 4 eV. Similarly, the conduction band of CS is at -0.63 eV and the valence band is at 1.73 eV. According to the valence band position, the heterojunction mechanism of ZO/CS nanocomposites is the classical Type-I model. In addition, ZO and CS are characterized by Mott–Schottky analysis in electrochemical methods (Figure 9a,b). The results show that the mechanism of the heterojunction formation between ZO and CS is the classical Type-I model, which verifies the experimental results.

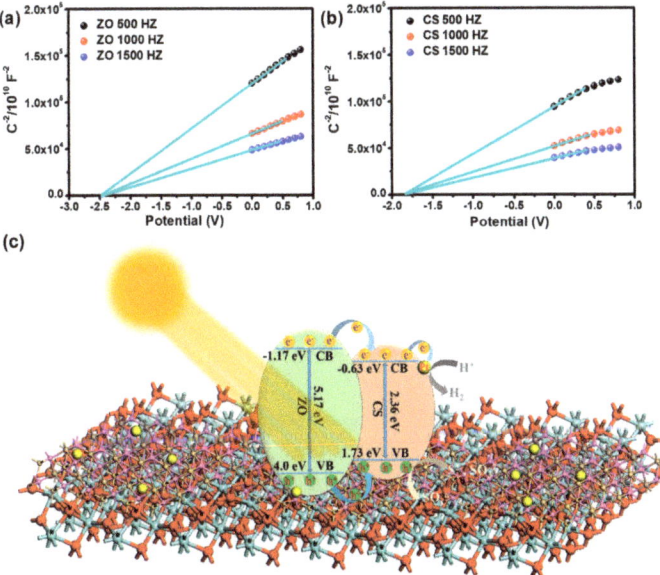

Figure 9. Mott–Schottky plots of (**a**) ZO and (**b**) CS. (**c**) The photocatalysis mechanism of ZO/CS nanocomposite under visible light.

As shown in Figure 9, ZO and CS produce photogenerated electron-hole pairs under visible light irradiation. After that, the photogenerated electron-hole pairs are separated rapidly, the photogenerated electrons gather in the conduction band of the semiconductor material, and the photogenerated holes gather in the valence band. At this time, many electrons on the surface of the ZO conduction band are transferred to the position of the CS conduction band. A large number of electrons on the surface of the ZO valence band are transferred to the CS valence band location. Abundant photogenerated electrons are gathered on the surface of the CS conduction band, which are continuously transferred to the surface of the co-catalyst Pt and combine with H^+ to produce hydrogen [39–41]. On the other hand, there are a large number of photogenerated holes on the surface of the CS valence band, which are consumed by the sacrificial agent [28]. In this way, the photogenerated electrons and holes in ZO/CS nanocomposites are continuously separated, which accelerates the photocatalytic hydrogen evolution reaction and alleviates the occurrence of photocorrosion. Therefore, ZO/CS nanocomposites show excellent photocatalytic activity and stability for hydrogen evolution.

3. Conclusions

In summary, we have successfully prepared ZO/CS nanocomposites, which effectively overcome the inherent defects of the single materials and elevate the photocatalytic hydrogen evolution activity and stability of the single materials. Among the ZO/CS nanocomposites, ZO/CS-0.6 nanocomposite showed the best photocatalytic hydrogen evolution activity (4.27 mmol g^{-1} h^{-1}), which is much higher than those of CS (1.75 mmol g^{-1} h^{-1}) and ZO (trace) individually. In addition, ZO/CS-0.6 nanocomposite showed excellent corrosion resistance and maintained excellent photocatalytic hydrogen evolution activity in four cycles. This is due to the Type-I mechanism at the heterojunction of ZO/CS nanocomposites, which effectively separates the photogenerated electron hole pairs, thus enabling efficient and stable photocatalytic hydrogen evolution. This work provides a way to change the inherent characteristics of a single material, which may be helpful for the development of high performance and stable photocatalysts.

Supplementary Materials: The following supporting information can be downloaded at: https://www.mdpi.com/article/10.3390/catal12111385/s1, Figure S1: HAADF and elemental mapping images of ZO/CS-0.6 nanocomposite; Figure S2: Optimized models of (a) ZrO_2 and (b) CdSe. Calculated energy band structures for the (c) ZrO_2 and (d) CdSe; Figure S3: (a) XRD patterns of ZO/CS-0.6 nanocomposite before and after cycling test. (b) TEM image ZO/CS-0.6 nanocomposite before cycling test. (c) TEM image ZO/CS-0.6 nanocomposite after cycling test; Table S1: The amounts of precursors in preparing CdSe-DETA, ZrO_2 and ZO/CS nanocomposites and the BET surface area, average pore size and total pore volume of above materials; Table S2: Comparison of photocatalytic H_2 production rate of the catalysts in references and this work.

Author Contributions: Conceptualization, Z.L. and Z.W.; methodology, Z.L.; software, Z.L.; validation, Z.L., L.Z. and T.M.; formal analysis, Z.W.; investigation, Z.W.; resources, Z.W. and J.Z.; data curation, Z.L.; writing—original draft preparation, Z.L.; writing—review and editing, Z.L.; visualization, Z.L.; supervision, Z.L.; project administration, Z.W. and J.Z; funding acquisition, Z.W. and J.Z. All authors have read and agreed to the published version of the manuscript.

Funding: This work was financial support from the National Natural Science Foundation of China (no. 51973078) and Natural Science Foundation of Anhui Province (2108085MB48) are gratefully acknowledged.

Data Availability Statement: The data presented in this study can be obtained from the first author.

Conflicts of Interest: The authors declare no conflict of interest.

References

1. Humayun, M.; Wang, C.; Luo, W. Recent Progress in the Synthesis and Applications of Composite Photocatalysts: A Critical Review. *Small Methods* **2022**, *6*, 2101395. [CrossRef] [PubMed]
2. Li, X.; Wu, X.; Liu, S.; Li, Y.; Fan, J.; Lv, K. Effects of fluorine on photocatalysis. *Chin. J. Catal.* **2020**, *41*, 1451–1467. [CrossRef]

3. Wang, J.; Liu, J.; Du, Z.; Li, Z. Recent advances in metal halide perovskite photocatalysts: Properties, synthesis and applications. *J. Energy Chem.* **2021**, *54*, 770–785. [CrossRef]
4. Yang, X.; Singh, D.; Ahuja, R. Recent Advancements and Future Prospects in Ultrathin 2D Semiconductor-Based Photocatalysts for Water Splitting. *Catalysts* **2020**, *10*, 1111. [CrossRef]
5. Zhang, Y.; Xu, J.; Zhou, J.; Wang, L. Metal-organic framework-derived multifunctional photocatalysts. *Chin. J. Catal.* **2022**, *43*, 971–1000. [CrossRef]
6. Zhang, L.; Zhang, J.; Yu, H.; Yu, J. Emerging S-Scheme Photocatalyst. *Adv. Mater.* **2022**, *34*, 2107668. [CrossRef]
7. Bie, C.; Wang, L.; Yu, J. Challenges for photocatalytic overall water splitting. *Chem* **2022**, *8*, 1567–1574. [CrossRef]
8. Wei, Y.; Qin, H.; Deng, J.; Cheng, X.; Cai, M.; Cheng, Q.; Sun, S. Semiconductor Photocatalysts for Solar-to-Hydrogen Energy Conversion: Recent Advances of CdS. *Curr. Anal. Chem.* **2021**, *17*, 573–589. [CrossRef]
9. Bao, Y.; Song, S.; Yao, G.; Jiang, S. S-Scheme Photocatalytic Systems. *Sol. RRL* **2021**, *5*, 2100118. [CrossRef]
10. Hayat, A.; Syed, J.A.S.G.; Al-Sehemi, A.S.; El-Nasser, K.; Taha, T.A.A.; Al-Ghamdi, A.A.; Amin, M.; Ajmal, Z.; Iqbal, W.; Palamanit, A.; et al. State of the art advancement in rational design of g-C_3N_4 photocatalyst for efficient solar fuel transformation, environmental decontamination and future perspectives. *Int. J. Hydrogen Energy* **2022**, *47*, 10837–10867. [CrossRef]
11. Li, Z.; Jin, D.; Wang, Z. $WO_3(H_2O)_{0.333}$/CdSe-diethylenetriamine nanocomposite as a step-scheme photocatalyst for hydrogen production. *Surf. Interfaces* **2022**, *29*, 101702. [CrossRef]
12. Fu, Y.; Zhang, K.; Zhang, Y.; Cong, Y.; Wang, Q. Fabrication of visible-light-active MR/NH_2-MIL-125(Ti) homojunction with boosted photocatalytic performance. *Chem. Eng. J.* **2021**, *412*, 128722. [CrossRef]
13. Chang, Y.-S.; Hsieh, P.-Y.; Chang, T.-F.M.; Chen, C.-Y.; Sone, M.; Hsu, Y.-J. Incorporating graphene quantum dots to enhance the photoactivity of CdSe-sensitized TiO_2 nanorods for solar hydrogen production. *J. Mater. Chem. A* **2020**, *8*, 13971–13979. [CrossRef]
14. Putri, L.K.; Ng, B.-J.; Ong, W.-J.; Lee, H.W.; Chang, W.S.; Mohamed, A.R.; Chai, S.-P. Energy level tuning of CdSe colloidal quantum dots in ternary 0D-2D-2D CdSe QD/B-rGO/O-gC_3N_4 as photocatalysts for enhanced hydrogen generation. *Appl. Catal. B Environ.* **2020**, *265*, 118592. [CrossRef]
15. Raheman, S.A.R.; Mane, R.S.; Wilson, H.M.; Jha, N. CdSe quantum dot/white graphene hexagonal porous boron nitride sheet (h-PBNs) heterostructure photocatalyst for solar driven H_2 production. *J. Mater. Chem. C* **2021**, *9*, 8524–8536. [CrossRef]
16. Xia, T.; Lin, Y.; Li, W.; Ju, M. Photocatalytic degradation of organic pollutants by MOFs based materials: A review. *Chin. Chem. Lett.* **2021**, *32*, 2975–2984. [CrossRef]
17. Guo, J.; Ma, D.; Sun, F.; Zhuang, G.; Wang, Q.; Al-Enizi, A.M.; Nafady, A.; Ma, S. Substituent engineering in g-C_3N_4/COF heterojunctions for rapid charge separation and high photo-redox activity. *Sci. China Chem.* **2022**, *65*, 1704–1709. [CrossRef]
18. Goktas, S.; Goktas, A. A comparative study on recent progress in efficient ZnO based nanocomposite and heterojunction photocatalysts: A review. *J. Alloys Compd.* **2021**, *863*, 158734. [CrossRef]
19. Zhang, K.J.; Fu, Y.J.; Hao, D.R.; Guo, J.Y.; Ni, B.J.; Jiang, B.Q.; Xu, L.; Wang, Q. Fabrication of CN75/NH_2-MIL-53(Fe) p-n heterojunction with wide spectral response for efficiently photocatalytic Cr(VI) reduction. *J. Alloys Compd.* **2022**, *891*, 161994. [CrossRef]
20. Padmanabhan, N.T.; Thomas, N.; Louis, J.; Mathew, D.T.; Ganguly, P.; John, H.; Pillai, S.C. Graphene coupled TiO_2 photocatalysts for environmental applications: A review. *Chemosphere* **2021**, *271*, 129506. [CrossRef]
21. Zhang, W.; Sun, A.; Pan, X.; Han, Y.; Zhao, X.; Yu, L.; Zuo, Z.; Suo, N. Magnetic transformation of Zn-substituted Mg-Co ferrite nanoparticles: Hard magnetism → soft magnetism. *J. Magn. Magn. Mater.* **2020**, *506*, 166623. [CrossRef]
22. Shah, N.R.A.M.; Yunus, R.M.; Rosman, N.N.; Wong, W.Y.; Arifin, K.; Minggu, L.J. Current progress on 3D graphene-based photocatalysts: From synthesis to photocatalytic hydrogen production. *Int. J. Hydrogen Energy* **2021**, *46*, 9324–9340. [CrossRef]
23. Mohamed, R.M.; Ismail, A.A. Mesoporous Ag_2O/ZrO_2 heterostructures as efficient photocatalyst for acceleration photocatalytic oxidative desulfurization of thiophene. *Ceram. Int.* **2022**, *48*, 12592–12600. [CrossRef]
24. Ma, T.; Li, Z.; Liu, W.; Chen, J.; Wu, M.; Wang, Z. Microwave hydrothermal synthesis of $WO_3(H_2O)_{0.333}$/CdS nanocomposites for efficient visible-light photocatalytic hydrogen evolution. *Front. Mater. Sci.* **2021**, *15*, 589–600. [CrossRef]
25. Wang, Y.; Zhu, B.; Cheng, B.; Macyk, W.; Kuang, P.; Yu, J. Hollow carbon sphere-supported Pt/CoO_x hybrid with excellent hydrogen evolution activity and stability in acidic environment. *Appl. Catal. B Environ.* **2022**, *314*, 121503. [CrossRef]
26. Fu, Y.; Tan, M.; Guo, Z.; Hao, D.; Xu, Y.; Du, H.; Zhang, C.; Guo, J.; Li, Q.; Wang, Q. Fabrication of wide-spectra-responsive NA/NH_2-MIL-125(Ti) with boosted activity for Cr(VI) reduction and antibacterial effects. *Chem. Eng. J.* **2023**, *452*, 139417. [CrossRef]
27. Cao, S.; Shen, B.; Tong, T.; Fu, J.; Yu, J. 2D/2D Heterojunction of Ultrathin MXene/Bi_2WO_6 Nanosheets for Improved Photocatalytic CO_2 Reduction. *Adv. Funct. Mater.* **2018**, *28*, 1800136. [CrossRef]
28. Li, Z.; Jin, D.; Wang, Z. ZnO/CdSe-diethylenetriamine nanocomposite as a step-scheme photocatalyst for photocatalytic hydrogen evolution. *Appl. Surf. Sci.* **2020**, *529*, 147071. [CrossRef]
29. Li, S.; Cai, M.; Liu, Y.; Wang, C.; Yan, R.; Chen, X. Constructing $Cd_{0.5}Zn_{0.5}S$/Bi_2WO_6 S-scheme heterojunction for boosted photocatalytic antibiotic oxidation and Cr(VI) reduction. *Adv. Powder Mater.* **2023**, *2*, 100073. [CrossRef]
30. Fu, J.; Xu, Q.; Low, J.; Jiang, C.; Yu, J. Ultrathin 2D/2D WO_3/g-C_3N_4 step-scheme H_2-production photocatalyst. *Appl. Catal. B Environ.* **2019**, *243*, 556–565. [CrossRef]

31. Liu, Z.R.; Ding, X.; Zhu, R.; Li, Y.A.; Wang, Y.Q.; Sun, W.; Wang, D.; Wu, L.; Zheng, L. Investigation on the Effect of Highly Active Ni/ZrO$_2$ Catalysts Modified by MgO-Nd$_2$O$_3$ Promoters in CO$_2$ Methanation at Low Temperature Condition. *Chemistryselect* **2022**, *7*, e202103774.
32. Li, S.; Cai, M.; Liu, Y.; Wang, C.; Lv, K.; Chen, X. S-Scheme photocatalyst TaON/Bi$_2$WO$_6$ nanofibers with oxygen vacancies for efficient abatement of antibiotics and Cr(VI): Intermediate eco-toxicity analysis and mechanistic insights. *Chin. J. Catal.* **2022**, *43*, 2652–2664. [CrossRef]
33. He, F.; Meng, A.; Cheng, B.; Ho, W.; Yu, J. Enhanced photocatalytic H2-production activity of WO$_3$/TiO$_2$ step-scheme heterojunction by graphene modification. *Chin. J. Catal.* **2020**, *41*, 9–20. [CrossRef]
34. Du, H.; Li, N.; Yang, L.; Li, Q.; Yang, G.; Wang, Q. Plasmonic Ag modified Ag$_3$VO$_4$/AgPMo S-scheme heterojunction photocatalyst for boosted Cr(VI) reduction under visible light: Performance and mechanism. *Sep. Purif. Technol.* **2023**, *304*, 122204. [CrossRef]
35. Wang, Z.; Chen, Y.; Zhang, L.; Cheng, B.; Yu, J.; Fan, J. Step-scheme CdS/TiO$_2$ nanocomposite hollow microsphere with enhanced photocatalytic CO$_2$ reduction activity. *J. Mater. Sci. Technol.* **2020**, *56*, 143–150. [CrossRef]
36. Lian, Z.; Sakamoto, M.; Kobayashi, Y.; Tamai, N.; Ma, J.; Sakurai, T.; Seki, S.; Nakagawa, T.; Lai, M.-W.; Haruta, M.; et al. Anomalous Photoinduced Hole Transport in Type I Core/Mesoporous-Shell Nanocrystals for Efficient Photocatalytic H2 Evolution. *ACS Nano* **2019**, *13*, 8356–8363. [CrossRef] [PubMed]
37. Martinez-Haya, R.; Miranda, M.A.; Marin, M.L. Type I vs Type II photodegradation of pollutants. *Catal. Today* **2018**, *313*, 161–166. [CrossRef]
38. Wang, Q.; Zhang, Y.; Li, J.; Liu, N.; Jiao, Y.; Jiao, Z. Construction of electron transport channels in type-I heterostructures of Bi$_2$MoO$_6$/BiVO$_4$/g-C$_3$N$_4$ for improved charge carriers separation efficiency. *J. Colloid Interf. Sci.* **2020**, *567*, 145–153. [CrossRef]
39. Liu, H.; Cheng, D.-G.; Chen, F.; Zhan, X. Porous lantern-like MFI zeolites composed of 2D nanosheets for highly efficient visible light-driven photocatalysis. *Catal. Sci. Technol.* **2020**, *10*, 351–359. [CrossRef]
40. Deng, L.; Fang, N.; Wu, S.; Shu, S.; Chu, Y.; Guo, J.; Cen, W. Uniform H-CdS@NiCoP core-shell nanosphere for highly efficient visible-light-driven photocatalytic H$_2$ evolution. *J. Colloid Interf. Sci.* **2022**, *608*, 2730–2739. [CrossRef]
41. Zhang, H.; Kong, X.; Yu, F.; Wang, Y.; Liu, C.; Yin, L.; Huang, J.; Feng, Q. Ni(OH)$_2$ Nanosheets Modified Hexagonal Pyramid CdS Formed Type II Heterojunction Photocatalyst with High-Visible-Light H$_2$ Evolution. *ACS Appl. Energy Mater.* **2021**, *4*, 13152–13160. [CrossRef]

Article

Construction of Novel Z-Scheme g-C$_3$N$_4$/AgBr-Ag Composite for Efficient Photocatalytic Degradation of Organic Pollutants under Visible Light

Xuefeng Hu *, Ting Luo, Yuhan Lin and Mina Yang *

School of Environmental Science and Engineering, Shaanxi University of Science and Technology, Xi'an 710021, China
* Correspondence: huxuefeng@sust.edu.cn (X.H.); 4318@sust.edu.cn (M.Y.)

Abstract: As a green and sustainable technology to relieve environmental pollution issues, semiconductor photocatalysis attracted great attention. However, most single-component semiconductors suffer from high carrier recombination rate and low reaction efficiency. Here, we constructed a novel visible-light-driven Z-scheme g-C$_3$N$_4$/AgBr-Ag photocatalyst (noted as CN-AA-0.05) using a hydrothermal method with KBr as the bromine source. The CN-AA-0.05 photocatalyst shows an excellent photocatalytic degradation performance, and a rhodamine B (RhB) degradation ratio of 96.3% in 40 min, and 2-mercaptobenzothiazole (MBT) degradation ratio of 99.2% in 18 min are achieved. Mechanistic studies show that the remarkable performance of CN-AA-0.05 is not only attributed to the enhanced light absorption caused by the Ag SPR effect, but also the efficient charge transfer and separation with Ag nanoparticles as the bridge. Our work provides a reference for the design and construction of efficient visible-light-responsive Z-scheme photocatalysts, and an in-depth understanding into the mechanism of Z-scheme photocatalysts.

Keywords: degradation of organic pollutants; heterogeneous catalysis; surface plasmon resonance; visible light; Z-scheme photocatalyst

Citation: Hu, X.; Luo, T.; Lin, Y.; Yang, M. Construction of Novel Z-Scheme g-C$_3$N$_4$/AgBr-Ag Composite for Efficient Photocatalytic Degradation of Organic Pollutants under Visible Light. Catalysts 2022, 12, 1309. https://doi.org/10.3390/catal12111309

Academic Editors: Yongming Fu and Qian Zhang

Received: 28 September 2022
Accepted: 20 October 2022
Published: 25 October 2022

Publisher's Note: MDPI stays neutral with regard to jurisdictional claims in published maps and institutional affiliations.

Copyright: © 2022 by the authors. Licensee MDPI, Basel, Switzerland. This article is an open access article distributed under the terms and conditions of the Creative Commons Attribution (CC BY) license (https://creativecommons.org/licenses/by/4.0/).

1. Introduction

Organic pollutants in environmental water bodies such as dyes and pesticides seriously endanger the ecological environment and human health [1]. Due to the requirements of green and low carbon, the greatest expectation for the treatment of organic pollutants in water is semiconductor photocatalytic degradation [2–4]. However, single-component semiconductors are commonly confronted with low efficiency due to the poor charge transfer and separation [5–8]. The construction of Z-scheme photocatalytic systems that mimic natural photosynthesis is a promising strategy to improve the photocatalytic efficiency of semiconductor photocatalysts [9–11]. A Z-scheme photocatalytic system generally consists of an oxidation reaction catalyst (PS II), a reduction reaction catalyst (PS I), and an electron mediator [5,6]. Under irradiation, both PS II and PS I catalysts of the Z-scheme system generate photo-generated charges [5,6]. The photo-generated electrons of PS II migrate to the electron mediator and recombine with the photo-generated holes of PS II, then the photo-generated electrons in PS I induce a reduction reaction while the photo-generated holes in PS II induce an oxidation reaction [5,6]. Since the reduction reaction and oxidation reaction occur at different sites, the Z-scheme system not only reduces the thermodynamic requirements of the photocatalytic reaction, providing a large space for the selection and design of photocatalytic materials, but also promotes the separation and transport of photo-generated carriers, greatly inhibiting the recombination of carriers [5,6].

Graphitic carbon nitride (g-C$_3$N$_4$) is a unique two-dimensional semiconductor photocatalyst without metal elements. It is regarded as one of the most likely semiconductors for large-scale applications in the future because of low cost, high stability, and visible-light-responsive activity [12,13]. Z-scheme systems based on g-C_3N_4, such as WO_3/g-C_3N_4 [14–16], Ag_3PO_4/g-C_3N_4 [17,18], TiO_2/g-C_3N_4 [19–21], and AgX/g-C_3N_4 [22], were reported to be widely used in various environmental remediation reactions. To further improve the performance of g-C_3N_4-based Z-scheme photocatalysts, researchers also tried to introduce a third component such as noble metal nanoparticles into the system [23–25]. On one hand, noble metal nanoparticles can act as electron acceptors to further promote the interfacial charge transfer and separation. On the one hand, noble metal nanoparticles can induce a surface plasmon resonance (SPR) effect and, effectively enhance the light absorption ability. For example, Shen [23] et al. constructed g-C_3N_4/Ag/Ag_3PO_4 composites by a simple in-situ deposition method. The g-C_3N_4/Ag/Ag_3PO_4 shows a phenol degradation kinetic constant of 1.13 min^{-1}, almost 60 and 2.5 times higher than that of pure g-C_3N_4 and Ag/Ag_3PO_4, respectively. In addition, the CdS/Ag/g-C_3N_4 Z-scheme photocatalyst reported by Qian [24] et al. has a high H_2 evolution rate of 1376.0 $\mu mol \cdot h^{-1} \cdot g^{-1}$ in lactic acid scavenger solution, which is 3.12 and 1.76 times that of CdS and CdS/g-C_3N_4, respectively.

In this work, we prepared a series of novel visible-light-driven Z-scheme g-C_3N_4/AgBr-Ag photocatalysts with different component ratios, through a hydrothermal method with KBr as the bromine source. Different from other methods [26–28] that use CTAB as the bromine source, our method with KBr as the bromine source avoids the surfactant contamination of water body. In addition, compared with the method of combining g-C_3N_4 and AgBr by direct physical means, our method of compounding different components by a hydrothermal process led to the formation of a new component metallic Ag. The g-C_3N_4/AgBr-Ag Z-scheme photocatalyst showed excellent photocatalytic degradation activities of RhB and MBT under visible light irradiation. Moreover, the stability and the possible photocatalytic mechanism of the g-C_3N_4/AgBr-Ag Z-scheme photocatalyst were also investigated in detail.

2. Results and Discussion

Figure 1a shows the XRD patterns of the CN, Ag/AgBr, and CN-AA-X catalysts (X = 0.03, 0.05, or 0.07), and CN-AA-0.05-D. The characteristic peaks at 13.1° and 27.5° for CN sample are clearly observed, which are attributed to the (1 0 0) in-plane of tris-triazine units and the (0 0 2) diffraction planes of g-C_3N_4, respectively [29]. The diffraction peaks at 26.8°, 31.0°, 44.4°, 55.1°, 64.6°, 73.3°, and 81.8° for the Ag/AgBr sample are assigned to the (1 1 1), (2 0 0), (2 2 0), (2 2 2), (4 0 0), (4 2 0), and (4 2 2) planes of AgBr crystal (JCPDS 06-0438), respectively [30], and the faint diffraction peak at 38.1° for Ag/AgBr sample corresponds to the metallic Ag. For all the CN-AA-X catalysts, the characteristic peaks present the coexistence of Ag, AgBr, and g-C_3N_4 phases, although the characteristic peaks ascribed to the metallic Ag are much weaker due to the low content. For the CN-AA-0.05-D catalyst, the characteristic peaks are similar to those of CN-AA-X catalysts, except that no peak ascribed to the metallic Ag is observed.

The microstructures and morphologies of prepared CN, CN-AA-0.05, and CN-AA-0.05-D catalysts are revealed by SEM, TEM, and HRTEM observations (Figure 1b–f). It can be seen that the pure g-C_3N_4 presents a compact lamellar structure with a rough surface (Figure 1b). Some slit-shaped pores appear in the CN-AA-0.05 sample due to the introduction of Ag/AgBr (Figure 1c). The CN-AA-0.05-D shows a microstructure similar to that of pure g-C_3N_4, but some irregular particles deposition is observed on the surface (Figure 1e). Figure 1d is the TEM of CN-AA-0.05, in which the Ag/AgBr nanoparticle is anchored on the surface of g-C_3N_4. The Ag/AgBr nanoparticle is confirmed by HRTEM image, as shown in Figure 1f, and the particle displays three distinct areas with different lattice fringes. The lattice with d spacing of 0.24 nm corresponds to the (1 1 1) plane of metallic Ag, while those of 0.28 and 0.33 nm can be indexed to the (2 0 0) plane and the (1 1 1) plane of AgBr, respectively. This undoubtedly shows the effectiveness of hydrothermal treatment to partially reduce Ag^+ to Ag^0. All of these confirm the formation of the contact interface between AgBr, g-C_3N_4, and metallic Ag.

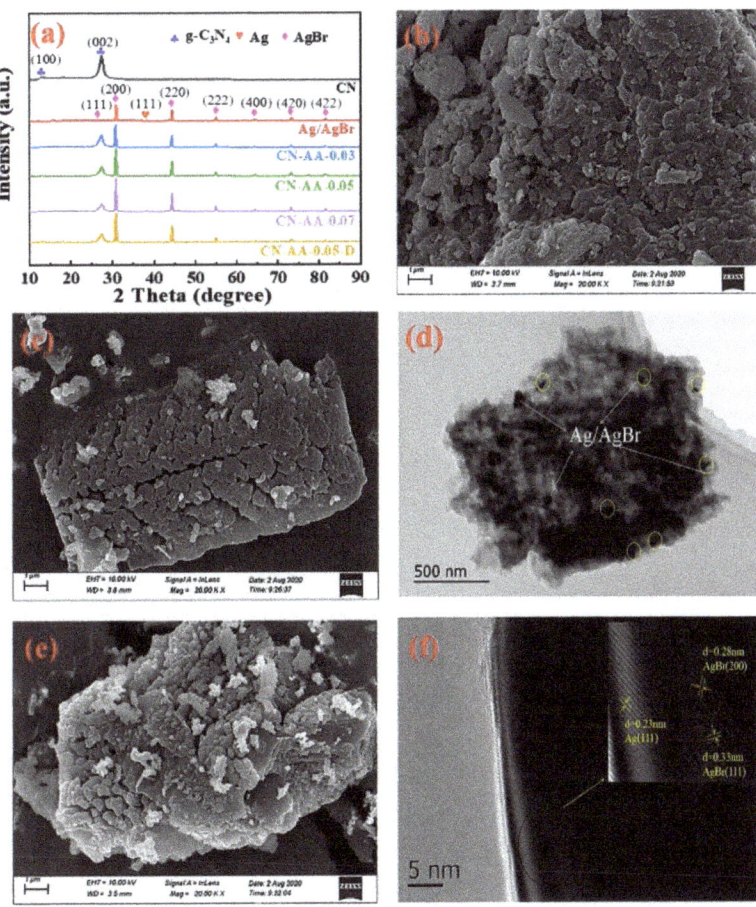

Figure 1. (a) XRD patterns of CN, Ag/AgBr, CN-AA-X(X = 0.03, 0.05 or 0.07), CN-AA-0.05-D. (b) The SEM image of CN. (c) SEM images of CN-AA-0.05 and (d) TEM images of CN-AA-0.05. (e) The SEM image of CN-AA-0.05-D. (f) The HRTEM image of CN-AA-0.05.

The elemental composition and chemical valence state of as-prepared CN, CN-AA-0.05, and CN-AA-0.05-D catalysts were investigated by XPS (Figure 2). It is clearly shown in Figure 2a that CN consists of C, N, and small amounts of adsorbed O elements, while both the CN-AA-0.05 and CN-AA-0.05-D samples consist of Ag, Br, C, N, and O elements. Figure 2b shows the C 1s XPS spectra of CN, CN-AA-0.05, and CN-AA-0.05-D samples. For all the three samples, the peaks at 284.62, 285.93, and 287.81 eV correspond to the surface adventitious carbon, sp^2 C atoms bonded to N in an aromatic ring (N-C=N), and sp^3 hybridized C atoms [C-(N)$_3$], respectively [31]. According to the N 1s XPS spectra of the CN, CN-AA-0.05, and CN-AA-0.05-D samples (Figure 2c), the peaks at 398.31, 398.93, 400.60, and 404.42 eV are attributed to the signals of sp^2 hybridized aromatic N atoms bonded to carbon atoms (C-N=C), tertiary N bonded to C atoms in the form of N-(C)$_3$, N–H structure, and charging effect, respectively [32]. Figure 2d displays the Ag 3d XPS spectra of the CN-AA-0.05 and CN-AA-0.05-D samples. For the CN-AA-0.05 sample, the two peaks corresponding to the Ag $3d_{5/2}$ and Ag $3d_{3/2}$, respectively, can be fitted to four peaks. The Ag$^+$ in AgBr is responsible for the peaks at 367.08 eV and 373.18 eV, while metallic Ag is responsible for the peaks at 367.80 eV and 374.36 eV [30,33]. For the CN-AA-0.05-D sample, only peaks ascribed to the Ag$^+$ can be observed, which is consistent with the XRD result. It implies that the hydrothermal treatment plays a key role in the formation of metallic Ag. For both the CN-AA-0.05 and CN-AA-0.05-D samples, the two peaks at 67.68 and 68.77 eV in Figure 2e are attributed to the Br $3d_{5/2}$ and Br $3d_{3/2}$, respectively [30].

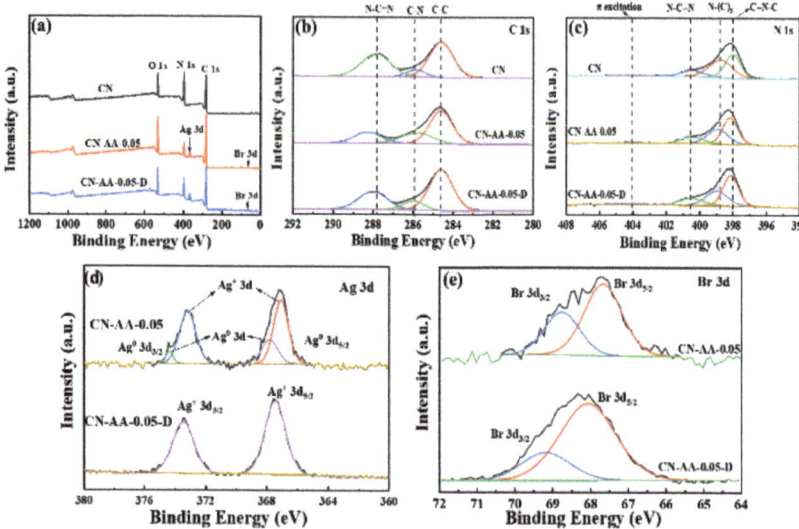

Figure 2. Full (**a**), C 1s (**b**), N 1s (**c**) XPS spectra of CN, CN-AA-0.05 and CN-AA-0.05-D; Ag 3d (**d**), Br 3d (**e**) XPS spectra of CN-AA-0.05 and CN-AA-0.05-D.

The optical property of CN, AgBr, CN-AA-X (X = 0.03, 0.05 or 0.07), and CN-AA-0.05-D were determined by UV–vis DRS test (Figure 3a). CN shows an absorption edge of about 460 nm, while AgBr exhibits an absorption edge of about 480 nm. For the CN-AA-X sample, the absorption intensity enhances compared with that of CN, and the absorption intensity increases with the increasing X value (the Ag/AgBr content), mainly due to the surface plasmon absorption of Ag and the interaction between Ag and AgBr. The estimated bandgaps of CN, AgBr, CN-AA-0.03, CN-AA-0.05, CN-AA-0.07, and CN-AA-0.05-D are 2.48, 2.36, 2.43, 2.38, 2.39, and 2.40 eV, respectively (Figure 3b) [34,35].

PL spectra (Figure 3c) and time-resolved fluorescence decay spectra (Figure 3d) were measured to study the transfer and annihilation processes of photo-generated carriers of CN-AA-0.05 and other comparative catalysts. As shown in Figure 3c, the PL intensity of CN is the highest, and the PL intensity decreases significantly after the introduction of Ag/AgBr. For CN-AA-0.05, the PL intensity is the lowest. These results indicate that an appropriate amount of Ag/AgBr introduction can effectively suppress the recombination of photo-generated carriers, thereby enhancing the catalytic performance. According to Figure 3d, the carrier lifetime of the CN-AA-0.05 catalyst is much shorter than that of CN, which indicates the efficient charge transfer among the components of the Z-scheme system.

Figure 3e shows the FTIR spectra of the pure g-C_3N_4 and CN-AA-X catalysts. The broadband at 3000–3400 cm^{-1} corresponds to the stretching modes of terminal NH_2 or NH groups. The absorption peaks at 1641 and 1567 cm^{-1} are attributed to C=N stretching, and 1406, 1330, and 1241 cm^{-1} are assigned to the aromatic C-N stretching [36,37]. Additionally, the sharp characteristic ring breath peak of the triazine units is found at 808 cm^{-1} [36,37]. It is noticed that the introduction of Ag/AgBr does not obviously affect the FTIR of g-C_3N_4.

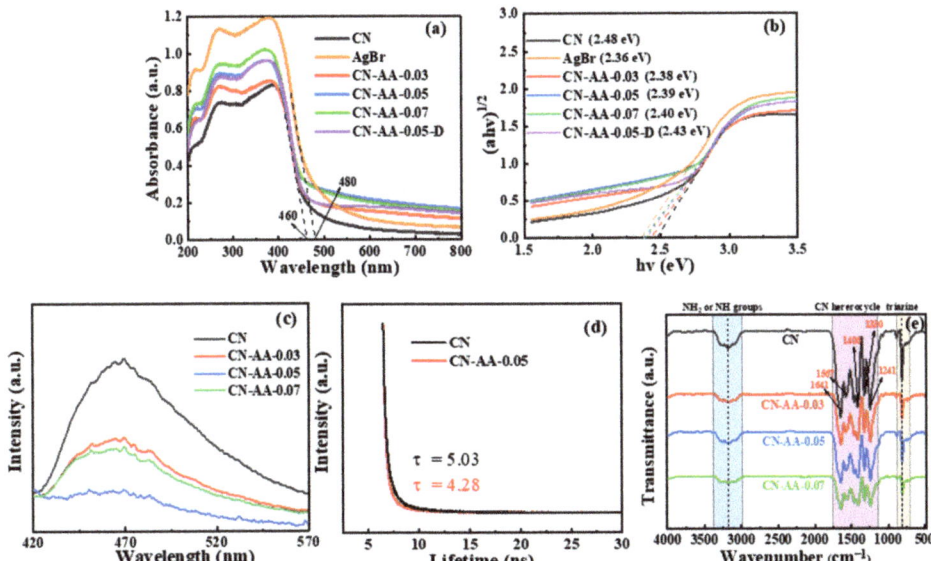

Figure 3. (a) UV–vis DRS and (b) plots of $(ahv)^{1/2}$ versus energy (hv) of CN, AgBr, CN-AA-X (X = 0.03, 0.05, 0.07), and CN-AA-0.05-D. (c) The PL spectra of CN and CN-AA-X (X = 0.03, 0.05, 0.07). (d) The time-resolved fluorescence decay spectra of CN and CN-AA-0.05. (e) The FTIR spectra of CN and CN-AA-X (X = 0.03, 0.05, 0.07).

The photocatalytic activities of CN, AgBr, CN-AA-X (X = 0.03, 0.05 or 0.07), and CN-AA-0.05-D materials were firstly evaluated by the degradation of dye RhB under visible light. As shown in Figure 4a, CN shows a photocatalytic degradation rate of 41.2% within 50 min, while AgBr or CN-AA-0.05-D exhibit a slightly better photocatalytic degradation performance than CN. For all CN-AA-X catalysts, the activities increase significantly compared with pure CN or AgBr, demonstrating the importance of the construction of g-C_3N_4/AgBr-Ag heterojunction photocatalysts for enhancing the photocatalytic performance. CN-AA-0.05 exhibits the best photocatalytic degradation performance of RhB among all CN-AA-X catalysts, with a degradation rate of 96.3% within 40 min, which indicates that there is an optimal Ag/AgBr introduction amount for the enhancement of photocatalytic degradation performance. Finally, after the first-order kinetic curve fitting, the calculated degradation rate constants are 0.008 min^{-1} for CN, 0.014 min^{-1} for AgBr, 0.030 min^{-1} for CN-AA-0.03, 0.061 min^{-1} for CN-AA-0.05, 0.032 min^{-1} for CN-AA-0.07, and 0.017 min^{-1} for CN-AA-0.05-D (Figure 4b). Subsequently, the activities of photocatalytic degradation MBT under visible light by these materials were further investigated. As shown in Figure 4c, CN-AA-0.05 exhibits the best photocatalytic degradation performance of MBT, with a degradation rate of 99.2% within 18 min. The activity trend of CN, AgBr, CN-AA-0.03, CN-AA-0.05, CN-AA-0.07, and CN-AA-0.05-D for photocatalytic degradation of MBT is the same as that of RhB, and the calculated degradation rate constants are 0.0234, 0.040, 0.128, 0.197, 0.157, and 0.049 min^{-1}, respectively (Figure 4d). The slope values of the linear fit corresponding to CN-AA-0.03, CN-AA-0.05, and CN-AA-0.07 seem to be affected especially by the very last irradiation point (i.e., 40 min) in Figure 4b. This may be due to degradation characterization of RhB. Degradation of RhB is accompanied by the blue shift of the absorption maximum due to the N-deethylation reaction, as we reported before [38], but the recorded value is still the maximum absorption position of RhB (553 nm), resulting in a deviation from first-order linear fitting.

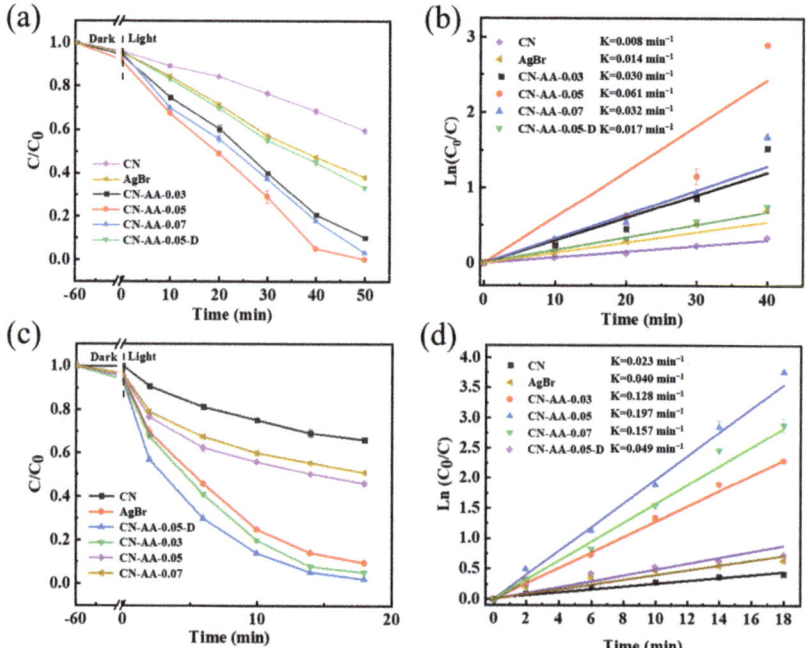

Figure 4. Photocatalytic (**a**) RhB and (**c**) MBT degradation activities of CN, AgBr, CN-AA-X (X = 0.03, 0.05, 0.07), and CN-AA-0.05-D. Kinetic curves of photocatalytic degradation of (**b**) RhB and (**d**) MBT by CN, AgBr, CN-AA-X (X = 0.03, 0.05, 0.07), and CN-AA-0.05-D.

In conclusion, the CN-AA-0.05 composite material prepared by hydrothermal method with the AgNO$_3$ addition amount of 0.05 g shows the best degradation performance, with the RhB degradation rate of 96.3% in 40 min and the MBT degradation rate of 99.2% in 18 min. As shown in Figure 5, the MBT degradation rate of CN-AA-0.05 remains above 90% after five cycles of 20 min per cycle, which demonstrates the excellent stability of the CN-AA-0.05 catalyst.

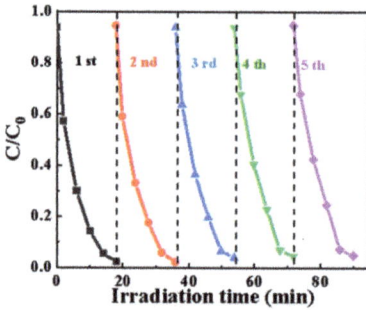

Figure 5. The cycling experiments of photocatalytic degradation of MBT by CN-AA-0.05 catalyst.

The reactive species that may be involved in a photocatalytic process mainly include ·OH, ·O$_2^-$, h$^+$, and e$^-$. To accurately infer the mechanism of a photocatalytic reaction, it is first necessary to determine the type of active species that plays the most critical role in the photocatalytic process. Here, the essential active species during the MBT degradation process were investigated by quenching experiments [39]. EA (10 mmol/L), BQ

(12 mmol/L), EDTA-2Na (12 mmol/L), and Cr(VI) (0.05 mmol/L) were applied to quench ·OH, ·O_2^-, h^+, and e^-, respectively. As shown in Figure 6a, the addition of BQ and EDTA-2Na greatly reduces the photocatalytic degradation performance of MBT, which indicates the important roles of ·O_2^- and h^+ in the photocatalytic degradation process. The addition of EA has little effect on MBT degradation, suggesting a negligible contribution of ·OH. Furthermore, the addition of Cr(VI) enhances the activity of photocatalytic degradation of MBT, which is attributed to the fact that more h^+ can react with MBT rather than recombine with electrons due to the quenching of electrons by Cr(VI). In detail, under visible light irradiation, CN-AA-0.05 can be photo-excited to yield electron (e^-) and hole (h^+). On one hand, organic pollutants that react irreversibly with photo-generated h^+ can enhance the photocatalytic electron–hole separation, which results in more CB electrons for Cr(VI) reduction. On the other hand, photoelectrons transfer to the conduction band and are captured by oxygen to form $O_2^{•-}$, or by Cr(VI) to form lower valent state chromium, which results in more holes for organic pollutants oxidation. The synergistic effect of Cr(VI) reduction and organic pollutants degradation over semiconductor photocatalysis was reported previously [40–42].

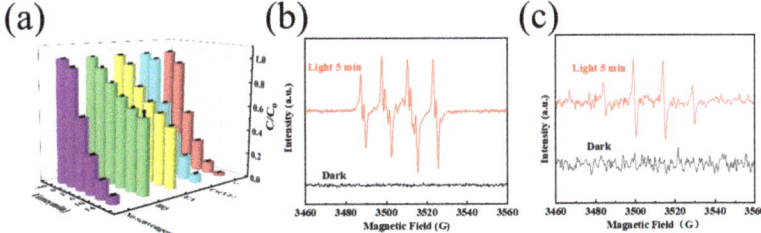

Figure 6. (**a**) Effects of different scavengers on photocatalytic degradation of MBT over CN-AA-0.05. (**b**) EPR spectra of CN-AA-0.05 in DMSO with DMPO as the capture agent. (**c**) EPR spectra of CN-AA-0.05 in deionized water with DMPO as the capture agent.

To identify the carrier transfer mechanism in the photocatalytic degradation process over CN-AA-0.05, radical spin-trapping experiments were further carried out. For the $O_2^{•-}$ spin-trapping test, there is no signal in dark, however, characteristic peaks of DMPO-OOH adduct (pointing to superoxide radical) emerge under visible light (Figure 6b). This result is coherent with the scavenging investigation, where $O_2^{•-}$ seems to have an important contribution to MBT degradation. For the ·OH spin-trapping test, the quartet ascribed to DMPO-OH spin-adduct (hydroxyl radical EPR fingerprint) is also observed under visible light (Figure 6c), but the signal disappears in an oxygen-free condition. It is very well-known that in aqueous solution, the DMPO-OOH spin-adduct has a lifetime of ca. 30 s that rapidly evolves into DMPO-OH spin-adduct [43]. This indicates that the observed DMPO-OH signal should be the transformation product of DMPO-OOH, which verifies the negligible contribution of ·OH to MBT degradation.

Based on the above results and the related literature [44–47], the possible photocatalytic mechanism of CN-AA-0.05 was proposed (Figure 7). Under visible light, both g-C_3N_4 and AgBr are excited to generate e_{CB}^- and h_{VB}^+. The e_{CB}^- of AgBr are quickly transferred to metallic Ag through the Schottky barrier, and, subsequently, the electrons in Ag are transferred to the VB of g-C_3N_4 and recombine with the h_{VB+} of g-C_3N_4. That is, Ag acts as an electron transfer bridge. Therefore, the e_{CB}^- of g-C_3N_4 with strong reduction capability and h_{VB}^+ of AgBr with strong oxidation capability remain, enabling the Z-scheme mechanism. The e_{CB}^- of g-C_3N_4 can further react with O_2 to form crucial active species $O_2^{·-}$, which, together with h_{VB}^+ of AgBr, can oxidize and degrade organic pollutants.

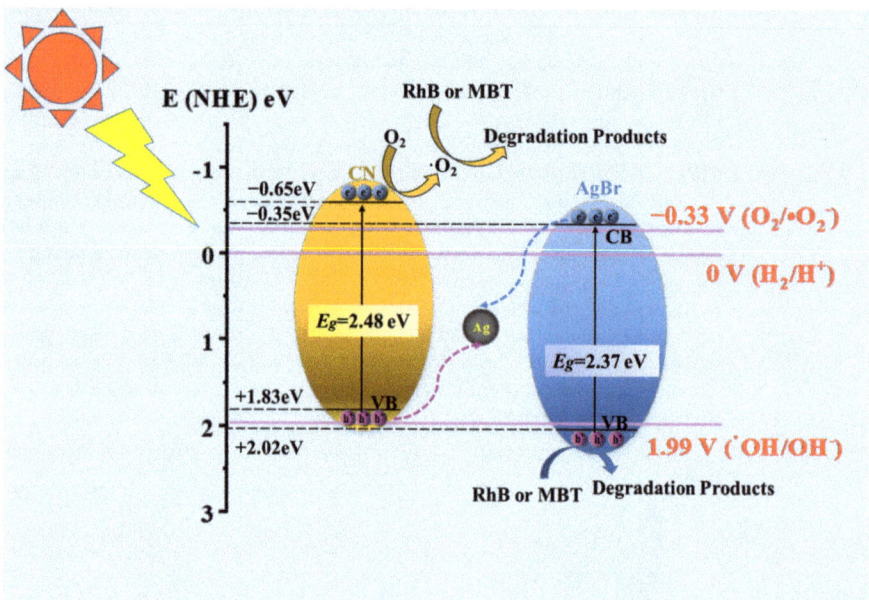

Figure 7. Schematic illustration of proposed photocatalysis mechanism of g-C$_3$N$_4$/AgBr-Ag composites under visible light irradiation.

3. Experimental Section

3.1. Catalyst Preparation

The g-C$_3$N$_4$ powders were synthesized by heating melamine in a muffle furnace. In a typical process, 10 g melamine was placed in a crucible with a cover. The crucible was heated to 550 °C at a heating rate of 5 °C/min and then kept for 3 h. After cooling to the room temperature, the yellow product g-C$_3$N$_4$ was obtained and noted as CN.

The g-C$_3$N$_4$/AgBr-Ag was synthesized as follows: X g of AgNO$_3$ (X = 0.03, 0.05, or 0.07) and 0.5 g of g-C$_3$N$_4$ were added to 80 mL of ethylene glycol, and the mixture was stirred at room temperature for 1 h. Then, 0.7X g KBr was added, and the mixture was stirred for another 6 h. Subsequently, the obtained mixture was transferred to an autoclave for the hydrothermal reaction at 180 °C for 10 h. Finally, the product was washed with deionized water 5 times, washed with ethanol once, and dried in an oven at 50 °C. The obtained sample was noted as CN-AA-X (X is the added mass of AgNO$_3$). For comparison, g-C$_3$N$_4$/AgBr-Ag without hydrothermal treatment was also prepared by the same method, which was marked as CN-AA-X-D.

3.2. Catalyst Characterization

The crystal structures of the samples were investigated on a Bruker D8 Advance X-ray diffractometer (XRD) using Cu Kα radiation source. The morphologies and microstructures of catalysts were observed by scanning electron microscopy (SEM) and transmission electron microscopy (HRTEM) on a Zeiss Sigma500 (Oberkochen, Baden-Württemberg, Germany) and a JEOL JEM 2100F electron microscope (Tokyo, Japan), respectively. X-ray photoelectron spectroscopy (XPS) was performed on a Kratos AXIS SUPRA spectrometer (Manchester, UK). The UV–vis diffuse reflectance spectra (DRS) were obtained by a Cary 5000 spectrophotometer (Santa Clara, CA, USA) using BaSO$_4$ as the reflectance standard. Photoluminescence (PL) spectra and time-resolved fluorescence emission decay spectra were recorded on an Edinburgh FS5 fluorescence spectrometer (Edinburgh, UK) with the excitation wavelength of 350 nm. Fourier-transform infrared (FTIR) spectra of synthesized samples were obtained on a spectrophotometer (Vertex70, Bruker, Saarbrucken, Germany) using the standard KBr disk method. Electron paramagnetic resonance (EPR) analyses were performed on a Bruker E500 spectrometer (Karlsruhe, Germany). Reactive radicals with short lifetimes are difficult to study directly by EPR spectroscopy. The spin-trapping approach allows us to identify the radical by causing them to react with trap molecules chosen so as to obtain relatively stable radical adducts [48]. DMPO (5,5-Dimethyl-1-pyrroline N-oxide) was used as spin-trap agent, $\cdot OH$ was detected in deionized water, but $\cdot O_2^-$ was detected in methyl sulfoxide (DMSO) solution in the present study. EPR tests were performed as follows: reaction solutions in 1 mm quartz capillary inside a 4 mm quartz tube were introduced into EPR cavity and tested before/after visible light irradiation. Oxygen-free spin-trapping investigation was performed through nitrogen bubbling of the solutions prior to the experiment. Sweep width of 100 G, microwave power of 0.2 mW, sweep time 5.24 s, microwave frequency of 9.41 GHz, and microwave attenuation of 30 dB were used during test.

3.3. Photocatalytic Tests

The photocatalytic activities of prepared photocatalysts were evaluated by the degradation of rhodamine B (RhB) and 2-mercaptobenzothiazole (MBT) under visible light. The light source used in the tests was a xenon lamp (300 W) with a 420 nm cut filter, and the distance between the reactor and the light source was 5 cm. For all photocatalysis experiments, 15 mg of photocatalyst was dispersed in MBT (30 mL, 20 mg/L) or RhB (30 mL, 50 mg/L) aqueous solution, and the suspension was stirred in the dark for 1 h. Then, the lamp was turned on to initiate the photocatalytic reaction. A total of 3 mL of suspension was taken at given time intervals, followed by centrifugation to remove the photocatalyst completely. The concentrations of RhB and MBT in the degradation process were determined by a UV–vis spectrometer (UV-2600) at wavelengths of 553 nm and 312 nm, respectively. Similar to aforementioned catalytic removal processes in the presence of sample CN-AA-0.05, EA (10 mmol/L), BQ (12 mmol/L), EDTA-2Na (12 mmol/L), and Cr(VI) (0.05 mmol/L) were added into reaction system to quench $\cdot OH$, $\cdot O_2^-$, h+, and e−, respectively. Each experiment was conducted three times.

MBT was selected as the target for the stability test experiment. The number of cycles was 5, and the reaction time for each cycle was 20 min.

4. Conclusions

Here, a novel visible-light-driven Z-scheme $g-C_3N_4$/AgBr-Ag photocatalyst was fabricated by a simple hydrothermal method with KBr as the bromine source. Compared with pure $g-C_3N_4$ or AgBr, the photocatalytic degradation activity of organic pollutants over $g-C_3N_4$/AgBr-Ag is significantly enhanced. Hydrothermal treatment is believed to transforms part of AgBr into metallic Ag. Metallic Ag initiates the SPR effect and acts as an electron transfer bridge, which finally improves the visible light absorption capacity and carrier separation efficiency of Z-scheme $g-C_3N_4$/AgBr-Ag. Our work not only provides an experimental basis for the design and construction of efficient visible-light-responsive Z-scheme photocatalysts, but also provides an in-depth understanding into the mechanism of Z-scheme photocatalytic degradation of organic pollutants.

Author Contributions: Conceptualization, X.H. and M.Y.; methodology, X.H., T.L. and M.Y.; investigation, T.L. and X.H.; writing—original draft preparation, Y.L.; writing—review and editing, X.H. and M.Y.; supervision, X.H. and M.Y.; project administration, X.H. and M.Y.; funding acquisition, X.H. and M.Y. All authors have read and agreed to the published version of the manuscript.

Funding: National Natural Science Foundation of China (no. 22176120, 22206118) and Shaanxi Thousand Talents Plan-Youth Program Scholars.

Data Availability Statement: Not applicable.

Conflicts of Interest: The authors declare no conflict of interest.

References

1. Alharbi, O.M.L.; Basheer, A.A.; Khattab, R.A.; Ali, I. Health and environmental effects of persistent organic pollutan. *J. Mol. Liq.* **2018**, *263*, 442–453. [CrossRef]
2. Chen, D.; Cheng, Y.; Zhou, N.; Chen, P.; Wang, Y.; Li, K.; Huo, S.; Cheng, P.; Peng, P.; Zhang, R.; et al. Photocatalytic degradation of organic pollutants using TiO_2-based photocatalysts: A review. *J. Clean. Prod.* **2020**, *268*, 121725. [CrossRef]
3. Zhao, Y.; Li, Y.; Sun, L. Recent advances in photocatalytic decomposition of water and pollutants for sustainable application. *Chemosphere* **2021**, *276*, 130201. [CrossRef] [PubMed]
4. Hassan, J.Z.; Raza, A.; Qumar, U.; Li, G. Recent advances in engineering strategies of Bi-based photocatalysts for environmental remediation. *Sustain. Mater. Technol.* **2022**, *33*, e00478. [CrossRef]
5. Li, H.; Tu, W.; Zhou, Y.; Zou, Z. Z-Scheme photocatalytic systems for promoting photocatalytic performance: Recent progress and future challenges. *Adv. Sci.* **2016**, *3*, 1500389. [CrossRef]
6. Xu, Q.; Zhang, L.; Yu, J.; Wageh, S.; Al-Ghamdi, A.A.; Jaroniec, M. Direct Z-scheme photocatalysts: Principles, synthesis, and applications. *Mater. Today* **2018**, *21*, 1042–1063. [CrossRef]
7. Li, S.; Cai, M.; Liu, Y.; Wang, C.; Lv, K.; Chen, X. S-scheme photocatalyst $TaON/Bi_2WO_6$ nanofibers with oxygen vacancies for efficient abatement of antibiotics and Cr (VI): Intermediate eco-toxicity analysis and mechanistic insights. *Chin. J. Catal.* **2022**, *43*, 2652–2664. [CrossRef]
8. Shi, Q.; Raza, A.; Xu, L.; Li, G. Bismuth oxyhalide quantum dots modified sodium titanate necklaces with exceptional population of oxygen vacancies and photocatalytic activity. *J. Colloid Interf. Sci.* **2022**, *625*, 750–760. [CrossRef]
9. Zhou, P.; Yu, J.; Jaroniec, M. All-solid-state Z-scheme photocatalytic systems. *Adv. Mater.* **2014**, *26*, 4920–4935. [CrossRef]
10. Zhang, W.; Mohamed, A.R.; Ong, W.-J. Z-Scheme photocatalytic systems for carbon dioxide reduction: Where are we now? *Angew. Chem. Int. Ed.* **2020**, *59*, 22894–22915. [CrossRef]
11. Huang, D.; Chen, S.; Zeng, G.; Gong, X.; Zhou, C.; Cheng, M.; Xue, W.; Yan, X.; Li, J. Artificial z-scheme photocatalytic system: What have been done and where to go? *Coord. Chem. Rev.* **2019**, *385*, 44–80. [CrossRef]
12. Shi, Q.; Zhang, X.; Liu, X.; Xu, L.; Liu, B.; Zhang, J.; Xu, H.; Han, Z.; Li, G. n-situ exfoliation and assembly of 2D/2D g-C_3N_4/TiO_2 (B) hierarchical microflower: Enhanced photo-oxidation of benzyl alcohol under visible light. *Carbon* **2022**, *196*, 401–409. [CrossRef]
13. Ong, W.-J.; Tan, L.-L.; Ng, Y.H.; Yong, S.-T.; Chai, S.-P. Graphitic carbon nitride (g-C_3N_4)-based photocatalysts for artificial photosynthesis and environmental remediation: Are we a step closer to achieving sustainability? *Chem. Rev.* **2016**, *116*, 7159–7329. [CrossRef]
14. Cadan, F.M.; Ribeiro, C.; Azevedo, E.B. Improving g-C_3N_4: WO_3 Z-scheme photocatalytic performance under visible light by multivariate optimization of g-C_3N_4 synthesis. *Appl. Surf. Sci.* **2021**, *537*, 147904. [CrossRef]
15. Zhou, S.; Wang, Y.; Zhou, K.; Ba, D.; Ao, Y.; Wang, P. In-situ construction of Z-scheme g-C_3N_4/WO_3 composite with enhanced visible-light responsive performance for nitenpyram degradation. *Chin. Chem. Lett.* **2021**, *32*, 2179–2182. [CrossRef]
16. Jing, H.; Ou, R.; Yu, H.; Zhao, Y.; Lu, Y.; Huo, M.; Huo, H.; Wang, X. Engineering of g-C_3N_4 nanoparticles/WO_3 hollow microspheres photocatalyst with Z-scheme heterostructure for boosting tetracycline hydrochloride degradation. *Sep. Purif. Technol.* **2021**, *255*, 117646. [CrossRef]
17. Du, J.; Xu, Z.; Li, H.; Yang, H.; Xu, S.; Wang, J.; Jia, Y.; Ma, S.; Zhan, S. Ag_3PO_4/g-C_3N_4 Z-scheme composites with enhanced visible-light-driven disinfection and organic pollutants degradation: Uncovering the mechanism. *Appl. Surf. Sci.* **2021**, *541*, 148487. [CrossRef]
18. Cheng, R.; Wen, J.; Xia, J.; Shen, L.; Kang, M.; Shi, L.; Zheng, X. Photo-catalytic oxidation of gaseous toluene by Z-scheme Ag_3PO_4-g-C_3N_4 composites under visible light: Removal performance and mechanisms. *Catal. Today* **2022**, *388–389*, 26–35. [CrossRef]
19. Xu, C.; Li, D.; Liu, X.; Ma, R.; Sakai, N.; Yang, Y.; Lin, S.; Yang, J.; Pan, H.; Huang, J.; et al. Direct Z-scheme construction of g-C_3N_4 quantum dots/TiO_2 nanoflakes for efficient photocatalysis. *Chem. Eng. J.* **2022**, *430*, 132861. [CrossRef]
20. Bi, X.; Yu, S.; Liu, E.; Liu, L.; Zhang, K.; Zang, J.; Zhao, Y. Construction of g-C_3N_4/TiO_2 nanotube arrays Z-scheme heterojunction to improve visible light catalytic activity. *Colloids Surf. A* **2020**, *603*, 125193. [CrossRef]
21. Hu, K.; Li, R.; Ye, C.; Wang, A.; Wei, W.; Hu, D.; Qiu, R.; Yan, K. Facile synthesis of Z-scheme composite of TiO_2 nanorod/g-C_3N_4 nanosheet efficient for photocatalytic degradation of ciprofloxacin. *J. Clean. Prod.* **2020**, *253*, 120055. [CrossRef]
22. Murugesan, P.; Narayanan, S.; Manickam, M.; Murugesan, P.K.; Subbiah, R. A direct Z-scheme plasmonic AgCl@g-C_3N_4 heterojunction photocatalyst with superior visible light CO_2 reduction in aqueous medium. *Appl. Surf. Sci.* **2018**, *450*, 516–526. [CrossRef]
23. Shen, Y.; Zhu, Z.; Wang, X.; khan, A.; Gong, J.; Zhang, Y. Synthesis of Z-scheme g-C_3N_4/Ag/Ag_3PO_4 composite for enhanced photocatalytic degradation of phenol and selective oxidation of gaseous isopropanol. *Mater. Res. Bull.* **2018**, *107*, 407–415. [CrossRef]
24. Qian, L.; Hou, Y.; Yu, Z.; Li, M.; Li, F.; Sun, L.; Luo, W.; Pan, G. Metal-induced Z-scheme CdS/Ag/g-C_3N_4 photocatalyst for enhanced hydrogen evolution under visible light: The synergy of MIP effect and electron mediator of A. *Mol. Catal.* **2018**, *458*, 43–51. [CrossRef]
25. Bao, Y.; Chen, K. AgCl/Ag/g-C_3N_4 hybrid composites: Preparation, visible light-driven photocatalytic activity and mechanism. *Nano-Micro Lett.* **2016**, *8*, 182–192. [CrossRef]

26. Yang, Y.; Guo, W.; Guo, Y.; Zhao, Y.; Yuan, X.; Guo, Y. Fabrication of Z-scheme plasmonic photocatalyst Ag@ AgBr/g-C$_3$N$_4$ with enhanced visible-light photocatalytic activity. *J. Hazard. Mater.* **2014**, *271*, 150–159. [CrossRef]
27. Li, Y.; Zhao, Y.; Fang, L.; Jin, R.; Yang, Y.; Xing, Y. Highly efficient composite visible light-driven Ag–AgBr/g-C$_3$N$_4$ plasmonic photocatalyst for degrading organic pollutants. *Mater. Lett.* **2014**, *126*, 5–8. [CrossRef]
28. Xu, Y.; Xu, H.; Yan, J.; Li, H.; Huang, L.; Xia, J.; Yin, S.; Shu, H. A plasmonic photocatalyst of Ag/AgBr nanoparticles coupled with g-C$_3$N$_4$ with enhanced visible-light photocatalytic ability. *Colloids Surf. A* **2013**, *436*, 474–483. [CrossRef]
29. Dong, F.; Wu, L.; Sun, Y.; Fu, M.; Wu, Z.; Lee, S.C. Efficient synthesis of polymeric gC 3 N 4 layered materials as novel efficient visible light driven photocatalysts. *J. Mater. Chem.* **2011**, *21*, 15171–15174. [CrossRef]
30. An, C.; Wang, J.; Jiang, W.; Zhang, M.; Ming, X.; Wang, S.; Zhang, Q. Strongly visible-light responsive plasmonic shaped AgX: Ag (X = Cl, Br) nanoparticles for reduction of CO$_2$ to methanol. *Nanoscale* **2012**, *4*, 5646–5650. [CrossRef]
31. Chai, B.; Peng, T.; Mao, J.; Li, K.; Zan, L. Graphitic carbon nitride (gC$_3$N$_4$)–Pt-TiO$_2$ nanocomposite as an efficient photocatalyst for hydrogen production under visible light irradiation. *Phys. Chem. Chem. Phys.* **2012**, *14*, 16745–16752. [CrossRef] [PubMed]
32. Ge, L.; Han, C. Synthesis of MWNTs/g-C$_3$N$_4$ composite photocatalysts with efficient visible light photocatalytic hydrogen evolution activity. *Appl. Catal. B* **2012**, *117–118*, 268–274. [CrossRef]
33. Luo, T.; Hu, X.; She, Z.; Wei, J.; Feng, X.; Chang, F. Synergistic effects of Ag-doped and morphology regulation of graphitic carbon nitride nanosheets for enhanced photocatalytic performance. *J. Mol. Liq.* **2021**, *324*, 114772. [CrossRef]
34. Li, S.; Cai, M.; Liu, Y.; Wang, C.; Yan, R.; Chen, X. Constructing Cd0. 5Zn0. 5S/Bi2WO6 S-scheme heterojunction for boosted photocatalytic antibiotic oxidation and Cr (VI) reduction. *Adv. Powder Mater.* **2023**, *2*, 100073. [CrossRef]
35. Li, S.; Cai, M.; Wang, C.; Liu, Y.; Li, N.; Zhang, P.; Li, X. Rationally designed Ta$_3$N$_5$/BiOCl S-scheme heterojunction with oxygen vacancies for elimination of tetracycline antibiotic and Cr (VI): Performance, toxicity evaluation and mechanism insight. *J. Mater. Sci. Technol.* **2022**, *123*, 177–190. [CrossRef]
36. Ji, H.; Chang, F.; Hu, X.; Qin, W.; Shen, J. Photocatalytic degradation of 2,4,6-trichlorophenol over g-C$_3$N$_4$ under visible light irradiation. *Chem. Eng. J.* **2013**, *218*, 183–190. [CrossRef]
37. Li, W.; Ma, Q.; Wang, X.; He, S.; Li, M.; Ren, L. Hydrogen evolution by catalyzing water splitting on two-dimensional g-C$_3$N$_4$-Ag/AgBr heterostructure. *Appl. Surf. Sci.* **2019**, *494*, 275–284. [CrossRef]
38. Hu, X.; Mohamood, T.; Ma, W.; Chen, C.; Zhao, J. Oxidative decomposition of rhodamine B dye in the presence of VO$_2^+$ and/or Pt(IV) under visible light irradiation: N-deethylation, chromophore cleavage, and mineralizatio. *J. Phys. Chem. B* **2006**, *110*, 26012–26018. [CrossRef]
39. Pan, C.; Zhu, Y. New type of BiPO4 oxy-acid salt photocatalyst with high photocatalytic activity on degradation of dye. *Environ. Sci. Technol.* **2010**, *44*, 5570–5574. [CrossRef]
40. Hu, X.; Ji, H.; Chang, F.; Luo, Y. Simultaneous photocatalytic Cr (VI) reduction and 2,4,6-TCP oxidation over g-C$_3$N$_4$ under visible light irradiation. *Catal. Today* **2014**, *224*, 34–40. [CrossRef]
41. Wang, C.; Li, S.; Cai, M.; Yan, R.; Dong, K.; Zhang, J.; Liu, Y. Rationally designed tetra (4-carboxyphenyl) porphyrin/graphene quantum dots/bismuth molybdate Z-scheme heterojunction for tetracycline degradation and Cr (VI) reduction: Performance, mechanism, intermediate toxicity appraisement. *J. Colloid Interface Sci.* **2022**, *619*, 307–321. [CrossRef] [PubMed]
42. Li, S.; Wang, C.; Cai, M.; Liu, Y.; Dong, K.; Zhang, J. Designing oxygen vacancy mediated bismuth molybdate (Bi$_2$MoO$_6$)/N-rich carbon nitride (C$_3$N$_5$) S-scheme heterojunctions for boosted photocatalytic removal of tetracycline antibiotic and Cr(VI): Intermediate toxicity and mechanism insight. *J. Colloid Interface Sci.* **2022**, *624*, 219–232. [CrossRef] [PubMed]
43. Finkelstein, E.; Rosen, G.M.; Rauchman, E.J.; Paxton, J. Spin trapping of superoxide. *Mol. Pharmacol.* **1979**, *16*, 676–685. [PubMed]
44. Wang, P.; Huang, B.; Dai, Y.; Whangbo, M.-H. Plasmonic photocatalysts: Harvesting visible light with noble metal nanoparticles. *Phys. Chem. Chem. Phys.* **2012**, *14*, 9813–9825. [CrossRef]
45. Chen, D.; Li, T.; Chen, Q.; Gao, J.; Fan, B.; Li, J.; Li, X.; Zhang, R.; Sun, J.; Gao, L. Hierarchically plasmonic photocatalysts of Ag/AgCl nanocrystals coupled with single-crystalline WO$_3$ nanoplates. *Nanoscale* **2012**, *4*, 5431–5439. [CrossRef]
46. Ye, L.; Liu, J.; Gong, C.; Tian, L.; Peng, T.; Zan, L. Two different roles of metallic Ag on Ag/AgX/BiOX (X = Cl, Br) visible light photocatalysts: Surface plasmon resonance and Z-scheme bridge. *ACS Catal.* **2012**, *2*, 1677–1683. [CrossRef]
47. Jiang, J.; Li, H.; Zhang, L. New insight into daylight photocatalysis of AgBr@Ag: Synergistic effect between semiconductor photocatalysis and plasmonic photocatalysis. *Chem. Eur. J.* **2012**, *18*, 6360–6369. [CrossRef]
48. Lauricella, R.; Tuccio, B. Detection and characterisation of free radicals after spin trapping. In *Electron Paramagnetic Resonance Spectroscopy*; Springer: Cham, Switzerland, 2020; pp. 51–82.

α-Fe$_2$O$_3$/Reduced Graphene Oxide Composites as Cost-Effective Counter Electrode for Dye-Sensitized Solar Cells

Lian Sun [1], Qian Zhang [1,*], Qijie Liang [2,*], Wenbo Li [2], Xiangguo Li [1], Shenghua Liu [1] and Jing Shuai [1]

[1] School of Materials, Shenzhen Campus of Sun Yat-sen University, No. 66, Gongchang Road, Guangming District, Shenzhen 518107, China; sunlian@mail2.sysu.edu.cn (L.S.); lixguo@mail.sysu.edu.cn (X.L.); liushengh@mail.sysu.edu.cn (S.L.); shuaij3@mail.sysu.edu.cn (J.S.)

[2] Songshan Lake Materials Laboratory, Room 425, C1 Building, University Innovation City, Songshan Lake, Dongguan 523000, China; liwb98@163.com

* Correspondence: zhangqian6@mail.sysu.edu.cn (Q.Z.); liangqijie@sslab.org.cn (Q.L.)

Abstract: The counter electrode (CE) is an important and vital part of dye-sensitized solar cells (DSSCs). Pt CEs show high-performance in DSSCs using iodide-based electrolytes. However, the high cost of Pt CEs restricts their large-scale application in DSSCs and the development of Pt-free CE is expected. Here, α-Fe$_2$O$_3$/reduced graphene oxide (α-Fe$_2$O$_3$/RGO) composites are prepared as the Pt-free CE materials for DSSCs. A simple hydrothermal technique was used to disseminate the α-Fe$_2$O$_3$ solid nanoparticles uniformly throughout the RGO surface. The presence of the α-Fe$_2$O$_3$ nanoparticles increases the specific surface area of RGO and allows the composites to be porous, which improves the diffusion of liquid electrolyte into the CE material. Then, the electrocatalytic properties of CEs with α-Fe$_2$O$_3$/RGO, α-Fe$_2$O$_3$, RGO, and Pt materials are compared. The α-Fe$_2$O$_3$/RGO CE has a similar electrocatalytic performance to Pt CE, which is superior to those of the pure α-Fe$_2$O$_3$ and RGO CEs. After being fabricated as DSSCs, the current–voltage measurements reveal that the DSSC based on α-Fe$_2$O$_3$/RGO CE has a power conversion efficiency (PCE) of 6.12%, which is 88% that of Pt CE and much higher than that of pure α-Fe$_2$O$_3$ and pure RGO CEs. All the results show that this work describes a promising material for cost-effective, Pt-free CEs for DSSCs.

Keywords: electrocatalytic; α-Fe$_2$O$_3$; reduced graphene oxide; counter electrode; dye-sensitized solar cells

1. Introduction

Due to their low price, greater energy conversion efficiency, and easy manufacturing technique, dye-sensitized solar cells (DSSCs) have received much interest [1,2]. The choice of counter electrode (CE) material is different for different electrolytes [3]. CE electron transmission from the external circuit to iodine and triiodide (I^-/I_3^-) in the redox electrolyte is crucial for developing DSSCs [4]. On the CE of DSSCs, thin films of Pt are often utilized as catalysts. However, large-scale production is not possible because Pt is a precious metal and costly. As a result, various attempts have been made to determine potential alternative materials for replacing Pt CE in DSSCs, such as transition metal oxides, which are cheaper, more conductive, and more chemically stable. Among the transitional metal oxides, iron oxides have the features of low cost, no toxicity, elemental abundance [5,6]. Fe$_2$O$_3$ is an important iron oxide found in various forms, such as α-Fe$_2$O$_3$, β-Fe$_2$O$_3$, γ-Fe$_2$O$_3$, and ε-Fe$_2$O$_3$ [7]. Each has its own unique characteristics. For example, α-Fe$_2$O$_3$ has very good electrochemical activity and is the most stable phase of iron oxide under ambient conditions [8,9], which makes it a promising candidate for cost-effective CE materials for DSSC [10]. Moreover, α-Fe$_2$O$_3$ is an n-type indirect semiconductor that is able to utilize about 40% of sunlight. Miao et al. reported that flower-shaped α-Fe$_2$O$_3$ could be used as the photoanode of DSSC and achieved 1.24% power conversion efficiency (PCE) [11]. However, the poor conductivity of the α-Fe$_2$O$_3$ restricts its further development and application [7].

Graphene is also a suitable material for the CE of DSSCs due to its long-term stability, with atoms organized in close-packed conjugated hexagonal lattices similar to graphite, but a one-atom-thick sheet [12,13]. Graphene has attracted considerable interest for energy convention owing to its superior electroconductibility, chemical stability, large specific area, and broad electrochemical window [14,15]. Graphene CEs possess a large surface area, defects, and oxygen-containing groups (such as reduced graphene oxide, RGO), suggesting the possibility of achieving comparable electrochemical performances to Pt CEs. [16] However, the poor electrocatalytic activity of RGO limits its application in DSSCs as a CE material.

Above all, the synergistic effect arising from interactions between α-Fe$_2$O$_3$ and RGO is essential because the presence of RGO can increase the electrical conductivity of α-Fe$_2$O$_3$ as well as α-Fe$_2$O$_3$ improving the electrocatalytic activity and reducing the cost of the CE materials. Furthermore, previous work found that the open spaces caused by α-Fe$_2$O$_3$ between graphene nanosheets may mitigate the effect of RGO volume change [17], which is beneficial to the electron transfer process in catalysis. Therefore, the composites based on α-Fe$_2$O$_3$ and RGO (α-Fe$_2$O$_3$/RGO) are expected to take advantage of the structural characteristics to yield improved CE performance in DSSCs. Chen et al. reported that a DSSC with 3D α-Fe$_2$O$_3$/GFs as the CE material displayed a superior PCE to that of Pt due to the positive synergistic effects of α-Fe$_2$O$_3$ and the 3D graphene frameworks (GFs) [18]. However, the synthesis process of the reported α-Fe$_2$O$_3$/GF materials not only requires a long time hydrothermal treatment, but also requires a high-temperature annealing process in Ar$_2$, which limits its large-scale application in DSSCs. Zhao et al. demonstrated 3D α-Fe$_2$O$_3$ hollow meso-microspheres on graphene sheets by applying a solvothermal strategy [19]. The synthesis process involves two hydrothermal treatments at 150 °C for hours. This structure delivers decent electrocatalytic performance in a dye-sensitized solar cell that is comparable to that of Pt. Employing α-Fe$_2$O$_3$, which has distinct morphology, as well as optimizing the energy and cost can further advance its potential in high-performance catalysts. In previous research, graphene oxide (GO) and Fe(OH)$_3$ were used to synthesize the α-Fe$_2$O$_3$/RGO composites by a facile and cost-effective process [20]. The Fe(OH)$_3$ sol contributed to the homogeneous size and excellent dispersity of α-Fe$_2$O$_3$ solid nanoparticles on RGO. The concentration of the composite was 73% α-Fe$_2$O$_3$/RGO, with the best performance exhibited when the volume ratio between the GO solution and the Fe(OH)$_3$ sol was 2:1.

Here, in this work, α-Fe$_2$O$_3$/RGO composites are developed to be used as a cost-effective CE in a DSSC system and the performances of DSSCs with α-Fe$_2$O$_3$/RGO, α-Fe$_2$O$_3$, RGO, and Pt CEs are compared under the same conditions. A facile synthesis process is used to develop α-Fe$_2$O$_3$ solid nanoparticles on RGO with outstanding homogeneity and dispersion by employing Fe(OH)$_3$ and a GO sol. As a CE material for DSSCs, the α-Fe$_2$O$_3$/RGO composites have a larger specific surface area than pure RGO, which induces the composites to have better electrocatalytic activity. The PCE of the DSSC using α-Fe$_2$O$_3$/RGO CE is increased by 30.5% compared to the pure α-Fe$_2$O$_3$ CE (4.69%), which itself is much higher than pure RGO (98.7% increase). This work presents a promising route for a cost-effective production way for CE materials for DSSCs.

2. Results and Discussion

By comparing the X-ray diffraction (XRD) patterns of α-Fe$_2$O$_3$/RGO composites, α-Fe$_2$O$_3$, RGO, and GO, it can seen that GO has been transformed to RGO (Figure 1a). Pure and composite α-Fe$_2$O$_3$ particles are all highly crystalline, which agrees with the reference (PDF#89-0597). The diffraction peaks at 2θ of 24.1°, 33.1°, 35.6°, 40.8°, 49.4°, 54.0°, 62.4°, and 64.0° correspond to (012), (104), (110), (113), (024), (116), (214), and (300). There are no significant variations in the XRD data of pure α-Fe$_2$O$_3$ and α-Fe$_2$O$_3$/RGO, except that the α-Fe$_2$O$_3$/RGO has a diffraction peak at 2θ = 25.6°, which corresponds to the graphene crystal faces (d-spacing of 3.35 Å) [21].

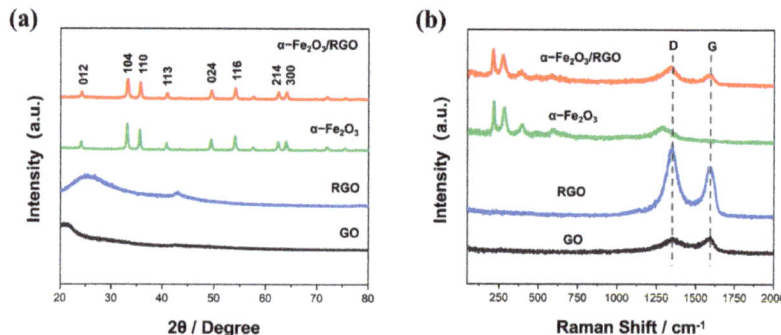

Figure 1. (a) XRD patterns and (b) Raman spectra of α-Fe$_2$O$_3$/RGO, α-Fe$_2$O$_3$, RGO, and GO.

Figure 1b shows the Raman spectra of the α-Fe$_2$O$_3$/RGO, α-Fe$_2$O$_3$, RGO, and GO samples. The D and G bands reveal typical peaks at 1350 cm^{-1} and 1589 cm^{-1} in all three graphene-related compounds. α-Fe$_2$O$_3$/RGO and RGO have higher I_D/I_G values than GO, indicating that GO has been reduced to RGO [22]. GO has an I_D/I_G ratio of 0.95, while α-Fe$_2$O$_3$/RGO has the most significant ratio at 1.57. Two peaks are presented for α-Fe$_2$O$_3$/RGO and pure α-Fe$_2$O$_3$ at 221 cm^{-1} and 285 cm^{-1}, which correspond to hematite's standard A_{1g} and E_g Raman modes, respectively, while the peak for pure α-Fe$_2$O$_3$ at 1299 cm^{-1} is due to two magnetic oscillator scattering of hematite [23]. From the XRD and Raman result, the RGO sheets are probably attached by α-Fe$_2$O$_3$ nanoparticles.

FE-SEM was utilized to examine the morphologies of the α-Fe$_2$O$_3$/RGO, α-Fe$_2$O$_3$, and RGO materials fabricated as films that were used as CEs in DSSCs. Figure 2 presents a top view of films of different materials on FTO at the same magnification. As shown in Figure 2a, the pure α-Fe$_2$O$_3$ nanoparticles exhibit uniformly regular solid shape with a size of about 20–50 nm. The pure RGO has a typical wrinkled and folded structure as observed in Figure 2b. Figure 2c illustrates the SEM image of the α-Fe$_2$O$_3$/RGO composites film. The graphene sheets are evenly coated with uniform α-Fe$_2$O$_3$ particles on both sides and the α-Fe$_2$O$_3$/RGO film has a rough surface and many more pores than pure RGO. This variation in shape indicates that the strong force between GO sheets and Fe^{3+} has a significant impact on the crystalline growth of α-Fe$_2$O$_3$ nanoparticles [24].

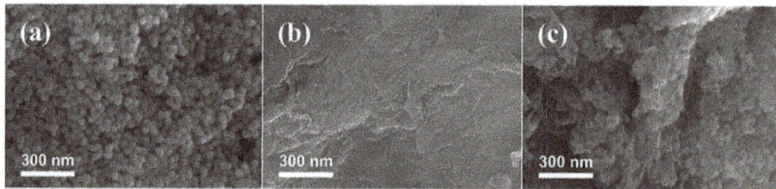

Figure 2. FE-SEM images of (a) α-Fe$_2$O$_3$, (b) RGO, (c) α-Fe$_2$O$_3$/RGO.

TEM studies of the α-Fe$_2$O$_3$/RGO composites were performed to define their microstructure further (Figure 3). Figure 3a shows that α-Fe$_2$O$_3$ solid nanoparticles ranging in size from 20 to 50 nm are evenly dispersed over RGO. This result also reveals efficient assembly of the α-Fe$_2$O$_3$ particles and graphene sheets during the hydrothermal treatment. The 0.22 nm lattice spacing in the (113) plane identifies the α-Fe$_2$O$_3$ particles (Figure 3b) [25], matching with the XRD data. The three images (Figure 3d–f) of the targeted region (Figure 3c) illustrate the TEM elemental mapping findings of the α-Fe$_2$O$_3$/RGO composites. Figure 3d demonstrates that carbon (C) atoms are numerous in the composites, but iron (Fe) and oxygen (O) atoms are scarce in Figure 3e,f. However, all three images demonstrate the distribution of C, Fe, and O elements in the α-Fe$_2$O$_3$/RGO composites are highly homogeneous.

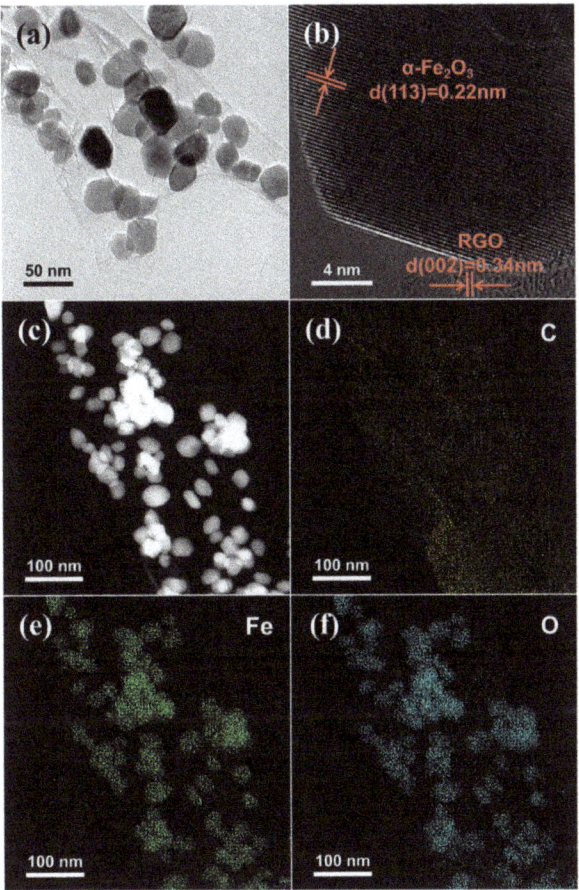

Figure 3. (a) TEM images of large-area and (b) high-resolution TEM image of α-Fe$_2$O$_3$/RGO composites; (c–f) EDS elemental mapping images of α-Fe$_2$O$_3$/RGO composites.

Figure 4a depicts the N$_2$ adsorption–desorption isotherms of the α-Fe$_2$O$_3$/RGO composites. The sample showed a curve pattern between Type IV (BDDT classification) that displays hysteresis loops predominantly of type H3 in the adsorption isotherms in Figure 4a [26]. The BET specific surface area of the α-Fe$_2$O$_3$/RGO samples are measured to be 136.16 m^2 g^{-1}. According to the previous study [4], the addition of α-Fe$_2$O$_3$ provides a larger specific surface area for the composites (the specific area of RGO prepared by the same synthesis method is 32.2 m^2 g^{-1}). In addition, obvious pressure hysteresis can be observed in the N$_2$ adsorption–desorption isotherms, revealing the porous nature of the α-Fe$_2$O$_3$/RGO nanostructure. This can be further identified by the BJH pore size distribution analysis which is shown in Figure 4b. The total pore volume of α-Fe$_2$O$_3$/RGO composites are 0.244 cm^3 g^{-1} according to the BJH method. The α-Fe$_2$O$_3$/RGO composites have a narrow pore size distribution centered at 3.7 nm, while the RGO sample was suggested in the before work to be a structure without large number of holes. The incorporation of α-Fe$_2$O$_3$ particles enhanced the specific surface area of α-Fe$_2$O$_3$/RGO composites with a porous architecture. The abovementioned results are also consistent with the result in

Figure 2. Liquid electrolyte may more readily permeate into this porous nanostructure, which significantly improves electrocatalytic performance.

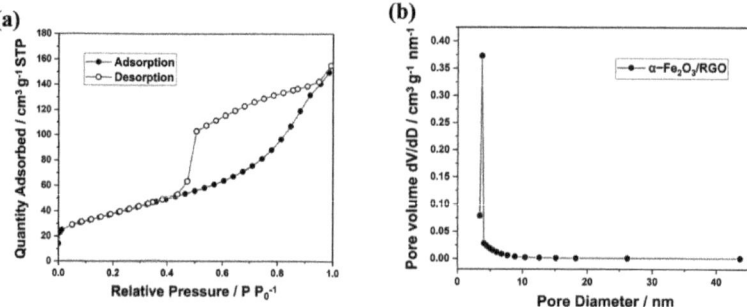

Figure 4. (a) N_2 adsorption and desorption isotherms for the α-Fe_2O_3/RGO samples; (b) pore size distribution of α-Fe_2O_3/RGO.

To evaluate the electrocatalytic activities of as-prepared α-Fe_2O_3/RGO, α-Fe_2O_3, RGO, and Pt, three-electrode cyclic voltammetry (CV) was performed (Figure 5). As previously reported, a representative curve with two couples of redox peaks was found for Pt. The I and I' peaks correspond to the oxidation and reduction peaks of I_3^-/I_2, whereas the II and II' peaks correspond to those of I^-/I_3^- [27]. During the electrochemical process in a DSSC, it is crucial that electrons from CE reduce I_3^- to I^-. Therefore, the II and II' peaks represent the electro-catalytic capabilities of the CEs. By analyzing the reduction peaks of all the CEs, we can see that the α-Fe_2O_3/RGO and Pt have sharper II' peaks, demonstrating that their catalytic activities are considerably higher than others. It is well known that peak-to-peak separation and peak current density are two essential metrics for assessing the electrocatalytic activities of CE [28]. The magnitude of peak current density is correlated to the capacity of the CE to decrease the I_3^- species. Compared with pure RGO and α-Fe_2O_3 CEs, the α-Fe_2O_3/RGO CE exhibits a greater peak current density (Figure 5), indicating that it has stronger electrocatalytic activity and a faster response rate.

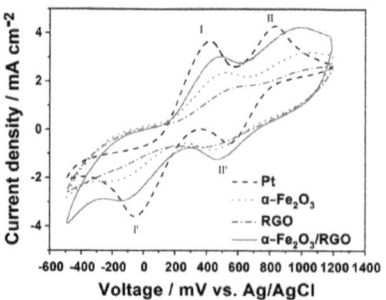

Figure 5. Cyclic voltammograms of α-Fe_2O_3/RGO, α-Fe_2O_3, RGO, and Pt CEs at a scan rate of 100 mV s^{-1}.

To further examine the liquid electrolyte diffusion rate into the material and interfacial charge transfer properties of the triiodide/iodide pair on the electrode surface, Tafel polarization experiments were performed in a virtual device made with two duplicate electrodes. Figure 6 depicts the logarithmic current density (log J_0) versus voltage (U) during the oxidation/reduction of triiodide to iodide. While the I_3^- is converted to I^- in the electrochemical cell, the impedance to charge transfer is inversely related to the exchange current density (J_0). Using the Equation (1), this may be computed from the intersection of the tangent line of the polarization curve and the prolongation of the linear section to zero bias.

$$J_0 = RT/nFR_{ct} \tag{1}$$

where R and F are constant, T is the room temperature, n is the amount of electrons participating in the reaction. The limiting diffusion exchange current density (J_{lim}) can also be computed from the Tafel curve using the Equation (2).

$$D = l\, J_{lim}/2nFC \tag{2}$$

where D is the diffusion coefficient of the triiodide, l is the thickness of the spacer, C is the triiodide concentration and n and F retain their defined meanings. J_{lim} depends on the diffusion rate of the I^-/I_3^- redox couple. An optimal auxiliary counter electrode should have high J_0, J_{lim}, and lower R_{ct} values.

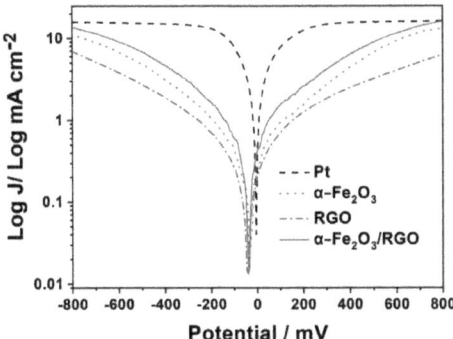

Figure 6. Tafel curves of the symmetrical cells fabricated with two identical α-Fe$_2$O$_3$/RGO, α-Fe$_2$O$_3$, RGO, and Pt electrodes.

Theoretically, the curve with a relatively low potential but higher than 0.1 V corresponds to the Tafel zone, in which the voltage is a linear function of the log of the current density (log J_0) (Equation (1)). It is possible to determine the exchange current density (J_0) in this region using Equation (1). In addition, the steeper the Tafel zone of the curve, the bigger the J_0 and the greater the material's catalytic activity. Pt has the greatest J_0 in the Tafel zone, followed by the α-Fe$_2$O$_3$/RGO, α-Fe$_2$O$_3$, and RGO. This reveals the α-Fe$_2$O$_3$/RGO composite CE has more electrocatalytic activity when compared to the pure α-Fe$_2$O$_3$ and RGO CEs. The Tafel curves of the α-Fe$_2$O$_3$/RGO, α-Fe$_2$O$_3$, and RGO Ces are asymmetric may due to the different diffusion time of the electrolyte on two electrodes of symmetric dummy cells. To test the Tafel curve, the electrolyte was firstly dripped on one electrode (always the cathode), and then covered with another electrode on the electrolyte. This process causes the electrolyte to diffuse first on the cathode, resulting in asymmetry between the two electrodes in the Tafel curve. For this reason, the Pt electrode is also slightly asymmetrical. However, the asymmetric degree of the Pt CE is lighter because of the better catalytic ability.

To investigate the effects of the α-Fe$_2$O$_3$/RGO composites as a CE material for DSSC, the DSSCs were assembled using α-Fe$_2$O$_3$/RGO as the CE. For comparison, the properties of DSSCs fabricated with α-Fe$_2$O$_3$, RGO, and Pt Ces were also investigated. Five parallel devices for each sample were tested. The photovoltaic properties results of the four different DSSCs are summarized in Table 1. The short-circuit photocurrent (j_{sc}), open-circuit voltage (V_{oc}), fill factor (FF), and the power conversion efficiency (PCE, η) calculated photovoltaic parameters are provided. Due to its weak electrocatalytic activity, the DSSC with RGO CE has a poor transfer efficiency, of 3.08%, whereas the DSSC with α-Fe$_2$O$_3$/RGO CE has an open-circuit voltage of 645 mV, a short-circuit current of 15.43 mA cm^{-2}, a fill factor of 0.61, and a cell efficiency of 6.12%, which is 88% of that of the DSSC with Pt CE (6.93%). In comparison to the α-Fe$_2$O$_3$ and RGO Ces, the α-Fe$_2$O$_3$/RGO CE has a substantially higher FF value. The photocurrent–voltage (I–V) curves of the DSSCs are shown in Figure 7a.

Table 1. PCE performances and EIS parameters of the DSSCs with different CEs.

CEs	R_{ct} (Ω)	j_{sc} (mA cm^{-2})	V_{oc} (mV)	FF	η (%)
Pt	1.78	16.33 ±0.06	635 ±0	0.67 ±0.00	6.93 ±0.05
α-Fe$_2$O$_3$	5.42	13.3 ±0.05	635 ±5	0.55 ±0.01	4.69 ±0.03
RGO	11.02	12.47 ±0.05	555 ±5	0.45 ±0.01	3.08 ±0.05
α-Fe$_2$O$_3$/RGO	3.81	15.43 ±0.00	645 ±0	0.61 ±0.00	6.12 ±0.004

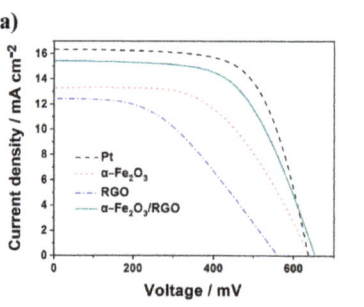
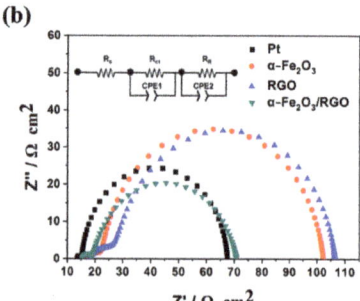

Figure 7. (a) Photocurrent density–voltage characteristics and (b) Nyquist plots of DSSCs with α-Fe$_2$O$_3$/RGO, α-Fe$_2$O$_3$, RGO, and Pt CEs measured at AM 1.5 G illumination (100 mW cm^{-2}). The inset in (b) is the equivalent circuit of the DSSCs.

Figure 7b illustrates the resulting Nyquist plots from the EIS measurements performed on the DSSCs using the CEs above to obtain insight into the variation in FF values. The spectra were simulated using the matching circuit shown in the inset of Figure 7b, which contains high-frequency series resistance (Rs). The interfacial charge transfer resistance R_{ct} and the capacitance of the electrical double layer (CPE1) are related to the RC processes of the CE/electrolyte interface in the intermediate frequency area. In addition, the interfacial charge transfer resistance (R_R) and capacitance of the depletion layer of the TiO$_2$ electrode (CPE2) are related to the RC processes of the TiO$_2$ electrode/electrolyte interface in the low frequency range. The electrocatalytic property of the CE for triiodide reduction may be assessed, which is defined from the first hemicycle of the EIS spectrum [29]. According to Table 1, the α-Fe$_2$O$_3$ CE exhibits a greater R_{ct}, suggesting weak electrocatalytic performance (as also proven by CV outcomes). When the RGO and the α-Fe$_2$O$_3$ are combined, the R_{ct} decreases from 5.42 Ω to 3.81 Ω. The R_{ct} value of the α-Fe$_2$O$_3$/RGO composites are much smaller than that of the RGO CE (11.02 Ω). This may be attributable to the integration of α-Fe$_2$O$_3$'s intrinsic high electrocatalytic activity on the RGO's highly active electric transport channel [30]. Nonetheless, the α-Fe$_2$O$_3$/RGO CE has a substantially bigger sum of R_s and R_{ct} than the Pt CE, leading to a lower FF and η values for the photovoltaic properties of the DSSCs [31].

3. Materials and Methods

3.1. Synthesis of Materials

First, a solution was prepared by adding 1.3 g FeCl$_3$ (Aldrich, 98%, Shanghai, China) in 4 mL distilled water. Second, 50 mL distilled water was boiled. Third, the as-prepared solution (0.5 mL) was added into the boiling distilled water dropwise. Then, the mixture was kept boiling for several minutes to prepare the Fe(OH)$_3$ sol. GO was manufactured from modified graphite oxide (GO), produced from natural scale graphite using a revised Hummers method technique [32,33]. In a typical synthesis, Fe(OH)$_3$ sol was added dropwise to the GO solution (1 mg mL^{-1}) at a volume ratio of 1:2 and then stirred for 30 min. Next, the mixture was heated to 85 °C in a water bath and hydrazine hydrate (Aldrich, 85%, Shanghai, China) was added into the mixture. Then, an ultrasonication process was applied for 30 min. After that, the solution was transferred into a Teflon-lined autoclave and heated at 150 °C for 6 h. The comparison mixture was prepared by centrifugation; α-Fe$_2$O$_3$ and RGO were prepared using the same technique, but without GO and FeCl$_3$, respectively.

3.2. Fabrication of CEs and DSSCs

The CEs were fabricated by a doctor blade technique [34]. In one batch, films of a certain thickness were made. Comparatively, platinized CEs were made by a thermal breakdown. The chloro-platinic acid hexahydrate (H$_2$PtCl$_6$·6H$_2$O, Pt $\geq$ 37.5%, AR, Aladdin, Shanghai, China) was mixed with isopropanol (C$_3$H$_8$O, $\geq$99.5%, AR, Aladdin, Shanghai, China) and the mixture was dropped on the cleaned fluorine-doped tin oxide (FTO) glass sheets (Nippon Sheet Glass Co., Osaka, Japan, surface resistance = 15 Ω/cm^2, transmittance = 90%) and then heated for 15 min at 390 °C.

TiO$_2$ was synthesized by using a sol–gel method [35]. Then, the TiO$_2$ films (ca. 11 μm) were coated on the FTO glass using the doctor blade technique. The FTO glass coated with TiO$_2$ was heated at 450 °C for 30 min. When the temperature dropped to around 90 °C, the electrodes were submerged for 24 h in a dye solution containing 0.5 mM cis-Ru (H$_2$dcbpy)$_2$ (NCS)$_2$ (H$_2$dcbpy = 4,4′-dicarboxy-2,2′-bipyridyl) (N3) dissolved in ethanol. After this, the TiO$_2$ photoanodes are complete. The DSSCs were constructed using the aforementioned TiO$_2$ photoanode, a CE, and a redox electrolyte comprising 0.5 M LiI (AR, Alfa, Zhengzhou, China), 0.05 M I$_2$ (AR, Alfa), 0.6 M 4-tert-butylpyridine (TBP, >96%, Tokyo Chemical Industry Co., Ltd. Tokyo, Japan), and 0.6 M 3-hexyl-1-methylimidazolium iodide (HMII, AR, Alfa, Zhengzhou, China) in 3-methoxypropionitrile (MPN, 99%, GC, Alfa, Zhengzhou, China). The tested DSSCs were masked to a working area of 0.2 cm^2.

3.3. Characterization

XRD (Cu Kα irradiation, D8-ADVANCE, Billerica, MA, USA) was used to analyze the crystalline structure and state of the samples. At room temperature, Raman spectra were acquired using a Raman spectroscope meter (Horiba LabRam HR Evolution, Tokyo, Japan) equipped with a 633 nm laser. Field emission scanning electron microscopy (FE-SEM, Gemini 300, Oberkochen, Germany) was used to examine the sample morphology (FE-SEM, Gemini 300, Oberkochen, Germany). The crystal structure was measured by high-resolution transmission electron microscopy (JEM-F200, Akishima, Japan). The specific surface area and total pore volume of samples were determined by the Brunauer–Emmett–Teller (BET) method, in which the N$_2$ adsorption at −195.8 °C was measured using an adsorption instrument (ASAP 2460, Norcross, GA, USA). The pore size distribution was estimated based on the desorption isotherm using the Barrett–Joyner–Halender (BJH) method [26].

Utilizing an electrochemical analyzer, cyclic voltammetry (CV) was performed in a three-electrode setup in an acetonitrile solution containing 0.1 M LiClO$_4$, 10 mM LiI, and 1 mM I$_2$ at a scan rate of 100 mV s^{-1} (Solartron SI 1287, Illinois, IL, USA). The Ag/Ag$^+$ combination acted as the comparison electrode, while platinum served as the counter electrode. The spectra were fitted by the Zview software. Using an electrochemical workstation system (Solartron SI 1287, Illinois, IL, USA) in symmetric dummy cells built with two identical CEs and a scan speed of 50 mV S^{-1}, Tafel polarization studies were performed. The photocurrent density–voltage (j$_{sc}$-V) performance of the DSSCs was measured with a Keithley digital source meter (Keithley 2410, Cleveland, OH, USA) and simulated under AM 1.5 illumination (100 mW cm^{-2}, Newport 69907). The incident light was calibrated with a power meter (model 350) and a detector (model 262). Electrochemical impedance spectroscopy (EIS) was performed on the Solartron SI 1260 frequency response analyzer and

Solartron SI 1287 electrochemical interface system. The frequency range was 0.1–100 kHz and the AC voltage was 10 mV.

4. Conclusions

In this work, we demonstrated a novel Pt-free CE made of α-Fe_2O_3/RGO composites for DSSCs. The α-Fe_2O_3/RGO composites were proven to be synthesized successfully via a facile method as the α-Fe_2O_3 nanoparticles were found to be dispersed over the RGO surface. Compared with pure α-Fe_2O_3 and RGO, the α-Fe_2O_3/RGO composites have a larger specific surface area and a more porous microstructure, which provides an advantage to catalytic reactions. The α-Fe_2O_3/RGO CE was prepared on an FTO glass-substrate via the doctor blade process. Comprehensive electrochemical investigations revealed that the α-Fe_2O_3/RGO CE exhibited Pt-like electrocatalytic activity for I_3^- reduction owing to the synergistic effect between the inherently high catalytic performance of α-Fe_2O_3 and the easy charge transfer properties of RGO. The PCE of the DSSCs constructed with the α-Fe_2O_3/RGO CE was 6.12%, up to 88% of that achieved with the Pt CE. Therefore, the α-Fe_2O_3/RGO CE is a promising candidate for application as a cost-effective CE material in Pt-free transparent DSSCs. This work presents a solution that can easily prepare large-scale DSSCs with low cost, which is expected to improve the possibility of commercialization of DSSCs in the future.

Author Contributions: Data curation, X.L.; Formal analysis, S.L.; Investigation, S.L.; Methodology, L.S., Q.Z. and Q.L.; Project administration, J.S.; Resources, Q.Z. and Q.L.; Software, X.L.; Writing—original draft, L.S. and Q.Z.; Writing—review & editing, L.S., Q.Z., Q.L., W.L. and J.S. All authors have read and agreed to the published version of the manuscript.

Funding: This research was funded by the open research fund of Songshan Lake Materials Laboratory (2021SLABFN21), the Hundreds of Talents Program of Sun Yat-sen University.

Acknowledgments: The authors would like to acknowledge financial support from the open research fund of Songshan Lake Materials Laboratory (2021SLABFN21), the Hundreds of Talents Program of Sun Yat-sen University.

Conflicts of Interest: The authors declare no conflict of interest.

References

1. O'regan, B.; Grätzel, M. A low-cost, high-efficiency solar cell based on dye-sensitized colloidal TiO_2 films. *Nature* **1991**, *353*, 737–740. [CrossRef]
2. Saranya, K.; Rameez, M.; Subramania, A. Developments in conducting polymer based counter electrodes for dye-sensitized solar cells–An overview. *Eur. Polym. J.* **2015**, *66*, 207–227. [CrossRef]
3. Kang, J.S.; Kim, J.; Kim, J.Y.; Lee, M.J.; Kang, J.; Son, Y.J.; Jeong, J.; Park, S.H.; Ko, M.J.; Sung, Y.E. Highly efficient bifacial dye-sensitized solar cells employing polymeric counter electrodes. *ACS Appl. Mater. Interfaces* **2018**, *10*, 8611–8620. [CrossRef]
4. Zhang, Q.; Liu, Y.; Duan, Y.; Fu, N.; Liu, Q.; Fang, Y.; Sun, Q.; Lin, Y. Mn_3O_4/graphene composite as counter electrode in dye-sensitized solar cells. *RSC Adv.* **2014**, *4*, 15091–15097. [CrossRef]
5. Singh, P.; Sharma, K.; Hasija, V.; Sharma, V.; Sharma, S.; Raizada, P.; Singh, M.; Saini, A.K.; Hosseini-Bandegharaei, A.; Thakur, V.K. Systematic review on applicability of magnetic iron oxides-integrated photocatalysts for degradation of organic pollutants in water. *Mater. Today Chem.* **2019**, *14*, 100186. [CrossRef]
6. Ye, H.; Wang, Y.; Liu, X.J.; Xu, D.D.; Yuan, H.; Sun, H.Q.; Wang, S.B.; Ma, X. Magnetically steerable iron oxides-manganese dioxide core-shell micromotors for organic and microplastic removals. *J. Colloid Interface Sci.* **2021**, *588*, 510–521. [CrossRef]
7. Kumar, Y.; Kumar, R.; Raizada, P.; Khan, A.A.P.; Singh, A.; Le, Q.V.; Nguyen, V.H.; Selvasembian, R.; Thakur, S.; Singh, P. Current status of hematite (α-Fe_2O_3) based Z-scheme photocatalytic systems for environmental and energy applications. *J. Environ. Chem. Eng.* **2022**, *10*, 107427. [CrossRef]
8. Gao, R.J.; Wang, J.; Huang, Z.F.; Zhang, R.R.; Wang, W.; Pan, L.; Zhang, J.F.; Zhu, W.K.; Zhang, X.W.; Shi, C.X.; et al. Pt/Fe_2O_3 with Pt-Fe pair sites as a catalyst for oxygen reduction with ultralow Pt loading. *Nat. Energy* **2021**, *6*, 614–623. [CrossRef]
9. Zhang, Q.; Liang, Q.; Liao, Q.; Ma, M.; Gao, F.; Zhao, X.; Song, Y.; Song, L.; Xun, X.; Zhang, Y. Amphiphobic hydraulic triboelectric nanogenerator for self-cleaning/charging power system. *Adv. Funct. Mater.* **2018**, *28*, 1803117. [CrossRef]
10. Hou, Y.; Wang, D.; Yang, X.H.; Fang, W.Q.; Zhang, B.; Wang, H.F.; Lu, G.Z.; Hu, P.; Zhao, H.; Yang, H. Rational screening low-cost counter electrodes for dye-sensitized solar cells. *Nat. Commun.* **2013**, *4*, 1583. [CrossRef]
11. Niu, H.; Zhang, S.; Ma, Q.; Qin, S.; Wan, L.; Xu, J.; Miao, S. Dye-sensitized solar cells based on flower-shaped α-Fe_2O_3 as a photoanode and reduced graphene oxide–polyaniline composite as a counter electrode. *RSC Adv.* **2013**, *3*, 17228–17235. [CrossRef]

12. Wang, M.; Huang, M.; Luo, D.; Li, Y.; Choe, M.; Seong, W.K.; Kim, M.; Jin, S.; Wang, M.; Chatterjee, S.; et al. Single-crystal, large-area, fold-free monolayer graphene. *Nature* **2021**, *596*, 519–524. [CrossRef] [PubMed]
13. Liu, L.; Qing, M.; Wang, Y.; Chen, S. Defects in graphene: Generation, healing, and their effects on the properties of graphene: A review. *J. Mater. Sci. Technol.* **2015**, *31*, 599–606. [CrossRef]
14. Fang, B.; Chang, D.; Xu, Z.; Gao, C. A review on graphene fibers: Expectations, advances, and prospects. *Adv. Mater.* **2020**, *32*, 1902664. [CrossRef]
15. Zhang, Q.; Liang, Q.; Liao, Q.; Yi, F.; Zheng, X.; Ma, M.; Gao, F.; Zhang, Y. Service behavior of multifunctional triboelectric nanogenerators. *Adv. Mater.* **2017**, *29*, 1606703. [CrossRef]
16. Kweon, D.H.; Baek, J.B. Edge-functionalized graphene nanoplatelets as metal-Free electrocatalysts for dye-sensitized solar cells. *Adv. Mater.* **2019**, *31*, 1804440. [CrossRef]
17. Zhu, X.; Zhu, Y.; Murali, S.; Stoller, M.D.; Ruoff, R.S. Nanostructured reduced graphene oxide/Fe$_2$O$_3$ composite as a high-performance anode material for lithium ion batteries. *ACS Nano* **2011**, *5*, 3333–3338. [CrossRef]
18. Yang, W.; Xu, X.W.; Li, Z.; Yang, F.; Zhang, L.Q.; Li, Y.F.; Wang, A.J.; Chen, S.L. Construction of efficient counter electrodes for dye-sensitized solar cells: Fe$_2$O$_3$ nanoparticles anchored onto graphene frameworks. *Carbon* **2016**, *96*, 947–954. [CrossRef]
19. Zhao, G.M.; Xu, G.J.; Jin, S. α-Fe$_2$O$_3$ hollow meso-microspheres grown on graphene sheets function as a promising counter electrode in dye-sensitized solar cells. *Rsc Adv.* **2019**, *9*, 24164–24170. [CrossRef]
20. Du, M.; Xu, C.; Sun, J.; Gao, L. Synthesis of α-Fe$_2$O$_3$ nanoparticles from Fe(OH)$_3$ sol and their composite with reduced graphene oxide for lithium ion batteries. *J. Mater. Chem. A* **2013**, *1*, 7154–7158. [CrossRef]
21. Wang, G. Preparation of α-Fe$_2$O$_3$/graphene composite and its electrochemical performance as an anode material for lithium ion batteries. *J. Alloys Compd.* **2011**, *509*, 216–220. [CrossRef]
22. Pradhan, G.K.; Padhi, D.K.; Parida, K.M. Fabrication of α-Fe$_2$O$_3$ nanorod/RGO composite: A novel hybrid photocatalyst for phenol degradation. *ACS Appl. Mater. Interfaces* **2013**, *5*, 9101–9110. [CrossRef]
23. De Faria, D.L.A.; Venâncio Silva, S.; De Oliveira, M.T. Raman microspectroscopy of some iron oxides and oxyhydroxides. *J. Raman. Spectrosc.* **1997**, *28*, 873–878. [CrossRef]
24. Wang, H.; Xu, Z.; Yi, H.; Wei, H.; Guo, Z.; Wang, X. One-step preparation of single-crystalline Fe$_2$O$_3$ particles/graphene composite hydrogels as high performance anode materials for supercapacitors. *Nano Energy* **2014**, *7*, 86–96. [CrossRef]
25. Sun, B.; Horvat, J.; Kim, H.S.; Kim, W.-S.; Ahn, J.; Wang, G. Synthesis of mesoporous α-Fe$_2$O$_3$ nanostructures for highly sensitive gas sensors and high capacity anode materials in lithium ion batteries. *J. Phys. Chem. C* **2010**, *114*, 18753–18761. [CrossRef]
26. Xu, F.; Zhang, J.; Zhu, B.; Yu, J.; Xu, J. CuInS$_2$ sensitized TiO$_2$ hybrid nanofibers for improved photocatalytic CO$_2$ reduction. *Appl. Catal. B Environ.* **2018**, *230*, 194–202. [CrossRef]
27. Zhang, T.-L.; Chen, H.-Y.; Su, C.-Y.; Kuang, D.-B. A novel TCO-and Pt-free counter electrode for high efficiency dye-sensitized solar cells. *J. Mater. Chem. A* **2013**, *1*, 1724–1730. [CrossRef]
28. Gong, F.; Wang, H.; Xu, X.; Zhou, G.; Wang, Z.-S. In situ growth of Co$_{0.85}$Se and Ni$_{0.85}$Se on conductive substrates as high-performance counter electrodes for dye-sensitized solar cells. *J. Am. Chem. Soc.* **2012**, *134*, 10953–10958. [CrossRef]
29. Fu, N.Q.; Xiao, X.; Zhou, X.W.; Zhang, J.; Lin, Y. Electrodeposition of platinum on plastic substrates as counter electrodes for flexible dye-sensitized solar cells. *J. Phys. Chem. C* **2012**, *116*, 2850–2857. [CrossRef]
30. Tai, S.-Y.; Liu, C.-J.; Chou, S.-W.; Chien, F.S.-S.; Lin, J.-Y.; Lin, T.-W. Few-layer MoS$_2$ nanosheets coated onto multi-walled carbon nanotubes as a low-cost and highly electrocatalytic counter electrode for dye-sensitized solar cells. *J. Mater. Chem.* **2012**, *22*, 24753–24759. [CrossRef]
31. Velten, J.; Mozer, A.J.; Li, D.; Officer, D.; Wallace, G.; Baughman, R.; Zakhidov, A. Carbon nanotube/graphene nanocomposite as efficient counter electrodes in dye-sensitized solar cells. *Nanotechnology* **2012**, *23*, 085201. [CrossRef]
32. Alkhouzaam, A.; Abdelrazeq, H.; Khraisheh, M.; AlMomani, F.; Hameed, B.H.; Hassan, M.K.; Al-Ghouti, M.A.; Selvaraj, R. Spectral and structural properties of high-quality reduced graphene oxide produced via a simple approach using tetraethylene-pentamine. *Nanomaterials* **2022**, *12*, 1240. [CrossRef]
33. Hummers, W.S., Jr.; Offeman, R.E. Preparation of graphitic oxide. *J. Am. Chem. Soc.* **1958**, *80*, 1339. [CrossRef]
34. Duan, Y.; Fu, N.; Zhang, Q.; Fang, Y.; Zhou, X.; Lin, Y. Influence of Sn source on the performance of dye-sensitized solar cells based on Sn-doped TiO$_2$ photoanodes: A strategy for choosing an appropriate doping source. *Electrochim. Acta* **2013**, *107*, 473–480. [CrossRef]
35. Duan, Y.; Fu, N.; Liu, Q.; Fang, Y.; Zhou, X.; Zhang, J.; Lin, Y. Sn-doped TiO$_2$ photoanode for dye-sensitized solar cells. *J. Phys. Chem. C* **2012**, *116*, 8888–8893. [CrossRef]

Article

Highly Efficient and Selective Carbon-Doped BN Photocatalyst Derived from a Homogeneous Precursor Reconfiguration

Qiong Lu [1,2], Jing An [3], Yandong Duan [3,*], Qingzhi Luo [3], Yunyun Shang [3], Qiunan Liu [4], Yongfu Tang [4], Jianyu Huang [4], Chengchun Tang [1,2,*], Rong Yin [3] and Desong Wang [1,3,4,*]

1. School of Materials Science and Engineering, Hebei University of Technology, Tianjin 300130, China; qiong253@163.com
2. Hebei Key Laboratory of Boron Nitride Micro and Nano Materials, Hebei University of Technology, Tianjin 300130, China
3. Hebei Key Laboratory of Photoelectric Control on Surface and Interface, School of Sciences, Hebei University of Science and Technology, Shijiazhuang 050018, China; anjinghebust@163.com (J.A.); lqz2004-1@163.com (Q.L.); shangyun1996@163.com (Y.S.); yinrong6868@163.com (R.Y.)
4. Applying Chemistry Key Laboratory of Hebei Province, State Key Laboratory of Metastable Materials Science and Technology, Yanshan University, Qinhuangdao 066000, China; 13731790498@163.com (Q.L.); tangyongfu@ysu.edu.cn (Y.T.); jyhuang8@hotmail.com (J.H.)
* Correspondence: ydduan@iccas.ac.cn (Y.D.); tangcc@hebut.edu.cn (C.T.); dswang06@126.com (D.W.)

Citation: Lu, Q.; An, J.; Duan, Y.; Luo, Q.; Shang, Y.; Liu, Q.; Tang, Y.; Huang, J.; Tang, C.; Yin, R.; et al. Highly Efficient and Selective Carbon-Doped BN Photocatalyst Derived from a Homogeneous Precursor Reconfiguration. Catalysts 2022, 12, 555. https://doi.org/10.3390/catal12050555

Academic Editors: Yongming Fu and Qian Zhang

Received: 27 April 2022
Accepted: 16 May 2022
Published: 18 May 2022

Publisher's Note: MDPI stays neutral with regard to jurisdictional claims in published maps and institutional affiliations.

Copyright: © 2022 by the authors. Licensee MDPI, Basel, Switzerland. This article is an open access article distributed under the terms and conditions of the Creative Commons Attribution (CC BY) license (https://creativecommons.org/licenses/by/4.0/).

Abstract: The modification of inert boron nitride by carbon doping to make it an efficient photocatalyst has been considered as a promising strategy. Herein, a highly efficient porous BCN (p-BCN) photocatalyst was synthesized via precursor reconfiguration based on the recrystallization of a new homogeneous solution containing melamine diborate and glucose. Two crystal types of the p-BCN were obtained by regulating the recrystallization conditions of the homogeneous solution, which showed high photocatalytic activities and a completely different CO_2 reduction selectivity. The CO generation rate and selectivity of the p-BCN-1 were 63.1 $\mu mol \cdot g^{-1} \cdot h^{-1}$ and 54.33%; the corresponding values of the p-BCN-2 were 42.6 $\mu mol \cdot g^{-1} \cdot h^{-1}$ and 80.86%. The photocatalytic activity of the p-BCN was significantly higher than those of equivalent materials or other noble metals-loaded nanohybrids reported in the literature. It was found that the differences in the interaction sites between the hydroxyl groups in the boric acid and the homolateral hydroxyl groups in the glucose were directly correlated with the structures and properties of the p-BCN photocatalyst. We expect that the developed approach is general and could be extended to incorporate various other raw materials containing hydroxyl groups into the melamine diborate solution and could modulate precursors to obtain porous BN-based materials with excellent performance.

Keywords: carbon-doped BN; homogeneous; precursor reconfiguration; photocatalysis; selectivity

1. Introduction

Hexagonal boron nitride (h-BN), called "white graphene", is a versatile material used in a number of diverse applications due to its thermal conductivity, mechanical strength, and chemical stability [1–4]. Although h-BN has so many advantages, it is a wide-band semiconductor (~5.5 eV) and is not suitable for use as a photocatalyst [4]. At present, h-BN is mainly used as a catalyst support, with its high specific surface areas and active edges examined in studies on its application as a photocatalyst [5].

Only a few studies have reported that the band gap of h-BN can be adjusted by doping to make it a suitable photocatalyst [6–10]. Among them, the doped h-BN structures with carbon atoms have been regarded with particular interest due to their simple preparation process and superior photocatalytic performance. The precursors of the reported carbon doping of h-BN (BCN) can be prepared by mechanically mixing a carbon source, boron source, and nitrogen source together using grinding mechanochemistry methods [8,11–14].

Manual grinding refines the size of the bulk raw materials into desired sizes; however, segregation often occurs in the mixing process due to the different densities of the raw materials, and an unevenly mixed precursor results [15,16]. Moreover, the influence of mechanical energy on the material properties in the grinding process is very complex and uncertain, resulting in the formation difficulty in controlling the crystal type and morphology of the product [17].

However, the chemical synthesis method can effectively avoid the above problems [18,19]. For instance, boric acid and melamine can be dissolved in water to form a homogeneous solution at high temperatures by this method, which are recrystallized to obtain a melamine diborate ($C_3N_6H_6 \cdot 2H_3BO_3$, M·2B) hydrogen-bond adduct with a uniform and definite structure [18,19]. Additionally, M·2B has been used as a highly promising precursor to porous h-BN nanosheets, with a desirable morphology and function [20,21]. In addition, glucose in the furanose form, as a polyhydroxy monosaccharide, can react with boric acid to form boronated complexes [22–24]. Therefore, it is expected that glucose as a carbon source can react with the boric acid in the M·2B solution to obtain a new precursor by the rational design, which, upon pyrolysis, could form BCN photocatalysts with a similar morphology as the porous h-BN nanosheets and with superior performance in the photocatalytic reaction. Such an insight makes it possible to tailor-make the synthesis of BCN photocatalysts with the desired properties and structures from the precursor [25].

In this work, a strategy was introduced for preparing precursors with a uniform and definite structure by recrystallization from a homogeneous solution containing M·2B and glucose, which was pyrolyzed to synthesize a highly efficient and selectively porous BCN (p-BCN) photocatalyst. By systematically studying the chemistry of the precursor, it was found that boric acid reacted with both melamine and a pair of adjacent hydroxyl groups on the anomeric carbon C(1) and C(2) or terminal carbon C(4) and C(6) of glucose to obtain the reconfigurated precursors by two cooling modes in the homogeneous solution. The resultant p-BCN photocatalysts showed two distinct structures with excellent photocatalytic activities and distinct selectivity for CO_2 conversion. This work is a new case of preparing a high-performance BCN photocatalyst by the recrystallization of a homogeneous solution, which will provide guidance to optimize precursor modulation so as to fully satisfy the application design of h-BN-based functional materials.

2. Results
2.1. Characterization of Precursors

The microstructures and composition of the P1 and P2 precursors were studied and Figure 1a shows the crystallographic character of the samples. M·2B was crystallized in a monoclinic P21/m space group by single crystal analysis (Table S1 and Figure S1 in Supplementary Materials) and the $I_{(031)}/I_{(033)}$ values of the XRD of P1 and P2 were lower than that of the M·2B [19,20]. These results imply that the addition of glucose had an evident effect on the crystal structure of the M·2B, which may have been caused by the preferred orientation between the boric acid and the melamine or glucose [19]. Figure 1b shows the Raman spectra of P1 and P2 in the region of 100~1100 cm^{-1}. Compared with the M·2B, the broader and stronger peaks for P1 and P2 at 593 cm^{-1} and 200 cm^{-1}, assigned to O−B−O bending and O−H twisting, originated from the interaction between the glucose, boric acid, and melamine, which resulted in an increase in the intermolecular distances and weaker intermolecular coupling [18]. Meanwhile, the lost intensity of the internal modes (200~1100 cm^{-1}) with the broadening peaks of both the precursors are attributed to a random arrangement and distortion within or between the molecules. The properties and variations of the N−H and O−H stretching modes of the precursors were further corroborated in the range of 3000~3600 cm^{-1} (Figure 1c). Since the melamine unit only utilized two N-donor sites to interact with the four boric acid units in the M·2B [20], the weak and wide Raman bands centered at 3362 cm^{-1} and 3298 cm^{-1} probably correspond to the stretching modes of the hydrogen-bonded N−H bonds of the melamine, while the strong bands observed at 3518 cm^{-1}, 3483 cm^{-1}, and 3410 cm^{-1} are assigned to the free

N−H bonds of the melamine. Additionally, the O−H stretching mode appeared as weak broad Raman bands around 3187 cm^{-1}. As displayed in Figure 1c, the relative intensity of the free N−H bonds in both P1 and P2 became weaker than that in the M·2B, suggesting that the surplus of the free N−H bonds of the melamine unit reacted with the O−H groups of the glucose through reconfiguration to form hydrogen bonds, resulting in less exposure of the free amino groups in the melamine. The N−H···O hydrogen bonds in P1 and P2, with an increase in the relative intensity of the Raman bands, verified that the O−H bonds of both the boric acid and glucose molecules could link with the N−H bonds of the melamine molecules via N−H···O hydrogen bonds. The results indicated that the addition of glucose had a great influence on the molecular interaction of the M·2B, and the new hydrogen bonds were formed in P1 and P2.

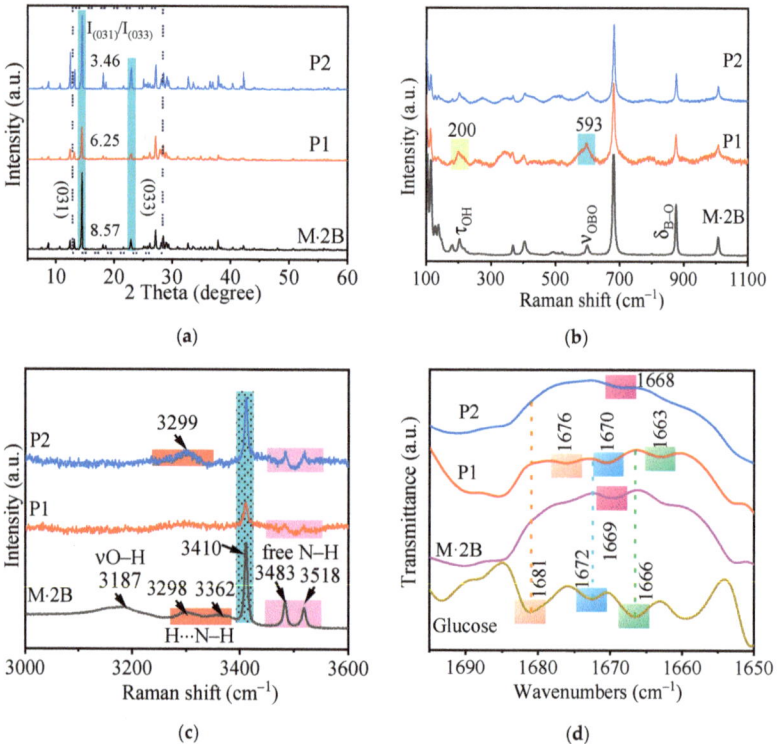

Figure 1. (**a**) Powder XRD patterns. (**b**,**c**) Raman spectra in the different wavenumber range. (**d**) An enlarged view of the FTIR spectra in the range of 1700~1650 cm^{-1} for glucose, M·2B, P1, and P2.

The partial enlargement of the FTIR spectra was performed in two different regions to explicitly investigate the changes in the precursor structures. For P1 and P2, the weak broad band observed at 3189 cm^{-1} corresponded to the O−H stretching mode (Figure S2) [18,26], which became broader with the addition of glucose. In addition to the peak at 1669 cm^{-1}, which was assigned to −NH$_2$ bending in the range of 1650~1695 cm^{-1}, several new absorption peaks were also detected in P1. The bands at 1681 cm^{-1}, 1672 cm^{-1}, and 1666 cm^{-1}, which were assigned to the C−OH vibrations of the hemiacetal groups in the glucose (Figure 1d), shifted to the lower wavenumbers of 1676 cm^{-1}, 1670 cm^{-1}, and 1663 cm^{-1} in P1, respectively [23]. These results indicate that glucose molecules existed in P1, and that there may have been hydrogen bonding interactions between the hydroxyl group of the glucose and the amino group of the melamine. The FTIR spectrum of P2 was similar to that of the M·2B; there were no C−OH vibrations of the hemiacetal groups in the glucose molecules. There were also some obvious differences in the FTIR spectra between P1 and P2 (Table S2). The results suggested that some of the peaks of P1 and P2 became broader, less intense, some of them were merged together, and some new peaks were even detected, inferring that the glucose was not simply mixed with the M·2B but interacted with the M·2B in P1 and P2 at a molecular level.

The ^{13}C and ^{11}B solid-state MAS NMR spectra are shown in Figure 2. Compared with the pure melamine, the ^{13}C chemical shifts of the melamine in P1 and P2 decreased to various degrees because of the hydrogen bonds formed between the melamine and glucose (Figure 2a). In addition, the ^{13}C chemical shifts of both C(1) and C(2) in P1 and C(4) and C(6) in P2 were lower than those in the glucose (Figure 2b and Table S3), probably because of the presence of the boron ester deriving from the reaction between the hydroxyl groups on C(1) and C(2) or the hydroxyl groups on C(4) and C(6) of the glucose and boric acid in P1 and P2, respectively [22,27]. Kennedy et al. proved that glucose is mainly complexed with ortho hydroxyl groups by ^{11}B NMR spectra [28]. This inference was further confirmed by ^{11}B MAS NMR spectra analysis. The ^{11}B chemical shift δ of the P1 and P2 samples were lower than that of the M·2B (Figure 2c). Because the electronegativity of the B atoms was lower than that of the H atoms and C atoms, the chemical shift of the ^{13}C moved to the high field due to the inductive effect, resulting in a decrease in the ^{13}C chemical shift. On the contrary, the chemical shift of the ^{11}B moved to the low field due to the inductive effect, resulting in an increases in the ^{11}B chemical shift. These results indicate that the boric acid could not only interact with the melamine to form a M·2B supramolecule, but could also interact with the glucose to form a boron ester. Therefore, the glucose molecules could combine with the boric acid and melamine through chemical bonding in P1 and P2 with large intermolecular distances, weak intermolecular coupling, random arrangement, and distortion, which are favorable for C element doping into the BN lattice to form p-BCN. Moreover, thermogravimetric (TG) analysis performed in N_2 confirmed that the temperatures of the endothermic peaks of P1 (150 °C) and P2 (167 °C) were different due to the difference in the hydrogen bonds of P1 and P2 in comparison with the M·2B (Figure S3) [29].

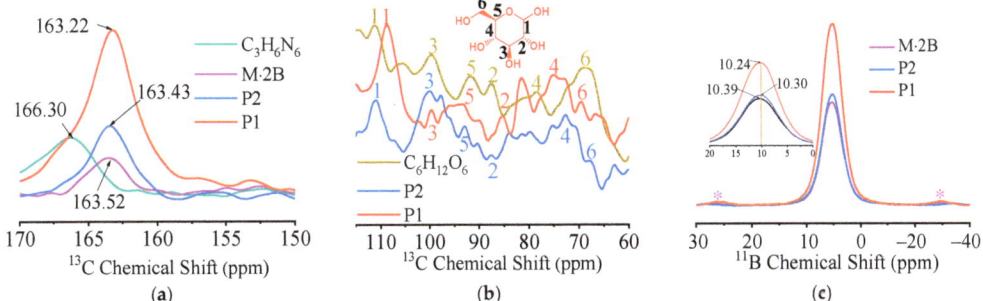

Figure 2. (a,b) ^{13}C Solid-state MAS NMR spectra of M·2B, P1, and P2. Arabic numerals in inset of (b) represent the location of carbon of glucose. (c) ^{11}B Solid-state MAS NMR spectra, * denotes spinning side band.

2.2. Characterization of p-BCN Photocatalysts

Figure 3a is a typical TEM image of the p-BCN-1 fibers with a diameter of 0.73 ± 0.40 μm (Figure S4a). It clearly discerns that the p-BCN-1 contained a high-density and uniform porous structure in sizes ranging from 10~20 nm (Figure S4b), which was similar to that of the p-BN (Figure S5a,b). The high-resolution TEM image (Figure 3b) presents the lattice fringes with a measured interspacing of 0.35 nm, accompanied by the selected area electron diffraction (SAED) pattern of the p-BCN-1 (inset of Figure 3a). It was found that the scarcely observed diffraction patterns suggested the poor crystallinity of the p-BCN-1. These results indicate that the p-BCN-1 had both a turbostratic and amorphous structure [12,25]. The p-BCN-2 displayed a non-uniform pore structure (Figure 3c) which was probably formed via bubbles blown by the gas released from the decomposition of the P2 precursor during calcination. Notably, the hexagonal phase (002) planes with a broader lattice fringe of 0.37 nm and the cubic phase (111) planes with a lattice fringe of 0.22 nm are clearly seen in Figure 3d, respectively [30]. It can be inferred that the composition and structure of P1 and P2 played an important role in the microstructure of the p-BCN. The p-BCN-1 and p-BCN-2 basically maintained the high porosity and amorphous structure of the p-BN, which facilitated the reactant diffusion and accommodated the linkage of the reactant transport channels to the catalytic active sites [31].

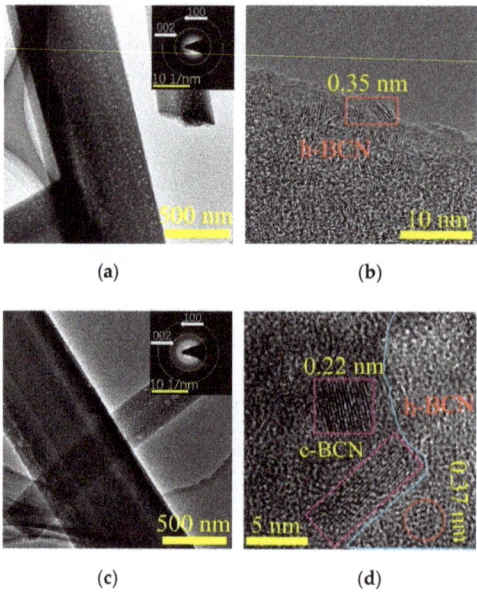

Figure 3. (a) TEM and (b) HRTEM images of p-BCN-1, inset of (a) selected area electron diffraction (scale bar: 500 nm and 10 nm). (c) TEM and (d) HRTEM image of p-BCN-2 (scale bar: 500 nm and 5 nm).

A N_2 adsorption/desorption isotherm and the corresponding pore size distribution of the samples were performed, and the results are summarized in Table S4. It was clearly found that the p-BN, p-BCN-1, and p-BCN-2 all showed a type IV isotherm with an H4 hysteresis loop (Figure 4a). The rapid growth at a low p/p_0 and the typical hysteresis loop observed at a higher p/p_0 region proved the existence of micropores and mesopores [25,32]. It is noteworthy that the p-BCN-1 and p-BCN-2 exhibited a high Brunauer–Emmett–Teller (BET) surface area of 918 $m^2 \cdot g^{-1}$ and 730 $m^2 \cdot g^{-1}$, respectively, which can provide abundant active adsorption sites [33]. Furthermore, the characteristic pore sizes of the p-BCN-1 and p-BCN-2 were ca. 22 nm, accompanied by a decrease in the amount of micropores at ca. 2 nm (Figure 4b). Therefore, the morphology of the p-BCN slightly changed in comparison with that of the p-BN. However, their pore size distributions were changed significantly, which further shows that the chemistry of the P1 and P2 precursors had an effect on the structure of the p-BCN.

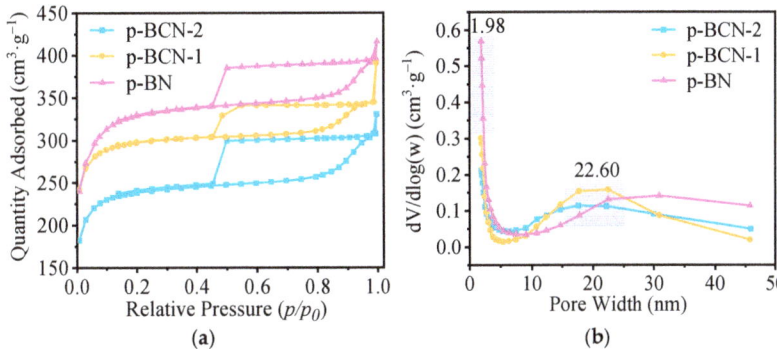

Figure 4. (a) N$_2$ adsorption–desorption isotherms. (b) The corresponding pore size distributions calculated by the BJH method of samples.

The X-ray diffraction (XRD) patterns of the samples are shown in Figure 5a. The diffraction peaks of the p-BN located at 25.6° and 42.9° can be ascribed to the (002) and (100) planes of the h-BN (PDF JCPDS No 34-0421), respectively. The p-BCN-1, obtained by P1 being calcinated at 900 °C, possessed the same crystal structure as the p-BN (Figure S6). The 2θ values of the (002) and (100) planes of the p-BCN-1 correspond to those of the p-BN, indicating that the basic characteristic structure of the h-BN was well maintained in the p-BCN-1 materials. The p-BCN-2 sample displayed a lower 2θ value of the (002) plane and a higher interlayer distance than those of the p-BN. These results confirmed that the precursor reconfiguration with the glucose had an effect on the crystal structure of the p-BCN. Figure 5b shows the Raman spectra of the p-BN, p-BCN-1, and p-BCN-2. Only one single peak at 1372 cm^{-1} was observed in the p-BN, assigned to the E_{2g} mode vibration of the h-BN [34,35]. Two strong peaks centered at 1347 and 1595 cm^{-1} in the p-BCN-1 were observed. Because the E_{2g} mode of the h-BN and the D band of the carbon were very close, we tended to think that the peak at 1347 cm^{-1} was the overlap of the above two peaks, and the peak at 1595 cm^{-1} can be ascribed to the G band of the carbon materials [36,37]. A red shift of the G band (1595 cm^{-1}) in the p-BCN-1 was observed compared with the pure graphene (1580 cm^{-1}), which originated from the structural distortion of the graphitic carbon with different bond lengths of B−N, C−B, and C−N [33]. The Raman peak located at 1855 cm^{-1} was related to the coalescence-inducing mode (CIM) vibration of the linear carbon chains, which could be detected in the carbon tube materials induced by the boron atoms [38]. The Raman spectrum of the p-BCN-1 also showed two bands at 2125 and 2283 cm^{-1}, corresponding to the C−C and C−N symmetric stretching modes [39]. These results show that the C atoms were incorporated within the h-BN network. The peaks at 1595, 1855, 2125, and 2283 cm^{-1} in the p-BCN-1 were the same as those in the p-BCN-2. However, two weak peaks around 1083 cm^{-1} and 1306 cm^{-1} in the p-BCN-2 may be ascribed to the cubic BN (c-BN) [40], inferring that in addition to the hexagonal phase, a new cubic phase emerged in the p-BCN-2 due to the reaction of the boric acid with the terminal carbon C(4) and C(6) hydroxyl groups of the glucose in P2.

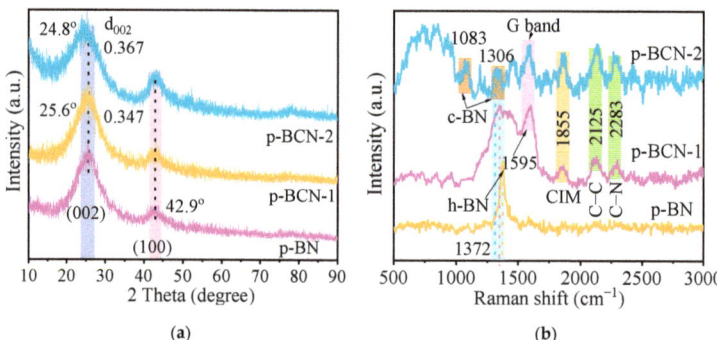

Figure 5. (a) Powder XRD patterns and (b) Raman spectra of p-BN, p-BCN-1, and p-BCN-2.

The FTIR spectra of the p-BN, p-BCN-1, and p-BCN-2 samples are shown in Figure 6. The FTIR spectra confirmed that the intrinsic structure of the BN could not be formed in the p-BCN-1 until the calcination temperature of P1 reached 900 °C (Figure S7). Compared with that of the p-BN, the position of the in-plane stretching band of the p-BCN-1 shifted from 792 cm^{-1} to 799 cm^{-1}, while the out-of-plane B−N bending band shifted from 1380 cm^{-1} to 1388 cm^{-1}, respectively (Figure 6a). It may be the disruption of the B−N−B bond by the C element due to the conjugative effect of the C−N−B bond [33,41] which lead to a vibration at a higher wavenumber. The position of the in-plane B−N stretching band of the p-BN was consistent with that of the p-BCN-2, while the peak assigned to the out-of-plane B−N bending band at 792 cm^{-1} for the p-BN blue shifted to 772 cm^{-1} for the p-BCN-2 (Figure 6b). Due to the typical overlap of the C–N with the B–N bands around 1100~1300 cm^{-1}, the peaks assigned to the C–N bonds faded. The FTIR spectra of the p-BCN-2 also exhibited absorption at 2850 cm^{-1} and 2922 cm^{-1}, which was assigned to C–H stretching vibrations [42], revealing the presence of an amorphous hydrogenated carbon (α-C:H) in the p-BCN-2. Those results probably led to the vibration of the in-plane B−N stretching bands at a lower wavenumber [43]. The possible chemical composition of the p-BCN-2 is illustrated in the inset of Figure 6b. Compared with the p-BCN-2, the sp^3 amorphous hydrogenated carbon band assigned to the C–H stretching vibrations was not found in the p-BCN-1 (Figure S7). The sp^2 hybridization of the carbon was in the p-BCN-1 while the sp^3 was in the p-BCN-2. The broad bands of the samples ranging from 3100 cm^{-1} to 3450 cm^{-1} could be assigned to B−OH and B−NH$_2$, indicating that there were the same B−OH and B−NH$_2$ active groups in the p-BCN-1 and the p-BCN-2 as the p-BN.

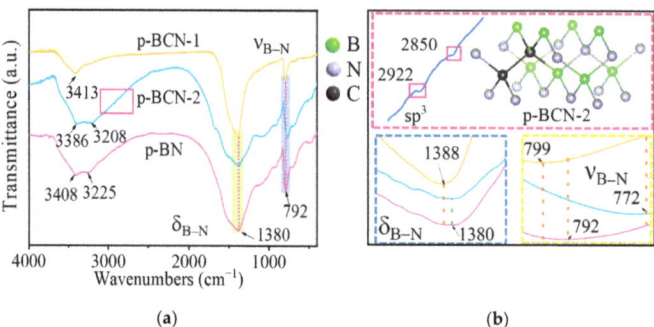

Figure 6. (a) FTIR spectra of p-BN, p-BCN-1, and p-BCN-2. (b) An enlarged view of the FTIR spectra of p-BCN-2 in the range of 3000~2400 cm^{-1} (rose rectangle), an enlarged view of the FTIR spectra of the light green rectangle in the range of 1500~1300 cm^{-1}, and the light purple rectangle in the range of 820~760 cm^{-1} in (a).

The XPS spectra are shown in Figure 7. The B 1s spectra show, for both samples, two components centered at 189.9 eV and 190.8 eV, assigned to the sp^2 of the B–C–N and B–N bonds (Figure 7a,b) [44]. The shoulder peak located at 192.1 eV corresponds to the edges or interfacial B atoms dangling bonds linked with –OH [45]. The deconvolution of the C 1s peaks shows that the p-BCN-1 exhibited a stronger preference of carbon to form new bonds with the B atoms than the p-BCN-2, as clearly observed from the intensity of the peaks at 284.1 eV assigned to C–B (Figure 7c,d) [8,11]. However, compared with the p-BCN-1 and p-BCN-2, for the deconvoluted spectra of the B 1s and C 1s, there were no B–C peaks in the p-BN (Figure S8a,b). Meanwhile, the presence of the B–C–N bonds suggests that hybridized atomic layers were formed in the p-BCN-1 and p-BCN-2. The relative atomic percentage calculated by the XPS data indicated that the contents of the B–N bonds were considerably higher than those of the B–C and C–N bonds (Table S5), indicating that the p-BCN-1 and p-BCN-2 retained the main structure of the p-BN.

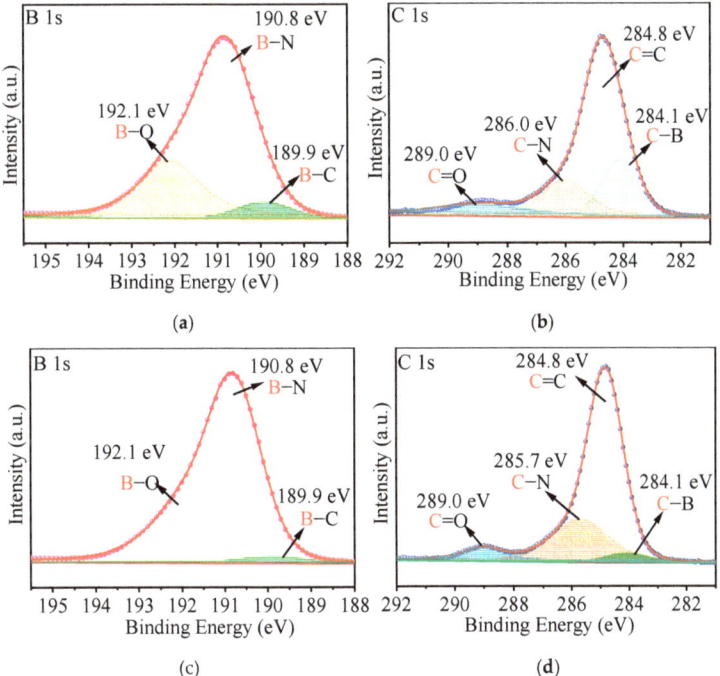

Figure 7. XPS high-resolved spectra of (**a**) B 1s and (**b**) C 1s of p-BCN-1 and (**c**) B 1s and (**d**) C 1s of p-BCN-2, respectively.

Solid-state MAS NMR spectroscopy was employed to obtain a closer insight into the chemical environment of the p-BN, p-BCN-1, and p-BCN-2 (Figure 8). As displayed in Figure 8a, the ^{1}H MAS NMR spectrum of the p-BN consisted of two distinct peaks centered at 6.9 and 4.1 ppm, which could be tentatively ascribed to –OH and –NH$_2$ groups, respectively [46]. After the precursor reconfiguration with glucose, the ^{1}H MAS NMR spectrum of the as-prepared p-BCN-1 showed three signal peaks; two peaks appeared at 6.7 and 4.9 ppm along with a small peak at 3.7 ppm. The signal at 3.7 ppm can be assigned to –NH$_2$ groups. The signals at 6.7 and 4.9 ppm, similar to those of the p-BN, can be attributed to –OH and –NH$_2$ groups due to the inductive effect stemming from the electronegativity of the C element [8]. When the N atoms in the BN were replaced by the C atoms, there were less electrons around the C–B–OH (6.7 ppm) bonds compared with those of the N–B–OH bonds in the BN. Similarly, after the B atoms in the BN were replaced by the C atoms, the electrons around C–NH$_2$ (4.9 ppm) were much richer than the B–NH$_2$ bonds in the BN. The only one peak that appeared at 6.9 ppm in the p-BCN-2 was assigned to the –OH group. Figure 8b shows the ^{11}B MAS NMR spectra of the samples. There was a single main signal at 17.7 ppm assigned to B in trigonal coordination (turbostratic B$_3$N$_3$) due to the B element's nature of *sp^2* hybridization in the p-BN, as already reported by Marchett's group [47]. In addition, a small resonance at ca. −4.0 ppm was detected, indicating the presence of a small quantity of 4-coordinate boron (BO$_4$) in the p-BN [47,48]. Compared with the p-BN, the ^{11}B position of the p-BCN-1 shifted slightly (0.2 ppm), which was probably because a small amount of C atoms reduced the chemical shielding of the boron atoms in the 3-coordinate boron compounds. In other words, some N atoms in the BN domains were partly substituted by C atoms, leading to a change in the chemical environment of the p-BCN-1. However, the ^{11}B spectrum of the p-BCN-2 showed mainly two signals at 16.6 ppm and 1.5 ppm, which correspond to the B in the h-BN and c-BN, respectively. This result indicates that both h-BN and c-BN were found in the p-BCN-2. The ^{13}C MAS NMR spectra of the samples are displayed in Figure 8c. The peak at ca. 158 ppm was assigned to the *sp^2* C–N bonds, which originated from C$_3$N$_3$ rings with a similar electron density [49] and were also detected in the p-BCN-1 and p-BCN-2. This may have been because there were still parts of the carbon atoms from the melamine remaining in the samples [50]. The new peaks centered at 187.9 and 176.8 ppm were detected in the p-BCN-1 and p-BCN-2, respectively, which could account for the *sp^2* B–C bonds [51]. The presence of the h-BN and c-BN phases in the p-BCN-2 can be connected with the difference in the chemical shift of the B–C bonds between the p-BCN-1 and p-BCN-2. It was also further confirmed that the C atoms were doped into the BN for the p-BCN, providing insights into the chemical bonding scheme within the two p-BCN materials. According to the above results, an illustration of the as-synthesized p-BCN-1 and p-BCN-2 is exhibited in Figure 8d.

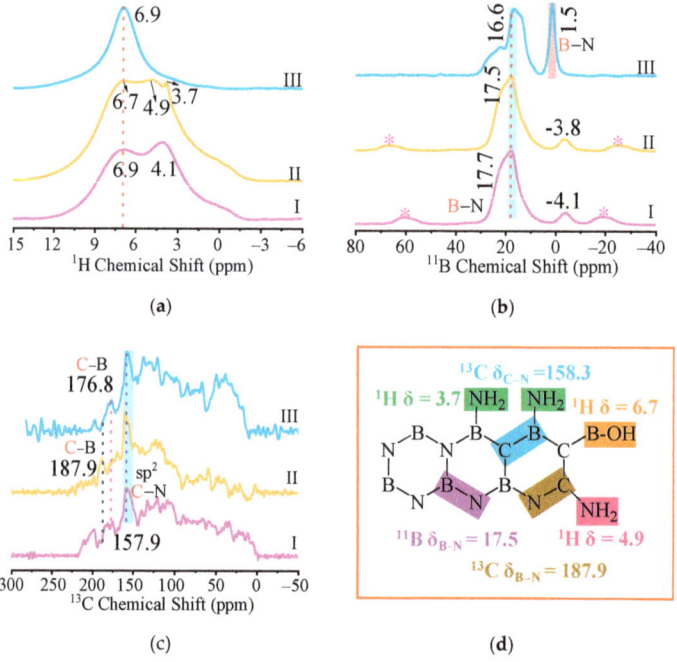

Figure 8. Solid-state MAS NMR spectra of p-BN (I), p-BCN-1 (II), and p-BCN-2 (III) samples: (**a**) ^{1}H MAS NMR spectra; (**b**) ^{11}B MAS NMR spectra. * denotes spinning side band; (**c**) ^{13}C MAS NMR spectra; and (**d**) the idealized structure determined by analysis of p-BCN-1.

XPS and NMR have often been used to validate the C doping of BCN, but the location of the C species on the plane has been vaguely described [8]. Herein, electron energy loss spectroscopy (EELS) was employed to probe the location of the C species in the p-BCN (Figure 9) since the light elements (B, C, N, and O) in p-BCN can be detected by EELS with the same resolution as TEM. The EELS of the p-BCN-1 obviously exhibited four distinct absorption peaks located at ~195, 295, 414, and 533 eV, which corresponds to the K-shell ionization edges of boron, carbon, nitrogen, and oxygen elements, respectively (Figure 9a). The sharp doublet of the 1s–π^* and 1s–σ^* K edges of the B and N are in accordance with those of the h-BN, which are characteristic peaks for the sp^2 hybridized B−N bonds, confirming that the intrinsic hexagonal structure of the h-BN still remained in the p-BCN-1 [8,52]. The C–K peak of the p-BCN-1 at ~295 eV in areas one, two, and three were observed, and the peak intensity along the edges of the p-BCN-1 was weak. However, the closer the area was to the center, the stronger the C–K peak intensity was. In addition, the signal C–K (π^*) and C–K (σ^*) also demonstrated rather perfect sp^2 bonding with the B and N for the C positions [8], confirming the introduction of C dopants into the BN lattice. The EELS of the p-BCN-2 obviously exhibited three distinct absorption peaks located at ~195, 295, and 414 eV, which correspond to the K-shell ionization edges of the boron, carbon, and nitrogen elements, respectively, and no O–K peak was detected (Figure 9b). The C–K peak of the p-BCN-2 at ~295 eV only in areas one and two were observed. The C–K peak intensity along the edges of the p-BCN-2 was strong. However, the closer the area was to the center, the weaker the C–K peak intensity was. The signal C–K (σ^*) demonstrated an sp^3 hybridization with the B and N for the C positions, which is consistent with the FTIR analysis. The 1s–π^* and 1s–σ^* peaks corresponding to the B–K edge and N–K edge in area three confirmed that the intrinsic hexagonal structure of the h-BN still remained in the p-BCN-2. Based on the results, the structures of the p-BCN-1 and p-BCN-2 changed dramatically, which is attributed to the unique structures of the P1 and P2 precursors.

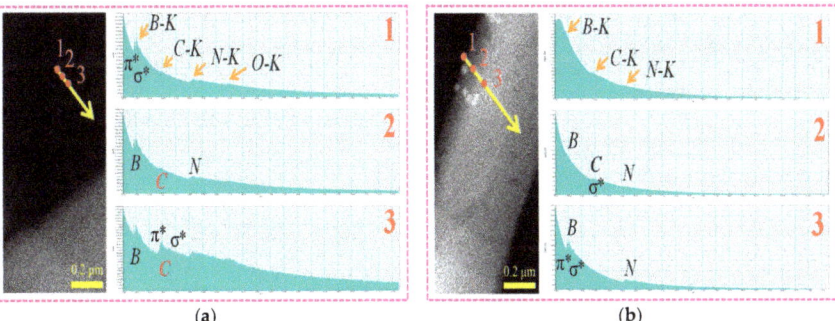

Figure 9. EELS mappings of (**a**) p-BCN-1 and (**b**) p-BCN-2. The Arabic digits on the yellow arrow represent the location of the tested sample, and the Arabic digits on the spectrograms represent the EELS mappings determined at the corresponding location of samples.

The UV–vis diffuse reflectance spectrum (DRS) of the as-prepared p-BN, p-BCN-1, and p-BCN-2 are shown in Figure 10. The obvious color variation of these samples corresponds to the absorbance curves (inset of Figure 10). It was observed that the p-BN sample was white and so nearly 100% transmittance throughout visible spectra occurred. The p-BCN-1 was dark gray and had enhanced absorption throughout the visible spectra by the modification of carbon atoms (Figure S9). The color of the p-BCN-2 was a mixture of white and dark grey; the absorption intensity was only slightly broadened compared with that of the p-BN. The absorption peak at 212 nm is attributed to the band gap transition absorption of the h-BN phase in the p-BN, which disappeared in the p-BCN-1 and p-BCN-2, indicating that bonding occurred among the B, C, and N atoms. Compared with the peak at 252 nm in the p-BN, the broad absorption peak at 243 nm in the p-BCN-1 and at 259 nm in the p-BCN-2 were observed, which can be attributed to the resonant exciton effects due to the π-π* transition that occurred for the samples [42]. Moreover, a new absorption peak appeared at 330 nm in the p-BCN-1, which is associated with defects such as vacancies and impurities [53].

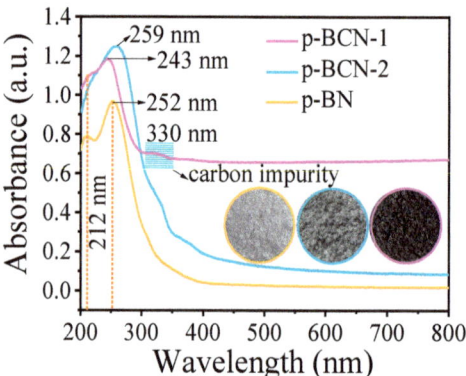

Figure 10. UV–vis diffuse reflectance spectra of p-BN, p-BCN-1, and p-BCN-2 samples.

As calculated by the Tauc plot method from the DRS spectra and the VB values (Figure S10), the energy levels (Figure S11) show that the bandgaps of the p-BCN-1 and p-BCN-2 narrowed with the addition of the C atoms from 3.57 eV of the p-BN to 2.97 eV and 3.28 eV. Considering that the electronegativity of the C element was lower than the N element but higher than the B element, the dopant atoms underwent a charge transfer with the atoms through the B–C and N–C bonds, which in turn altered their electronic structure [54,55]. Furthermore, as shown in Figure 11a, the higher transient photocurrent response was observed for the p-BCN-1 and p-BCN-2 in contrast to the p-BN. This result demonstrated the more efficient separation of the photogenerated electron holes at the p-BCN interface, owing to the donation of the carbon doping in the p-BCN [7]. The electrochemical impedance spectroscopy (EIS) in Figure 11b displayed that the p-BCN-1 and p-BCN-2 composites exhibited a much smaller diameter of the semicircular Nyquist plots compared with that of the p-BN, revealing that the decreased electron transfer resistance in the p-BCN-1 and p-BCN-2 could effectively promote the interface charge transport compared with the p-BN.

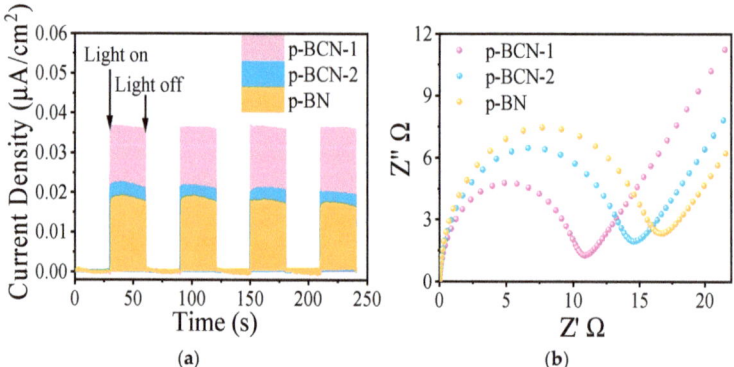

Figure 11. (a) Transient photocurrent density responses and (b) EIS Nyquist plots of p-BN, p-BCN-1, and p-BCN-2 samples.

2.3. Photocatalytic CO_2 Reduction Performance of p-BCN

The photocatalytic CO_2 reduction performances of the p-BN, p-BCN-1, and p-BCN-2 were studied as displayed in Figures 12 and S12. The p-BN sample exhibited a very poor photocatalytic performance of a CO_2 reduction reaction with a H_2 generation rate of 0.78 $\mu mol \cdot g^{-1} \cdot h^{-1}$, and H_2 was the only product. However, for the p-BCN-1, the products included not only 69.4% CO but also 21.4% H_2 and 9.2% CH_4, and their generation rates were 63.1 $\mu mol \cdot g^{-1} \cdot h^{-1}$, 19.4 $\mu mol \cdot g^{-1} \cdot h^{-1}$, and 8.4 $\mu mol \cdot g^{-1} \cdot h^{-1}$, respectively. Remarkably, CO instead of H_2 became the main product of the CO_2 conversion. The selectivity of the CO, CH_4, and H_2 was 54.33%, 28.97%, and 16.70%, respectively, which means that CO_2 was more likely to reduce into CO than CH_4 and H_2. With the coexistence of the two-phase structure in the p-BCN-2, the products of the CO_2 reduction reaction were 94.4% CO and 5.60% CH_4 over the p-BCN-2, and their generation rates were 42.6 and 2.52 $\mu mol \cdot g^{-1} \cdot h^{-1}$, respectively. Moreover, no H_2 products were observed. The selectivity of the p-BCN-2 for CO was improved significantly (80.86% for p-BCN-2 and 54.33% for p-BCN-1, Figure 12), indicating that the p-BCN-2 had a higher catalytic selectivity for CO_2 reduction to CO in comparison with the p-BCN-1. Meanwhile, as shown in Table 1, the photocatalytic CO_2 reduction performances of the p-BCN-1 and p-BCN-2 were significantly higher than those of equivalent materials or other noble metals-loaded nanohybrids reported in the literature. These results indicate that the formation of p-BCN-1 with a hexagonal (h) type can significantly promote CO_2 conversion, while the formation of p-BCN-2 with an h type and cubic (c) type can not only improve CO_2 conversion, but also remarkably improve the CO selectivity.

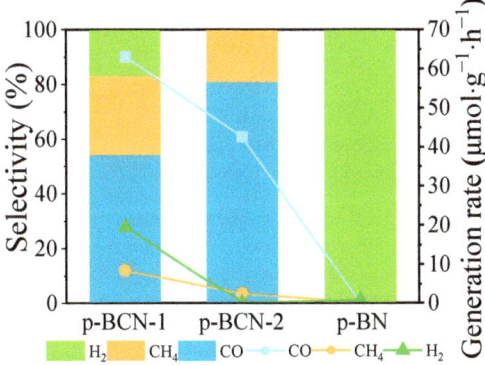

Figure 12. Catalytic performance of p-BN and p-BCN photocatalyzed CO_2 conversion.

Table 1. Photocatalytic CO_2 reduction performance of different photocatalysts in the presence of H_2O.

Photocatalysts	Formation Rate ($\mu mol \cdot g^{-1} \cdot h^{-1}$)		
	CO	CH_4	H_2
BN [56]	0	-	0.7
O/BN [56]	12.5	-	3.3
g-C_3N_4 [57]	0.1	0.07	1.0
TiO_2 [58]	1.2	0.38	2.1
Pt-TiO_2 [58]	1.1	5.2	33
Pd-TiO_2 [58]	1.1	4.3	25
Rh-TiO_2 [58]	0.62	3.5	18
Au-TiO_2 [58]	1.5	3.1	20
Ag-TiO_2 [58]	1.7	2.1	16
Porous BN [59]	1.17	-	-
p-BN [a]	0	0	0.78
p-BCN-1 [a]	63.1	8.4	19.4
p-BCN-2 [a]	42.6	2.52	-

[a] this work.

Based on the above characterization and analysis results, as shown in Figure 13, p-BCN with different crystal structures can be prepared by regulating the recrystallization conditions of the homogeneous precursor solution. Condition (I): The P1 precursor was prepared when boric acid reacted with both melamine and the hydroxyl groups on the C(1) and C(2) of glucose according to cooling in bath ice. The resultant p-BCN-1 sample with a homogeneous h-type crystal was successfully developed by the calcination of P1. Condition (II): The P2 precursor was prepared when boric acid reacted with both melamine and the hydroxyl groups on the C(4) and C(6) of glucose according to natural cooling. A p-BCN-2 sample with an h-type and c-type coexistent crystal was developed by the calcination of P2. Furthermore, the differences between P1 and P2 were found to be directly correlated with the photocatalytic performance of the p-BCN.

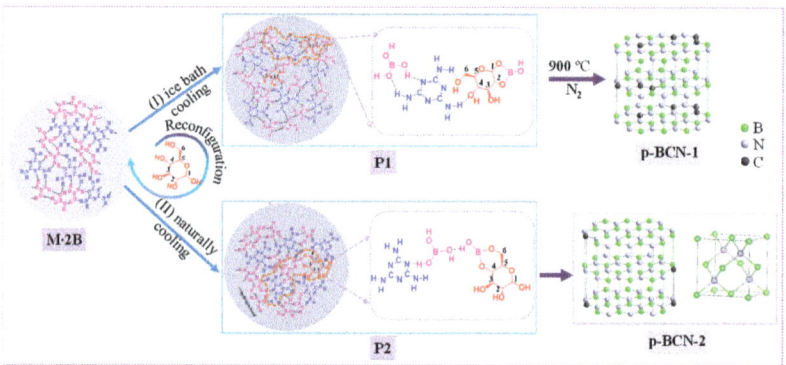

Figure 13. Schematic illustration of p-BCN with different crystal forms prepared by regulating recrystallization conditions of the homogeneous precursor solution: (**I**) P1-induced h-type p-BCN formation; (**II**) P2-induced h-type and c-type p-BCN formation.

A Strong Electron paramagnetic resonance (EPR) signal was detected at a magnetic field of ~3503 G with a calculated g-value of 2.0033 in the p-BCN-1 and 2.0034 in the p-BCN-2 (Figure 14). This implies the presence of free electrons along with the substitution of B or N with C atoms in a BN lattice in the p-BCN and an increase in the density state of conduction band electrons from the electron donation by C atoms [7,60]. The N atom had one extra electron in comparison with the substituted C atom, which resulted in the presence of unpaired electrons after the in-plane bonding with the B atoms in the p-BCN. The free electrons promoted electron separation to improve photocatalytic CO_2 reduction activity. As a comparison, no such resonance peak was observed for the p-BN. In addition, the difference in the photocatalytic activity and selectivity between the p-BCN-1 and p-BCN-2 under simulated solar irradiation depends on the energy band structure [61]. It was reported that the required energy for the conversion of CO_2 into CO, CH_4, and H_2 production is −0.52 eV, −0.24 eV, and −0.41 eV (vs. NHE), respectively [62,63]. The main factors influencing the selectivity of photocatalytic CO_2 reduction reactions include light-excitation attributes, band structures, charge separation efficiencies, and so on. Here, the position of the CB of the p-BCN-1 (−1.65 eV) was more negative than that of the p-BCN-2 (−0.77 eV), resulting in the p-BCN-1 having a higher reduction potential to drive charge transfer. Moreover, the band gap of the p-BCN-1 (2.97 eV) was narrower than that of the p-BCN-2 (3.28 eV), making the p-BCN-1 absorb a wider wavelength of light and produce more photogenerated electrons. Additionally, more free electrons caused by C doping in the p-BCN-1 improved the charge separation efficiency. Therefore, the higher surface density of the photogenerated electrons in the p-BCN-1 was more conducive to the occurrence of CO_2 reduction reactions, producing higher reduced state products. When the p-BCN-1 and p-BCN-2 photocatalysts were excited by the same incident light, on one hand, the faster and more photogenerated electrons transfer made a more complete reduction reaction in the p-BCN-1, while on the other hand, the greater energy level difference between the CB of the p-BCN-1 and the reduction potential of the reaction product (CO, CH$_4$, and H$_2$) provided a stronger driving force for the charge transfer between them. However, there were not enough electrons in the p-BCN-2 to reduce the H$^+$ to H$_2$ and to, in turn, improve the selectivity of the CO. As a result, photocatalytic performance tests indicated that the CO and CH$_4$ production yields of the p-BCN-1 were significantly higher than those of the p-BCN-2, but the selectivity of the CO of the p-BCN-1 was lower than that of the p-BCN-2. Furthermore, the p-BCN provided a high surface area for CO$_2$ adsorption and abundant active sites at the edges of the p-BCN for the reduction reaction. A possible photocatalytic CO$_2$ reduction mechanism of the p-BCN photocatalysts was proposed and is shown in Figure 14. "*" is the adsorption state at the surface of the photocatalysts. The surfaces of the catalysts adsorbed the CO$_2$ molecules. Firstly, the p-BCN-1 effectively absorbed the UV–vis light and generated electron holes. Secondly, the holes on the VB of the p-BCN-1 oxidized H$_2$O to generate OH and protons (H$^+$). Thirdly, CO$_2$* reacted with photogenerated electrons and H$^+$ to reduce into CO*. Fourthly, CO* combined with photogenerated electrons and H$^+$ to form CH$_4$. Lastly, the photogenerated electrons and H$^+$ were further reduced to H$_2$. The process of the photocatalytic reduction of the CO$_2$ of the p-BCN-2 is similar to that of the p-BCN-1. However, there were not enough photogenerated electrons in the p-BCN-2 to reduce H$^+$ to H$_2$. Therefore, H$^+$ could not be reduced to H$_2$. Furthermore, the catalytic performance and selectivity of the CO$_2$ reduction of the p-BCN were also correlated with its phase structure, which will be the focus of our future research.

$$\text{p-BCN-1} + h\nu \longrightarrow h^+ + e^- \tag{1-1}$$

$$H_2O + h^+ \longrightarrow H^+ + OH^- \tag{1-2}$$

$$CO_2^* + H^+ + e^- \longrightarrow COOH^* \longrightarrow CO^* + OH^* \tag{1-3}$$

$$CO^* + H^+ + e^- \longrightarrow CH_4 \tag{1-4}$$

$$H^* + H^+ + e^- \longrightarrow H_2 \tag{1-5}$$

$$\text{p-BCN-2} + h\nu \longrightarrow h^+ + e^- \tag{2-1}$$

$$H_2O + h^+ \longrightarrow H^+ + OH^- \tag{2-2}$$

$$CO_2^* + H^+ + e^- \longrightarrow COOH^* \longrightarrow CO^* + OH^* \tag{2-3}$$

$$CO^* + H^+ + e^- \longrightarrow CH_4 \tag{2-4}$$

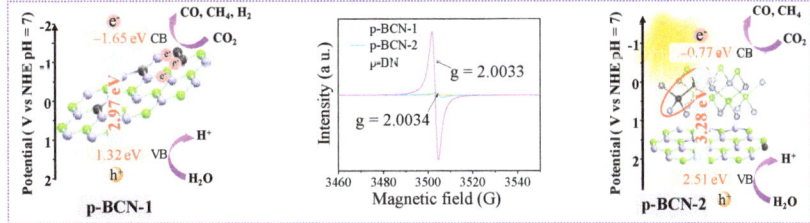

Figure 14. The possible scheme of p-BCN-1 and p-BCN-2 for photoreduction of CO$_2$ under simulated solar irradiation.

3. Materials and Methods

3.1. Materials

Melamine (M), boric acid (B), and D-glucose (G) were obtained from Shanghai Aladdin Biochemical Technology Co., Ltd., Shanghai, China. All chemicals were used as received without further purification. Deionized water was used in the experiment.

3.2. Synthesis of C-Doped BN (p-BCN)

The p-BCN samples were prepared by a two-step-synthesis method using melamine, boric acid, and glucose as the source materials, which is based on the previous preparation method of p-BNNS [25]. Firstly, the solution was prepared at 95 °C upon mixing 18.92 g of melamine (M, N source), 18.55 g of boric acid (B, B source), and 13.5 g of D-glucose (G, C source) in 1000 mL deionized water. Subsequently, the transparent and homogeneous solution was heated at 90 °C for 6 h. The mixture precursor solution was labeled as M·2B·G. The solid precursor named P1 was obtained by the prompt recrystallization from the homogeneous solution in an ice bath (Figures S13 and S14a) then was filtered and dried at 70 °C for 24 h. Secondly, P1 was pyrolyzed at 300 °C for 1 h, 550 °C for 1 h, and 900 °C for 2 h under a N_2 atmosphere (a heat rate of 2 °C·min^{-1}) to obtain a sample named p-BCN-1. Meanwhile, the solid precursor named P2 was obtained by recrystallization from the homogeneous solution via natural cooling to room temperature (Figures S13 and S14b). The p-BCN-2 was prepared by calcining P2 under the same conditions. Figure 15 and Figure S13 describe the preparation process of the p-BCN-1 and p-BCN-2 samples by different cooling modes. As a comparison, when the glucose was removed, the p-BN sample was further prepared by calcining the M·2B precursor under the same condition as P2.

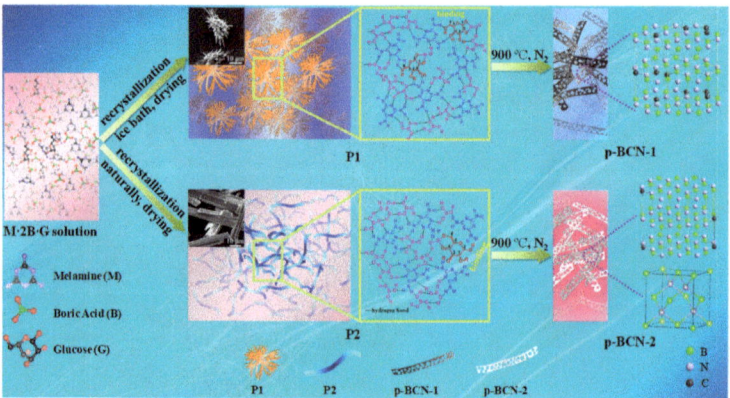

Figure 15. Schematic synthesis of p-BCN-1 and p-BCN-2 samples.

4. Conclusions

In summary, p-BCN photocatalysts were synthesized by the pyrolysis of new precursors prepared by the recrystallization of a homogeneous solution with M·2B and glucose. This experimental study suggests that the two precursors were obtained from the reactions between boric acid and specific hydroxyl groups (on anomeric carbons C(1) and C(2) or on terminal carbons C(4) and C(6)) of glucose during the recrystallization of the homogeneous solution. The chemical composition and structure of the precursors could be remarkably changed by adjusting the recrystallization conditions of the homogeneous solution, leading to the resultant p-BCN products with distinctly different structures and properties. The resultant p-BCN-1 sample with a homogeneous h-type crystal and a p-BCN-2 sample with an h-type and c-type crystal exhibited excellent photocatalytic activity and high selectivity for CO_2 conversion, respectively, accompanying with the changes in the structure and properties of the p-BCN, such as narrowing the band gap, broadening the range of visible light absorption, promoting the interface charge transport, and efficiently separating photogenerated electron holes. This work provides a novel strategy to modulate precursors to design and develop BN-based materials with superior performance and to broaden the potential applications of BN-based materials.

Supplementary Materials: The following supporting information can be downloaded at: https://www.mdpi.com/article/10.3390/catal12050555/s1, Experimental details for the apparatus and characterization, and photocatalytic performance test for CO2 reduction. Table S1. Crystal data and structure refinement for M·2B; Figure S1. Arrangements of melamine and boric acid molecules in the crystal of M·2B; Figure S2. (a) Raman and FTIR spectra of M·2B (black line), P1 (red line), and P2 (blue line): (b) in the 500~4000 cm^{-1} region: (c) in the 3000~3600 cm^{-1} region; Table S2. Spectral data (cm^{-1}) and band assignments of M·2B, P1 and P2; Table S3. Chemical shifts of 13C of glucose, P1, and P2; Figure S3. TGA thermograms of M·2B, P1 and P2 samples: (a) from 40 to 1000 °C in N2; (b) from 40 to 300 °C in N2; and (c) DSC curves of M·2B, P1, and P2 samples from 40 °C to 1000 °C in N2; Figure S4. (a) The SEM image of p-BCN-1, inset: the diameter distribution histogram. (b) HRTEM image of p-BCN-1; Figure S5. Morphologies of the p-BN: (a) TEM (scale bar: 500 nm); (b) HRTEM (scale bar: 10 nm); and (c) Measured the layer distance of the red rectangle of (b); Table S4. Physical and textural properties of p-BN, p-BCN-1, and p-BCN-2; Figure S6. Powder XRD patterns of calcined product p-BCN-1 of the P1 at different temperatures; Figure S7. FTIR spectra of the calcined product p-BCN-1 of P1 at different temperatures; Figure S8. XPS high-resolved spectra of p-BN, (a) B 1s and (b) C 1s; Table S5. The peak position and relative atomic percentage of various functional groups in p-BN, p-BCN-1, and p-BCN-2 samples; Figure S9. UV–vis diffuse reflectance spectra of the calcined product p-BCN-1 of P1 at different temperatures; Figure S10. (a) Plots of $\varepsilon 0.5/\lambda$ versus $1/\lambda$ based on the optical absorption data from p-BN, p-BCN-1, and p-BCN-2 samples, respectively. VB-XPS spectra of (b) p-BN, (c) p-BCN-1, and (d) p-BCN-2; Figure S11. Schematic illustration of the band structures of p-BN, p-BCN-1, and p-BCN-2 samples; Figure S12. Photocatalytic activities over the p-BN, p-BCN-1, and p-BCN-2 samples: (a) CO production, (b) H2 production, and (c) CH4 production; Figure S13. Photographic synthesis of the p-BCN samples by two different cooling modes; Figure S14. Cooling curves of (a) P1 and (b) P2. [18,20,33,61,64]

Author Contributions: Conceptualization, Q.L. (Qiong Lu), J.A., Y.D., Q.L. (Qingzhi Luo), C.T. and D.W.; data curation, Q.L. (Qiong Lu); investigation, Q.L. (Qiong Lu); methodology, Q.L. (Qiong Lu), J.A., Y.D., Q.L. (Qingzhi Luo), Y.S., C.T. and D.W.; resources, Q.L. (Qingzhi Luo), Q.L. (Qiunan Liu), Y.T., J.H. and R.Y.; writing—original draft, Q.L. (Qiong Lu); writing—review and editing, Q.L. (Qiong Lu), J.A., Y.D., C.T. and D.W. All authors have read and agreed to the published version of the manuscript.

Funding: This work was financially supported by the Natural Science Foundation of Hebei Province (No. B2021208007), the Science and Technology Research Project of Higher Education of Hebei Province (No. ZD2021321), the Science and Technology Project of Hebei Education Department (ZD2022122) and the S&T Program of Hebei (No. 199676242H).

Conflicts of Interest: The authors declare no conflict of interest.

References

1. Gong, Y.; Xu, Z.-Q.; Li, D.; Zhang, J.; Aharonovich, I.; Zhang, Y. Two-dimensional hexagonal boron nitride for building Next-generation energy-efficient devices. *ACS Energy Lett.* **2021**, *6*, 985–996. [CrossRef]
2. Kim, K.K.; Lee, H.S.; Lee, Y.H. Synthesis of hexagonal boron nitride heterostructures for 2D van der Waals electronics. *Chem. Soc. Rev.* **2018**, *47*, 6342–6369. [CrossRef] [PubMed]
3. Zhang, J.; Tan, B.; Zhang, X.; Gao, F.; Hu, Y.; Wang, L.; Duan, X.; Yang, Z.; Hu, P. Atomically thin hexagonal boron nitride and its heterostructures. *Adv. Mater.* **2020**, *33*, 2000769. [CrossRef] [PubMed]
4. Weng, Q.; Wang, X.; Wang, X.; Bando, Y.; Golberg, D. Functionalized hexagonal boron nitride nanomaterials: Emerging properties and applications. *Chem. Soc. Rev.* **2016**, *45*, 3989–4012. [CrossRef] [PubMed]
5. Zhou, C.; Lai, C.; Zhang, C.; Zeng, G.; Huang, D.; Cheng, M.; Hu, L.; Xiong, W.; Chen, M.; Wang, J.; et al. Semiconductor/boron nitride composites: Synthesis, properties, and photocatalysis applications. *Appl. Catal. B* **2018**, *238*, 6–18. [CrossRef]
6. Mendelson, N.; Chugh, D.; Reimers, J.R.; Cheng, T.S.; Gottscholl, A.; Long, H.; Mellor, C.J.; Zettl, A.; Dyakonov, V.; Beton, P.H.; et al. Identifying carbon as the source of visible single-photon emission from hexagonal boron nitride. *Nat. Mater.* **2020**, *20*, 321–328. [CrossRef]
7. Chen, L.; Zhou, M.; Luo, Z.; Wakeel, M.; Asiri, A.M.; Wang, X. Template-free synthesis of carbon-doped boron nitride nanosheets for enhanced photocatalytic hydrogen evolution. *Appl. Catal. B* **2019**, *241*, 246–255. [CrossRef]
8. Huang, C.; Chen, C.; Zhang, M.; Lin, L.; Ye, X.; Lin, S.; Antonietti, M.; Wang, X. Carbon-doped BN nanosheets for metal-free photoredox catalysis. *Nat. Commun.* **2015**, *6*, 7698. [CrossRef]
9. Zheng, M.; Shi, J.; Yuan, T.; Wang, X. Metal-free dehydrogenation of N-heterocycles by ternary h-BCN nanosheets with visible light. *Angew. Chem. Int. Ed.* **2018**, *57*, 5487–5491. [CrossRef]

10. Shi, J.; Yuan, T.; Zheng, M.; Wang, X. Metal-free heterogeneous semiconductor for visible-light photocatalytic decarboxylation of carboxylic acids. *ACS Catal.* **2021**, *11*, 3040–3047. [CrossRef]
11. Guo, F.; Yang, P.; Pan, Z.; Cao, X.-N.; Xie, Z.; Wang, X. Carbon-doped BN nanosheets for the oxidative dehydrogenation of ethylbenzene. *Angew. Chem. Int. Ed.* **2017**, *56*, 8231–8235. [CrossRef] [PubMed]
12. Zheng, M.; Yuan, T.; Shi, J.; Cai, W.; Wang, X. Photocatalytic oxygenation and deoxygenation transformations over BCN nanosheets. *ACS Catal.* **2019**, *9*, 8068–8072. [CrossRef]
13. Wang, B.; Anpo, M.; Lin, J.; Yang, C.; Zhang, Y.; Wang, X. Direct hydroxylation of benzene to phenol on h-BCN nanosheets in the presence of $FeCl_3$ and H_2O_2 under visible light. *Catal. Today* **2019**, *324*, 73–82. [CrossRef]
14. Zhou, M.; Wang, S.; Yang, P.; Huang, C.; Wang, X. Boron carbon nitride semiconductors decorated with CdS nanoparticles for photocatalytic reduction of CO_2. *ACS Catal.* **2018**, *8*, 4928–4936. [CrossRef]
15. Garay, A.L.; Pichon, A.; James, S.L. Solvent-free synthesis of metal complexes. *Chem. Soc. Rev.* **2007**, *36*, 846–855. [CrossRef]
16. Ambika, S.; Devasena, M.; Nambi, I.M. Synthesis, characterization and performance of high energy ball milled meso-scale zero valent iron in Fenton reaction. *J. Environ. Manag.* **2016**, *181*, 847–855. [CrossRef]
17. Blanco, M.C.; Cámara, J.; Gimeno, M.C.; Laguna, A.; James, S.L.; Lagunas, M.C.; Villacampa, M.D. Synthesis of gold-silver luminescent honeycomb aggregates by both solvent-based and solvent-free methods. *Angew. Chem. Int. Ed.* **2012**, *51*, 9777–9779. [CrossRef]
18. Wang, K.; Duan, D.; Wang, R.; Lin, A.; Cui, Q.; Liu, B.; Cui, T.; Zou, B.; Zhang, X.; Hu, J.; et al. Stability of hydrogen-bonded supramolecular architecture under high pressure conditions: Pressure-induced amorphization in melamine-boric acid adduct. *Langmuir* **2009**, *25*, 4787–4791. [CrossRef]
19. Kawsaki, T.; Kuroda, Y.; Nishikawa, H. The crystal structure of melamine diborate. *J. Ceram. Soc. Jpn.* **1996**, *104*, 935–938. [CrossRef]
20. Roy, A.; Choudhury, A.C.; Rao, N.R. Supramolecular hydrogen-bonded structure of a 1:2 adduct of melamine with boric acid. *J. Mol. Struct.* **2002**, *613*, 61–66. [CrossRef]
21. Atalay, Y.; Avcı, D.; Başoğlu, A.; Okur, İ. Molecular structure and vibrational spectra of melamine diborate by density functional theory and ab initio Hartree-Fock calculations. *J. Mol. Struc. THEOCHEM* **2005**, *713*, 21–26. [CrossRef]
22. Norrild, J.C.; Eggert, H. Evidence for mono-and bisdentate boronate complexes of glucose in the furanose form. Application of $^1J_{C-C}$ coupling constants as a structural probe. *J. Am. Chem. Soc.* **1995**, *117*, 1479–1484. [CrossRef]
23. Franco, A.; Ascenso, J.R.; Ilharco, L.; Diogo, H.P.; André, V.; da Silva, J.A.L. Ribose-borate esters as potential components for prebiological evolution. *J. Mol. Struct.* **2019**, *1184*, 281–288. [CrossRef]
24. Cordes, D.B.; Gamsey, S.; Singaram, B. Fluorescent quantum dots with boronic acid substituted viologens to sense glucose in aqueous solution. *Angew. Chem. Int. Ed.* **2006**, *45*, 3829–3832. [CrossRef]
25. Lu, Q.; An, J.; Duan, Y.; Luo, Q.; Yin, R.; Li, X.; Tang, C.; Wang, D. A strategy for preparing efficient Ag/p-BNNS nanocatalyst with a synergistic effect between Ag and p-BNNS. *J. Catal.* **2021**, *395*, 457–466. [CrossRef]
26. Panicker, C.Y.; Varghese, H.T.; John, A.; Philip, D.; Nogueira, H.I.S. Vibrational spectra of melamine diborate, $C_3N_6H_6 2H_3BO_3$. *Spectrochim. Acta A* **2002**, *58*, 1545–1551. [CrossRef]
27. Bagno, A.; Rastrelli, F.; Saielli, G. Prediction of the ^1H and ^{13}C NMR spectra of α-D-glucose in water by DFT methods and MD simulations. *J. Org. Chem.* **2007**, *72*, 7373–7381. [CrossRef]
28. Henderson, W.G.; How, M.J.; Kennedy, G.R.; Mooney, E.F. The interconversion of aqueous boron species and the interaction of borate with diols: A ^{11}B NMR study. *Carbohydr. Res.* **1973**, *28*, 1–12. [CrossRef]
29. Zhang, S.; Mandai, T.; Ueno, K.; Dokko, K.; Watanabe, M. Hydrogen-bonding supramolecular protic salt as an "all-in-one" precursor for nitrogen-doped mesoporous carbons for CO_2 adsorption. *Nano Energy* **2015**, *13*, 376–386. [CrossRef]
30. Pascallon, J.; Stambouli, V.; Ilias, S.; Bouchier, D.; Nouet, G.; Silva, F.; Gicquel, A. Deposition of c-BN films on diamond: Influence of the diamond roughness. *Mater. Sci. Eng. B* **1999**, *59*, 239–243. [CrossRef]
31. Taguchi, A.; Schüth, F. Ordered mesoporous materials in catalysis. *Microporous Mesoporous Mater.* **2005**, *77*, 1–45. [CrossRef]
32. Song, Q.; Fang, Y.; Liu, Z.; Li, L.; Wang, Y.; Liang, J.; Huang, Y.; Lin, J.; Hu, L.; Zhang, J.; et al. The performance of porous hexagonal BN in high adsorption capacity towards antibiotics pollutants from aqueous solution. *Chem. Eng. J.* **2017**, *325*, 71–79. [CrossRef]
33. Wang, J.; Hao, J.; Liu, D.; Qin, S.; Portehault, D.; Li, Y.; Chen, Y.; Lei, W. Porous boron carbon nitride nanosheets as efficient metal-free catalysts for the oxygen reduction reaction in both alkaline and acidic solutions. *ACS Energy Lett.* **2017**, *2*, 306–312. [CrossRef]
34. Weng, Q.H.; Wang, X.B.; Zhi, C.Y.; Bando, Y.; Golberg, D. Boron nitride porous microbelts for hydrogen storage. *ACS Nano* **2013**, *7*, 1558–1565. [CrossRef] [PubMed]
35. Wang, L.; Wu, B.; Liu, H.; Huang, L.; Li, Y.; Guo, W.; Chen, X.; Peng, P.; Fu, L.; Yang, Y.; et al. Water-assisted growth of large-sized single crystal hexagonal boron nitride grains. *Mater. Chem. Front.* **2017**, *1*, 1836–1840. [CrossRef]
36. Liu, A.C.Y.; Arenal, R.; Montagnac, G. In situ transmission electron microscopy observation of keV-ion irradiation of single-walled carbon and boron nitride nanotubes. *Carbon* **2013**, *62*, 248–255. [CrossRef]
37. Qin, L.; Yu, J.; Kuang, S.; Xiao, C.; Bai, X. Few-atomic-layered boron carbonitride nanosheets prepared by chemical vapor deposition. *Nanoscale* **2012**, *4*, 120–123. [CrossRef]

38. Endo, M.; Kim, Y.A.; Hayashi, T.; Muramatsu, H.; Terrones, M.; Saito, R.; Villalpando-Paez, F.; Chou, S.G.; Dresselhaus, M.S. Nanotube coalescence-inducing mode: A novel vibrational mode in carbon systems. *Small* **2006**, *2*, 1031–1036. [CrossRef]
39. Dargelos, A.; Karamanis, P.; Pouchan, C. Ab-initio calculations of the IR spectra of dicyanodiacetylene (C_6N_2) beyond the harmonic approximation. *Chem. Phys. Lett.* **2019**, *723*, 155–159. [CrossRef]
40. Yu, L.L.; Gao, B.; Chen, Z.; Sun, C.T.; Cui, D.L.; Wang, C.J.; Wang, Q.L.; Jiang, M.H. In situ FTIR investigation on phase transformations in BN nanoparticles. *Chin. Sci. Bull.* **2005**, *50*, 2827–2831.
41. Zeng, X.; Chen, H.; He, X.; Zhang, H.; Fang, W.; Du, X.; Li, W.; Huang, Z.; Zhao, L. In-situ synthesis of non-phase-separated boron carbon nitride for photocatalytic reduction of CO_2. *Environ. Res.* **2022**, *207*, 112178. [CrossRef] [PubMed]
42. López-Salas, N.; Ferrer, M.L.; Gutiérrez, M.C.; Fierro, J.L.G.; Cuadrado-Collados, C.; Gandara-Loe, J.; Silvestre-Albero, J.; del Monte, F. Hydrogen-bond supramolecular hydrogels as efficient precursors in the preparation of freestanding 3D carbonaceous architectures containing BCNO nanocrystals and exhibiting a high CO_2/CH_4 adsorption ratio. *Carbon* **2018**, *134*, 470–479. [CrossRef]
43. Lei, W.; Portehault, D.; Liu, D.; Qin, S.; Chen, Y. Porous boron nitride nanosheets for effective water cleaning. *Nat. Commun.* **2013**, *4*, 1777. [CrossRef] [PubMed]
44. Giusto, P.; Cruz, D.; Heil, T.; Tarakina, N.; Patrini, M.; Antonietti, M. Chemical vapor deposition of highly conjugated, transparent boron carbon nitride thin films. *Adv. Sci.* **2021**, *8*, 2101602. [CrossRef] [PubMed]
45. Weng, Q.; Zeng, L.; Chen, Z.; Han, Y.; Jiang, K.; Bando, Y.; Golberg, D. Hydrogen storage in carbon and oxygen Co-doped porous boron nitrides. *Adv. Funct. Mater.* **2020**, *31*, 2007381. [CrossRef]
46. Torii, S.; Jimura, K.; Hayashi, S.; Kikuchi, R.; Takagaki, A. Utilization of hexagonal boron nitride as a solid acid-base bifunctional catalyst. *J. Catal.* **2017**, *355*, 176–184. [CrossRef]
47. Marchetti, P.S.; Kwon, D.; Schmidt, W.R.; Interrante, L.V.; Maciel, G.E. High-field boron-11 magic-angle spinning NMR characterization of boron nitrides. *Chem. Mater.* **2002**, *3*, 482–486. [CrossRef]
48. Bawari, S.; Sharma, K.; Kalita, G.; Madhu, P.K.; Narayanan, T.N.; Mondal, J. Structural evolution of BCN systems from graphene oxide towards electrocatalytically active atomic layers. *Mater. Chem. Front.* **2020**, *4*, 2330–2338. [CrossRef]
49. Portehault, D.; Giordano, C.; Gervais, C.; Senkovska, I.; Kaskel, S.; Sanchez, C.; Antonietti, M. High-surface-area nanoporous boron carbon nitrides for hydrogen storage. *Adv. Funct. Mater.* **2010**, *20*, 1827–1833. [CrossRef]
50. Chen, S.; Li, P.; Xu, S.; Pan, X.; Fu, Q.; Bao, X. Carbon doping of hexagonal boron nitride porous materials toward CO_2 capture. *J. Mater. Chem. A* **2018**, *6*, 1832–1839. [CrossRef]
51. Wrackmeyer, B. Carbon-13 NMR spectroscopy of boron compounds. *Prog. Nucl. Magn. Reson. Spectrosc.* **1979**, *12*, 227–259. [CrossRef]
52. Tay, R.Y.; Li, H.; Tsang, S.H.; Zhu, M.; Loeblein, M.; Jing, L.; Leong, F.N.; Teo, E.H.T. Trimethylamine borane: A new single-source precursor for monolayer h-BN single crystals and h-BCN thin films. *Chem. Mater.* **2016**, *28*, 2180–2190. [CrossRef]
53. Liu, F.; Yu, J.; Ji, X.; Qian, M. Nanosheet-structured boron nitride spheres with a versatile adsorption capacity for water cleaning. *ACS Appl. Mater. Inter.* **2015**, *7*, 1824–1832. [CrossRef] [PubMed]
54. Liu, Q.; Chen, C.; Du, M.; Wu, Y.; Ren, C.; Ding, K.; Song, M.; Huang, C. Porous hexagonal boron nitride sheets: Effect of hydroxyl and secondary amino groups on photocatalytic hydrogen evolution. *ACS Appl. Nano Mater.* **2018**, *1*, 4566–4575. [CrossRef]
55. Bhattacharya, A.; Bhattacharya, S.; Das, G.P. Band gap engineering by functionalization of BN sheet. *Phys. Rev. B* **2012**, *85*, 035415. [CrossRef]
56. Cao, Y.; Zhang, R.; Zhou, T.; Jin, S.; Huang, J.; Ye, L.; Huang, Z.; Wang, F.; Zhou, Y. B–O bonds in ultrathin boron nitride nanosheets to promote photocatalytic carbon dioxide conversion. *ACS Appl. Mater. Interfaces* **2020**, *12*, 9935–9943. [CrossRef]
57. Zhao, L.; Ye, F.; Wang, D.; Cai, X.; Meng, C.; Xie, H.; Zhang, J.; Bai, S. Lattice engineering on metal cocatalysts for enhanced photocatalytic reduction of CO_2 into CH_4. *ChemSusChem* **2018**, *11*, 3524–3533. [CrossRef]
58. Xie, S.; Wang, Y.; Zhang, Q.; Deng, W.; Wang, Y. MgO- and Pt-promoted TiO_2 as an efficient photocatalyst for the preferential reduction of carbon dioxide in the presence of water. *ACS Catal.* **2014**, *4*, 3644–3653. [CrossRef]
59. Shankar, R.; Sachs, M.; Francàs, L.; Lubert-Perquel, D.; Kerherve, G.; Regoutz, A.; Petit, C. Porous boron nitride for combined CO_2 capture and photoreduction. *J. Mater. Chem. A* **2019**, *7*, 23931–23940. [CrossRef]
60. Weng, Q.; Kvashnin, D.G.; Wang, X.; Cretu, O.; Yang, Y.; Zhou, M.; Zhang, C.; Tang, D.-M.; Sorokin, P.B.; Bando, Y.; et al. Tuning of the optical, electronic, and magnetic properties of boron nitride nanosheets with oxygen doping and functionalization. *Adv. Mater.* **2017**, *29*, 1700695. [CrossRef]
61. Fu, J.; Jiang, K.; Qiu, X.; Yu, J.; Liu, M. Product selectivity of photocatalytic CO_2 reduction reactions. *Mater. Today* **2020**, *32*, 222–243. [CrossRef]
62. Park, H.-a.; Choi, J.H.; Choi, K.M.; Lee, D.K.; Kang, J.K. Highly porous gallium oxide with a high CO_2 affinity for the photocatalytic conversion of carbon dioxide into methane. *J. Mater. Chem.* **2012**, *22*, 5304. [CrossRef]
63. White, J.L.; Baruch, M.F.; Pander, J.E., III; Hu, Y.; Fortmeyer, I.C.; Park, J.E.; Zhang, T.; Liao, K.; Gu, J.; Yan, J.; et al. Light-driven heterogeneous reduction of carbon dioxide: Photocatalysts and photoelectrodes. *Chem. Rev.* **2015**, *115*, 12888–12935. [CrossRef] [PubMed]
64. Wu, C.; Wang, B.; Wu, N.; Han, C.; Zhang, X.; Shen, S.; Tian, Q.; Qin, C.; Li, P.; Wang, Y. Molecular-scale understanding on the structure evolution from melamine diborate supramolecule to boron nitride fibers. *Ceram. Int.* **2020**, *46*, 1083–1090. [CrossRef]

MDPI
St. Alban-Anlage 66
4052 Basel
Switzerland
www.mdpi.com

Catalysts Editorial Office
E-mail: catalysts@mdpi.com
www.mdpi.com/journal/catalysts

Disclaimer/Publisher's Note: The statements, opinions and data contained in all publications are solely those of the individual author(s) and contributor(s) and not of MDPI and/or the editor(s). MDPI and/or the editor(s) disclaim responsibility for any injury to people or property resulting from any ideas, methods, instructions or products referred to in the content.

www.ingramcontent.com/pod-product-compliance
Lightning Source LLC
LaVergne TN
LVHW070413100526
838202LV00014B/1450